Advances in Intelligent Systems and Computing

Volume 860

Series editor

Janusz Kacprzyk, Polish Academy of Sciences, Warsaw, Poland
e-mail: kacprzyk@ibspan.waw.pl

The series "Advances in Intelligent Systems and Computing" contains publications on theory, applications, and design methods of Intelligent Systems and Intelligent Computing. Virtually all disciplines such as engineering, natural sciences, computer and information science, ICT, economics, business, e-commerce, environment, healthcare, life science are covered. The list of topics spans all the areas of modern intelligent systems and computing such as: computational intelligence, soft computing including neural networks, fuzzy systems, evolutionary computing and the fusion of these paradigms, social intelligence, ambient intelligence, computational neuroscience, artificial life, virtual worlds and society, cognitive science and systems, Perception and Vision, DNA and immune based systems, self-organizing and adaptive systems, e-Learning and teaching, human-centered and human-centric computing, recommender systems, intelligent control, robotics and mechatronics including human-machine teaming, knowledge-based paradigms, learning paradigms, machine ethics, intelligent data analysis, knowledge management, intelligent agents, intelligent decision making and support, intelligent network security, trust management, interactive entertainment, Web intelligence and multimedia.

The publications within "Advances in Intelligent Systems and Computing" are primarily proceedings of important conferences, symposia and congresses. They cover significant recent developments in the field, both of a foundational and applicable character. An important characteristic feature of the series is the short publication time and world-wide distribution. This permits a rapid and broad dissemination of research results.

More information about this series at http://www.springer.com/series/11156

Radek Silhavy · Petr Silhavy
Zdenka Prokopova
Editors

Intelligent Systems in Cybernetics and Automation Control Theory

Springer

Editors
Radek Silhavy
Faculty of Applied Informatics
Tomas Bata University in Zlin
Zlin, Czech Republic

Zdenka Prokopova
Faculty of Applied Informatics
Tomas Bata University in Zlin
Zlin, Czech Republic

Petr Silhavy
Faculty of Applied Informatics
Tomas Bata University in Zlin
Zlin, Czech Republic

ISSN 2194-5357 ISSN 2194-5365 (electronic)
Advances in Intelligent Systems and Computing
ISBN 978-3-030-00183-4 ISBN 978-3-030-00184-1 (eBook)
https://doi.org/10.1007/978-3-030-00184-1

Library of Congress Control Number: 2018953177

This Springer imprint is published by the registered company Springer Nature Switzerland AG
The registered company address is: Gewerbestrasse 11, 6330 Cham, Switzerland

Preface

This book constitutes the refereed proceedings of the Computational Methods in Systems and Software 2018 (CoMeSySo 2018), held in September 2018.

CoMeSySo 2018 conference intends to provide an international forum for the discussion of the latest high-quality research results in all areas related to cybernetics and intelligent systems. The addressed topics are the theoretical aspects and applications of software engineering in intelligent systems, cybernetics and automation control theory, econometrics, mathematical statistics in applied sciences, and computational intelligence.

CoMeSySo 2018 has received (all sections) 117 submissions, and 63 of them were accepted for publication.

The volume Intelligent Systems in Cybernetics and Automation Control Theory brings new approaches and methods to real-world problems and exploratory research that describes novel approaches in the cybernetics, algorithms, and software engineering in the scope of the intelligent systems.

The editors believe that readers will find the following proceedings interesting and useful for their research work.

July 2018

Radek Silhavy
Petr Silhavy
Zdenka Prokopova

Organization

Program Committee

Program Committee Chairs

Petr Silhavy	Department of Computers and Communication Systems, Faculty of Applied Informatics, Tomas Bata University in Zlin, Czech Republic
Radek Silhavy	Department of Computers and Communication Systems, Faculty of Applied Informatics, Tomas Bata University in Zlin, Czech Republic
Zdenka Prokopova	Department of Computers and Communication Systems, Tomas Bata University in Zlin, Czech Republic
Krzysztof Okarma	Faculty of Electrical Engineering, West Pomeranian University of Technology, Szczecin, Poland
Roman Prokop	Department of Mathematics, Tomas Bata University in Zlin, Czech Republic
Viacheslav Zelentsov	Doctor of Engineering Sciences, Chief Researcher of St. Petersburg Institute for Informatics and Automation of Russian Academy of Sciences (SPIIRAS), Russian Federation
Lipo Wang	School of Electrical and Electronic Engineering, Nanyang Technological University, Singapore
Silvie Belaskova	Head of Biostatistics, St. Anne's University Hospital Brno, International Clinical Research Center, Czech Republic

International Program Committee Members

Pasi Luukka	North European Society for Adaptive and Intelligent Systems & School of Business and School of Engineering Sciences Lappeenranta University of Technology, Finland
Ondrej Blaha	Louisiana State University Health Sciences Center New Orleans, New Orleans, USA
Izabela Jonek-Kowalska	Faculty of Organization and Management, the Silesian University of Technology, Poland
Maciej Majewski	Department of Engineering of Technical and Informatic Systems, Koszalin University of Technology, Koszalin, Poland
Alena Vagaska	Department of Mathematics, Informatics and Cybernetics, Faculty of Manufacturing Technologies, Technical University of Kosice, Slovak Republic
Boguslaw Cyganek	Department of Computer Science, University of Science and Technology, Krakow, Poland
Piotr Lech	Faculty of Electrical Engineering, West Pomeranian University of Technology, Szczecin, Poland
Monika Bakosova	Institute of Information Engineering, Automation and Mathematics, Slovak University of Technology, Bratislava, Slovak Republic
Pavel Vaclavek	Faculty of Electrical Engineering and Communication, Brno University of Technology, Brno, Czech Republic
Miroslaw Ochodek	Faculty of Computing, Poznan University of Technology, Poznan, Poland
Olga Brovkina	Global Change Research Centre Academy of Science of the Czech Republic, Brno, Czech Republic
Elarbi Badidi	College of Information Technology, United Arab Emirates University, Al Ain, United Arab Emirates
Gopal Sakarkar	Shri. Ramdeobaba College of Engineering and Management, Republic of India
V. V. Krishna Maddinala	GD Rungta College of Engineering and Technology, Republic of India
Anand N Khobragade	Maharashtra Remote Sensing Applications Centre, Republic of India
Abdallah Handoura	Computer and Communication Laboratory, Telecom Bretagne, France

Organizing Committee Chair

Radek Silhavy — Tomas Bata University in Zlin, Faculty of Applied Informatics
email: comesyso@openpublish.eu

Conference Organizer (Production)

OpenPublish.eu s.r.o.
Web: http://comesyso.openpublish.eu
Email: comesyso@openpublish.eu

Conference Website, Call for Papers

http://comesyso.openpublish.eu

Contents

Efficient Framework to Secure Communication in IoT Using Novel Finite Field Encryption

Eisha Akanksha(✉)

Department of Electronics & Communication, CMRIT, Bengaluru, India
eisha.a@cmrit.ac.in

Abstract. Offering resiliency against major threats to the exponentially massive number of physical devices in Internet-of-Things (IoT) is one of the most challenging problems in Ubiquitous Computing. Most recently, researchers have evolved up with many security solutions to find that they do not offer proper performance equilibrium between security performance and communication. Therefore, this problem is addressed by proposed a novel analytical framework that uses highly enhanced finite field encryption to ensure that messages exchanged between IoT nodes are highly secured irrespective of any forms of threat level. The study outcome of proposed system shows that it offer reduced energy, reduced delay, and maximized data delivery performance with higher degree of data integrity and privacy preservation performance in contrast to frequently used security protocols in IoT.

Keywords: Internet-of-Things · Cloud computing · Security · Attack Vulnerability · Public key cryptography · Cyber-physical security

1 Introduction

Internet-of-Things (IoT) comprises of a hugely connected physical machines or devices for acquisition as well as exchange of information [1]. One of the biggest problem in IoT is that it keeps on adding exponentially higher number of devices that goes beyond a limited protected environment e.g. firewall system. Owing to addition of number of different forms of cyber-physical system, identifying an accurate threat is one of the highly complex processes till date. The proliferation of the mobile devices is another root cause of vulnerability in IoT [2, 3]. At the same time, securing IoT over cloud infrastructure is quite a challenging task, which has a long way to go with some preliminary research work [4, 5]. The Application Programming Interface (APIs) used over IoT devices in cloud is extremely poorly built even compared to normal API on other cloud-based application [6]. Owing to usage of poor security features in IoT devices, the private information of the users as well as device are at stake. With the increase of network span of IoT, offering equivalent protection to smart cities will be inevitable requirement in IoT [7]. However, it is highly a tedious task to do so. The prime reason behind this is a lack of updating process for both the devices as well as service by the organization. The safety of the device lies on how fast it can get its updates through certain gateway. Although, certain studies towards gateways has been carried out in IoT [8–10]; however, they are found to not focus on identifying threat

R. Silhavy et al. (Eds.): CoMeSySo 2018, AISC 860, pp. 1–11, 2019.
https://doi.org/10.1007/978-3-030-00184-1_1

level. Therefore, this manuscript offers a novel contribution to identify and resist threats over IoT devices. Unlike existing approach, the proposed system also targets to offer better communication services along with security in IoT. The organization of the proposed paper is as follows: Sect. 2 discusses about the existing research work followed by problem identification in Sect. 3. Section 4 discusses about proposed methodology followed by elaborated discussion of algorithm implementation in Sect. 5. Comparative analysis of accomplished result is discussed under Sect. 6 followed by conclusion in Sect. 7.

2 Related Work

There are various research work carried out towards securing IoT infrastructure. This section briefs of some of the recent work being carried out towards IoT security. The work carried out by Aslam and Curry [11] have presented machine learning based approach to utilize feature extraction technique and classifier selection method in order to enhance performance of event processing in IoT. Bhattrai and Wang [12] have studied various security and privacy challenges and designed a security model to achieve end-to-end trust factor in IoT applications. Chen et al. [13] have designed energy-efficient scheduling schemes to overcome the implementation problem of user scheduling in IoT data transmission. An optimization framework is constructed by Farooq and Zhu [14] to reconfigure the setting of physical network according to need in order to obtain real-time transmission in the internet of battlefield things. Gong et al. [15] have presented a dynamic mechanism that analyzes the behavior of sensing nodes of IoT for realizing the trust factor of nodes. He et al. [16] have addressed the security issues in wearable devices and designed strong password selector against password breaching attacks. Hudson [17] has discussed various security principles for maintaining security and privacy in IoT. Lee et al. [18] have presented new security model based on scaling approach that aim to detect the clone attacks in IoT. The improved version of DHCPv6 protocol is claimed to offer privacy as reported in work of Li et al. [19]. Liu et al. [20] have emphasized on the usage of lattice-based encryption mechanism for securing IoT. Privacy problem has been also addressed by Ma et al. [21] that deal with autonomous service identification in IoT. The work of Li et al. [22] discusses about an assessment technique for different threats using neural network. Apart from above literatures, there are various other approaches viz. traffic obfuscation (Liu et al. [23]), experimental-based approach (Premsankar et al. [24]), heuristic-based approach (Qian et al. [25]), exploration-analysis (Sen et al. [26]), fog computing (Li et al. [27]), multi-level based approach (Yan et al. [28]), similarity-based identification in software defined network (Yin et al. [29]), and hardware cryptographic approach (Zhang et al. [30]). Therefore, there are good numbers of research work being carried out most recently in order to secure IoT system. The next section briefs about issues that are yet left unsolved.

3 Problem Description

The recent security approaches lacks sufficient performance equilibrium between robust security and computational demands of IoT devices. Moreover, the usage of cryptographic approach is not assessed for their practical viability as they are highly recursive in their operation without checking how they affect non-security performance. It was also noticed that existing solution is highly attack specific and hence they are less likely to be functional if the attack scenario is changed. Apart from this, there are very much less studies focusing on public key cryptographic usage as they are quite suitable for sensors networked in IoT. Hence, there is a need of a system that offers comprehensive security with less involvement of computational/network resources and is resistive against maximum level of threats.

4 Proposed Methodology

The prime goal of the proposed system is to offers a cost-effective security features in communication system over an IoT devices. An analytical modeling has been considered for this purpose. The block diagram of the proposed system is shown in Fig. 1.

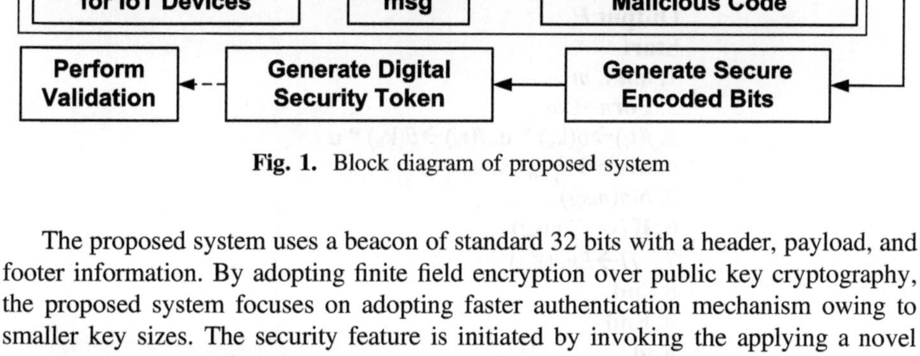

Fig. 1. Block diagram of proposed system

The proposed system uses a beacon of standard 32 bits with a header, payload, and footer information. By adopting finite field encryption over public key cryptography, the proposed system focuses on adopting faster authentication mechanism owing to smaller key sizes. The security feature is initiated by invoking the applying a novel encryption algorithm in order to generate a ciphered message. The next step focuses on identifying the presence of malicious codes by using process of generating a secured encoded bits followed by generation of digital security token. This phenomenon results in generating a hash value of the encoded bits for compressing the size as well as for faster authentication purpose. Finally, the digital signature token is another revised form of encoded bits that are required to be validated for its legitimacy. The complete design methodology involved very less iterative operations in order to obtain better response time and lowering the dependencies of resources from constrained IoT devices. The next section illustrates algorithm design.

5 Algorithm Implementation

The core design principle of the proposed algorithm is to formulate a very light weight self-authentication mechanism among the IoT devices. The algorithm aims for improving the performance of public key encryption that could assist in authorizing as well as authenticating the IoT devices. The study also considers such IoT devices to possess highly constrained resource characteristics for assisting in secure communication. The proposed system mainly implements two core security algorithms that run over both IoT devices and the gateway system in IoT.

5.1 Encryption Algorithm for IoT Devices

The prime role of this algorithm is to ensure the safety of the security principle running within an IoT environment by not disclosing its operational pattern. For this purpose, the proposed system uses finite field encryption mechanism. Inspired from the work of Chen et al. [31], the study considers three points X_1, X_2, and X_3 to be lying within a finite field, where these points represents secret key, unlike the legacy work. The prime motive of this algorithm is to render random selection of prime fields that directly represent private key and perform encryption of ciphered contents in order to protect it from being threat of disclosure of any specific patterns. The steps of the algorithm are as follows:

Encryption Algorithm for IoT Devices

Input:*i*, *m*

Output:*H*

Start

1. *init* i, *m*

2. **For***n*=1:*m*

3. $f(t_r) \rightarrow \theta(k_a) * \alpha, f(r_c) \rightarrow \theta(k_b) * \alpha$

4. Mem$\rightarrow \theta_{point}$(R. α)

5. *bin*(*msg*)

6. **If** ($\psi \subseteq$ *msg*)

7. $H \rightarrow$ *Enc*(K_1)

8.**End**

9. **End**

End

The algorithm takes the input of *i* (number of IoT devices), *m* (number of communicating IoT devices) that after processing results in *H* (Encrypted message). For all the communicating nodes (Line-2), consider that source and destination node selects an arbitrary secret key $[k_a k_b] = (k \rightarrow [1\ X_3])$. A new function is created that allows the gateway node to forwards secret key to both source (t_r) and destination node (r_c) (Line-3). The secret key is formed by scalar multiplication of random secret key where θ can be any form of random function and α will represents the prime value within the finite field. The arbitrary point θ_{point} is selected from the range of smart city R and prime field α (Line-4) that is stored in memory. The message to be forwarded is binarized (Line-5). The next part of the implementation is all about assessing the similarity pattern residing within the forwarded beacons (Line-6). A function ψ is designed that represents a matrix where *Mem* is stored. Although, there are 99% probability that this *Mem* matrix will never be equal to any broadcasted beacon, but there is also 1% positive chance for this. Hence, this 1% of probability of error is addressed by performing unique encoding mechanism that calls for substitution of all iterative codes present in diverse string values. This operation will result in absolutely two different ciphered broadcasted values. Although, this is a very simplified solution but there is no research papers ever reported to perform such operation in IoT. The next step is to carry out ciphering considering secret key, unique message, prime fields X_3 and its modulus i.e. $|X_3|$ (Line-7). The next process is to further ensure if any malicious codes are running within the IoT devices.

5.2 Algorithm for Identifying Malicious Code

Unlike adoption of symmetric keys in IoT devices, the proposed system uses asymmetric keys. This algorithm mainly focuses on exploring the legitimacy of the nodes involved in communication by performing two unique operation i.e. (i) generating a dynamic security token d_c followed by validating it. The steps of the algorithm are as discussed below:

Algorithm for Identifying Malicious Code
Input: *m*
Output: O_1/O_2 (flag message for identification of malicious codes)
Start
1. **For** $n = 1$: *m*
2. **If** $\phi \rightarrow 0$
3. $\theta(1, \gamma\text{-}1)$, Enc(msg, ϕ)
4. SecEnc→f(ϕ, Enc_{nodeA}, k_{nodeA}, mod(γ))
5. **If** SecEnc==0
6. Re-Compute ϕ
7. Generate d_c→f(ϕ, SecEnc)
8. Enc_{nodeA}, (v_1, v_2)= f(SecEnc, ϕ, Enc_{nodeA}, K_{pu}
9. **If** ($\phi = \phi_1$)
10. O_1→flag dc as legitimate
11. **Else**
12. O_2→flag dc as illegitimate
13. **End**
14. **End**
15. **End**

End

For all the communicating IoT devices m (Line-1), the algorithm initially computes ϕ by performing scalar multiplication of any random values of source node and modulus of any natural number (Line-2). In case, the value of the new variable ϕ is found to be equivalent to 0 (Line-2) than it performs a random selection of any integer value within a lower range of 1 and higher range of $(1 - \gamma)$ (Line-3). Along with this, the algorithm also extracts encrypted value of message *msg* and new variable ϕ (Line-3). The proposed algorithm contributes to perform computation of the dynamic security token that is uses the new variable ϕ and a secured encoded bits *SecEnc* (Line-4). A function is designed that considers following arguments e.g. new variable ϕ, *secured* encoded bits of nodes A message i.e. Enc_{nodeA}, secret key of node-A i.e. k_{nodeA}, and modulus of new variable ϕ (Line-4). The next part of the algorithm performs generation of digital secure token that uses nearly same parameters as used for generating secured encoded bits except it excludes the new variable ϕ and modulus of γ (Line-8). Once the digital secure token d_c is generated, the algorithm performs validation of it. It is ensured that the algorithm uses a public key K_{pu} (Line-8). Prior to performing validation of digital security token being generated, the algorithm computes another temporary variable only for that session of validation. This temporary variable is obtained by scalar multiplication of random value of source node and modulus of natural number. Hence, upon comparing new variable ϕ with temporary variable ϕ_1, it represents a condition where only the gateway has a control over the authenticated nodes. Hence, this condition represents presence of legitimate node and so digital security token d_c is considered authentic (Line-10) or else it is considered to be a part of compromised node (Line-12). Figure 2 highlights the operational flow of proposed algorithm.

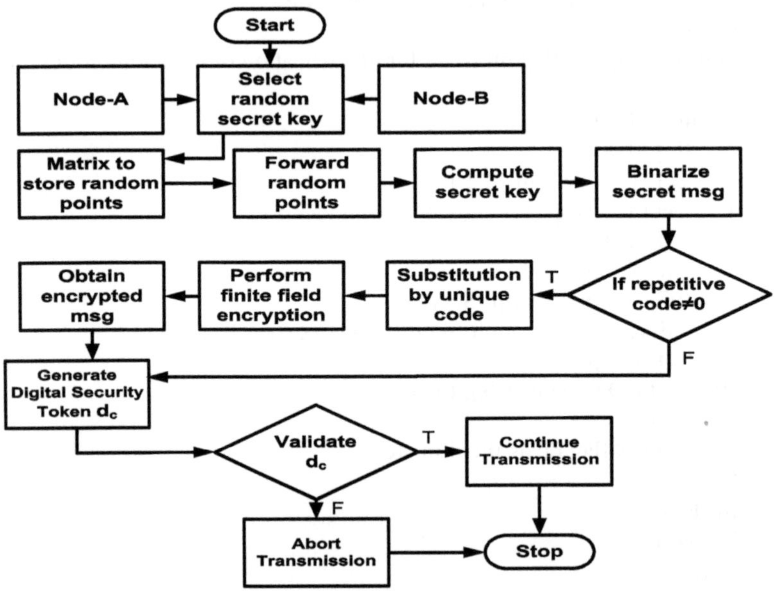

Fig. 2. Operational flow of Algorithm-1

6 Results Discussion

As proposed system is basically a security algorithm that will be executed over IoT devices, there are good chances that they could offer security as the cost of computational complexity. It is because the IoT devices cannot perform higher computational processing. Running any security algorithm can exhaust the battery and many induce latency that can directly impact communication process. Therefore, the proposed study considers energy consumption, end-to-end delay, and packet delivery ratio is the prime performance parameters. The implementation is carried out considering 1000 × 1200 m^2 simulation area considering channel sensing capacity of 0.2 s and transmission energy initialized to 0.05 Joules. For an effective analysis, the proposed results are also compared with existing security protocols in IoT e.g. RPL security [32], DTLS [33], and IPSec [34].

The computation of the energy is carried out by implementing the first-order radio energy model over the IoT devices as it represents direct energy consumption for RF circuitry involved in it. A closer look into the outcome in Fig. 3 shows that energy consumption of proposed system *ProP* is highly reduced as compared to others. The prime reasons behind this are as follows:-DTLS is basically meant for transport layer security and it offers comprehensive resistance in it. However, the handshaking mechanism of DTLS is quite dependent iteratively on many network-based entities, time factors, etc. Hence, the resource consumption increases with the increase in traffic

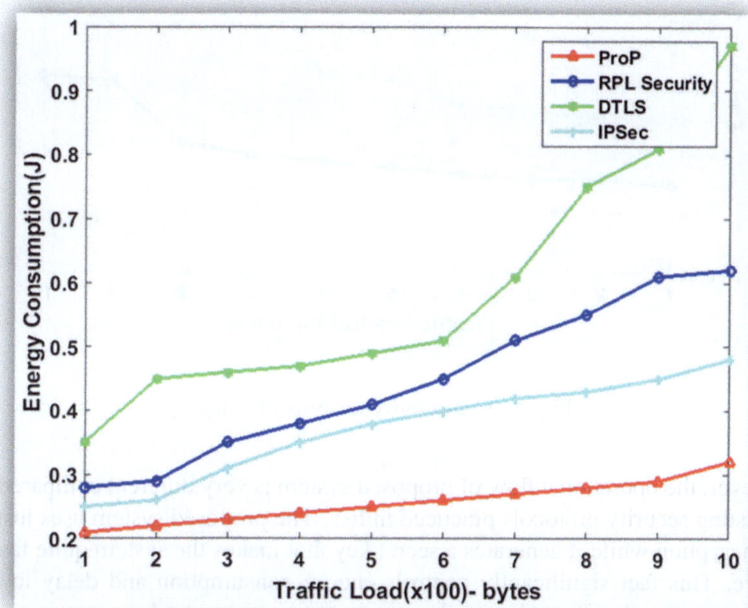

Fig. 3. Comparative analysis of energy

loads leading to excessive energy depletion. This problem doesn't existing in RPL security scheme and hence it offers slightly improved energy dissipation compared to DTLS. It offers a comprehensive security against potential threats of sinkhole attack, denial of service attack, wormhole attack, etc. An ICMPv6 message with 4 byte header is used to offer this security; however, RPL security cannot be considered as a security protocol but rather it is considered as a framework for which reason it becomes massively heavy process to run within constrained IoT devices. To some extent, IPSec offers lesser dependencies for it to be executed as its architecture is highly simplified compared to RPL and DTLS. However, the problem of heavyweight encryption still affects its performance as IPSec is designed on the foundation of using heavier security protocol e.g. RSA and DES that has higher resource dependencies towards computational resources. Because of the similar reasons, the performance of all these protocols are found to exhibit nearly similar trend s even in delay analysis (Fig. 4).

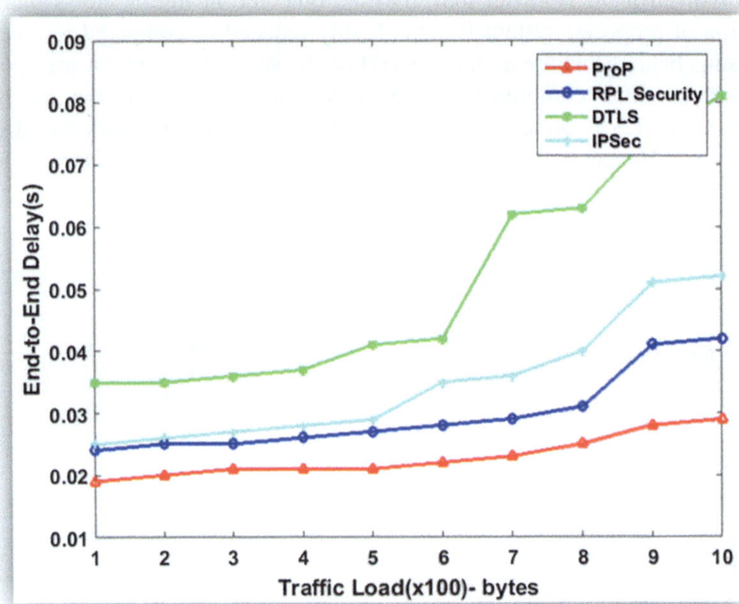

Fig. 4. Comparative analysis of delay

However, the operational flow of proposed system is very different compared to any of the existing security protocols practiced in IoT. The proposed system uses just single step of encryption while it generates a secret key that makes the system quite faster and responsive. This fact significantly controls energy consumption and delay to a large extent. The proposed system also implements a unique process where a secure encoded bits results in minimization of size of the routing beacons. Hence, forwarding the data will not only be easier but with extremely lesser dependency of computational as well as network resources. This fact is also cross checked for packet delivery system. The assessment of packet delivery ratio shows that proposed system offers significantly improved data forwarding performance as compared to any existing IoT security protocols.. The concept of digital secure token is another contribution that minimizes the iterative steps in encryption and further improves the performance of public key protocol by choosing the best and unique key for a specific session of authentication. Because of this faster operation, the proposed system offers faster packet delivery ratio. The declining trend of packet delivery ratio is due to progressive energy consumption with increasing simulation trials. Figure 5 highlights the performance of packet delivery ratio.

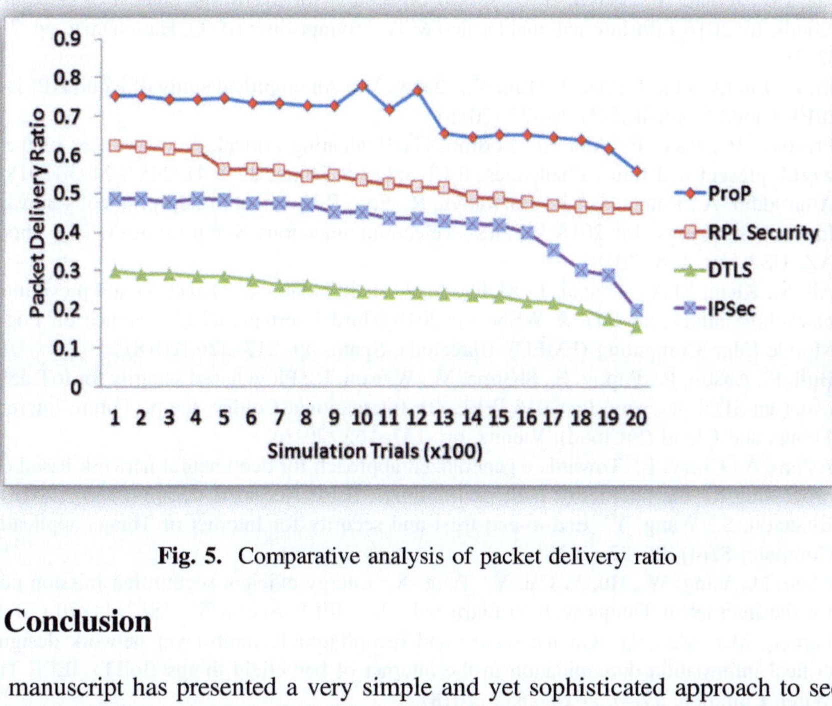

Fig. 5. Comparative analysis of packet delivery ratio

7 Conclusion

The manuscript has presented a very simple and yet sophisticated approach to secure the communication system in IoT. The contributions of the proposed system are: (i) it offers a less iterative algorithmic approach that not only offers faster response time but also offers lesser dependencies on computational/network related resources. The computational time recorded is approximately 0.26659 s which is very less as compared to other e.g. RPL Security (8.6621 s), DTLS (15.7922 s), and IPSec (6.01411 s). (ii) the proposed system not only identifies but also performs prevention as it is not feasible for intruder to decrypt secure encoded bits or digital security token owing to multiple dependencies of variable that are highly dynamic in nature. (iii) The proposed system significantly promotes data integrity and privacy. Hence, even if the data from any intermediate node is captured by attacker, they will never be able to decrypt it back, (iv) The most significant contribution of this paper is its substitution process of every encoded bits. Hence, it always ensures that none of the beacons bear any form of pattern among each other.

References

1. Bouhaï, N., Saleh, I.: Internet of Things: Evolutions and Innovations. Wiley, New York (2017)
2. Chen, L.W., Chung, J.J.: Mobility-aware and congestion-relieved dedicated path planning for group-based emergency guiding based on internet of things technologies. IEEE Trans. Intell. Transp. Syst. **18**(9), 2453–2466 (2017)
3. Bi, Y., Han, G., Lin, C., Deng, Q., Guo, L., Li, F.: Mobility support for fog computing: an SDN approach. IEEE Commun. Mag. **56**(5), 53–59 (2018)
4. Chen, L., et al.: Robustness, security and privacy in location-based services for future IoT: a survey. IEEE Access **5**, 8956–8977 (2017)
5. Ida, I.B., Jemai, A., Loukil, A.: A survey on security of IoT in the context of eHealth and clouds. In: 2016 11th International Design & Test Symposium (IDT), Hammamet, pp. 25–30 (2016)
6. Li, Z., Lu, Q., Zhu, L., Xu, X., Liu, Y., Zhang, W.: An empirical study of cloud API issues. IEEE Cloud Comput. **5**(2), 58–72 (2018)
7. Frustaci, M., Pace, P., Aloi, G., Fortino, G.: Evaluating critical security issues of the IoT world: present and future challenges. IEEE Internet Things J. **5**(4), 2483–2495 (2018)
8. Amiruddin, A., Ratna, A.A.P., Harwahyu, R., Sari, R.F.: Secure multi-protocol gateway for Internet of Things. In: 2018 Wireless Telecommunications Symposium (WTS), Phoenix, AZ, USA, pp. 1–8 (2018)
9. Ali, S., Khan, M.A., Ahmad, J., Malik, A.W., ur Rehman, A.: Detection and prevention of black hole attacks in IOT & WSN. In: 2018 Third International Conference on Fog and Mobile Edge Computing (FMEC), Barcelona, Spain, pp. 217–226 (2018)
10. Bull, P., Austin, R., Popov, E., Sharma, M., Watson, R.: Flow based security for IoT devices using an SDN gateway. In: 2016 IEEE 4th International Conference on Future Internet of Things and Cloud (FiCloud), Vienna, pp. 157–163 (2016)
11. Aslam, A., Curry, E.: Towards a generalized approach for deep neural network based event processing for the internet of multimedia things. IEEE Access **6**, 25573–25587 (2018)
12. Bhattarai, S., Wang, Y.: End-to-end trust and security for Internet of Things applications. Computer **51**(4), 20–27 (2018)
13. Chen, D., Yang, W., Hu, J., Cai, Y., Tang, X.: Energy-efficient secure transmission design for the Internet of Things with an untrusted relay. IEEE Access **6**, 11862–11870 (2018)
14. Farooq, M.J., Zhu, Q.: On the secure and reconfigurable multi-layer network design for critical information dissemination in the internet of battlefield things (IoBT). IEEE Trans. Wirel. Commun. **17**(4), 2618–2632 (2018)
15. Gong, B., Wang, Y., Liu, X., Qi, F., Sun, Z.: A trusted attestation mechanism for the sensing nodes of Internet of Things based on dynamic trusted measurement. China Commun. **15**(2), 100–121 (2018)
16. He, D., Ye, R., Chan, S., Guizani, M., Xu, Y.: Privacy in the Internet of Things for smart healthcare. IEEE Commun. Mag. **56**(4), 38–44 (2018)
17. Hudson, F.D.: Enabling trust and security: TIPPSS for IoT. IT Prof. **20**(2), 15–18 (2018)
18. Lee, P.Y., Yu, C.M., Dargahi, T., Conti, M., Bianchi, G.: MDSClone: multidimensional scaling aided clone detection in Internet of Things. IEEE Trans. Inf. Forensics Secur. **13**(8), 2031–2046 (2018)
19. Li, L., Ren, G., Liu, Y., Wu, J.: Secure DHCPv6 mechanism for DHCPv6 security and privacy protection. Tsinghua Sci. Technol. **23**(1), 13–21 (2018)
20. Liu, Z., Choo, K.K.R., Grossschadl, J.: Securing edge devices in the post-quantum Internet of Things using lattice-based cryptography. IEEE Commun. Mag. **56**(2), 158–162 (2018)

21. Ma, Y., Wu, Y., Ge, J., Li, J.: An architecture for accountable anonymous access in the Internet-of-Things network. IEEE Access **6**, 14451–14461 (2018)
22. Li, S., Bi, F., Chen, W., Miao, X., Liu, J., Tang, C.: An improved information security risk assessments method for cyber-physical-social computing and networking. IEEE Access **6**, 10311–10319 (2018)
23. Liu, J., Zhang, C., Fang, Y.: EPIC: a differential privacy framework to defend smart homes against internet traffic analysis. IEEE Internet Things J. **5**(2), 1206–1217 (2018)
24. Premsankar, G., Di Francesco, M., Taleb, T.: Edge computing for the Internet of Things: a case study. IEEE Internet Things J. **5**(2), 1275–1284 (2018)
25. Qian, Y., Chen, M., Chen, J., Hossain, M.S., Alamri, A.: Secure enforcement in cognitive internet of vehicles. IEEE Internet Things J. **5**(2), 1242–1250 (2018)
26. Sen, S., Koo, J., Bagchi, S.: TRIFECTA: security, energy efficiency, and communication capacity comparison for wireless IoT devices. IEEE Internet Comput. **22**(1), 74–81 (2018)
27. Li, G., Wu, J., Li, J., Guan, Z., Guo, L.: Fog computing-enabled secure demand response for internet of energy against collusion attacks using consensus and ACE. IEEE Access **6**, 11278–11288 (2018)
28. Yan, Q., Huang, W., Luo, X., Gong, Q., Yu, F.R.: A multi-level DDoS mitigation framework for the industrial Internet of Things. IEEE Commun. Mag. **56**(2), 30–36 (2018)
29. Yin, D., Zhang, L., Yang, K.: A DDoS attack detection and mitigation with software-defined Internet of Things framework. IEEE Access **6**, 24694–24705 (2018)
30. Zhang, L.X., Dong, Q., Wang, J., Blaauw, D., Sylvester, D.: Recryptor: a reconfigurable cryptographic cortex-M0 processor with In-memory and near-memory computing for IoT security. IEEE J. Solid State Circ. **53**(4), 995–1005 (2018)
31. Chen, Y., et al.: A programmable Galois field processor for the Internet of Things. In: 2017 ACM/IEEE 44th Annual International Symposium on Computer Architecture (ISCA), Toronto, ON, pp. 55–68 (2017)
32. A security threat analysis for the routing protocol for low-power and lossy networks (RPLs). https://tools.ietf.org/html/rfc7416. Accessed 05 June 2018
33. Datagram transport layer security version 1.2. https://tools.ietf.org/html/rfc6347. Accessed 05 June 2018
34. IP Security (IPsec) and internet key exchange (IKE) document roadmap. https://tools.ietf.org/html/rfc6071. Accessed 05 June 2018

A Novel Template - Based Data Structurization Scheme for Normalizing and Analyzing Medical Data

A. S. Chandru[1(✉)] and K. Seetharam[2]

[1] Department of Information Science & Engineering, NIE Institute of Technology, Mysore, India
chandru.as.research@gmail.com

[2] Department of Computer Science & Engineering, N.M.A.M. Institute of Technology, Nitte, Karkala Taluk, Udupi, India

Abstract. With the ever-increasing size of the data in recent times, big data approaches are widely initiated to be adopted. This phenomenon can also be seen in the area of healthcare sector where there are generation of massive medical data that is not only bigger in dimension but also highly unstructured. After reviewing existing approaches, it was found that existing system doesn't offer comprehensive and practical solution against data structurization problem. Therefore, this paper presents a novel template-based data structurization scheme that takes the unstructured medical data as input and subjects it to a novel data structurization scheme. The system also contributes in obtaining knowledge related to criticality of disease associated with the dataset. The study outcome shows that proposed system offers reduced response time for both data structurization and knowledge discovery.

Keywords: Big data · Analytics · Knowledge discovery
Electronic health record · Analysis · Data structurization

1 Introduction

The concept of big data has successfully attracted an attention among the research community [1]. Although, there is no standard specification of big data at present, but it is said that when a data is associated with five explicit problems (also known as 5V) all at same time than it turn out to take the shape of big data [2, 3]. Therefore, big data approach essentially deals with normalizing such complex data so that analytical operation can be carried out over such data. There are various research work e.g. [4–6] that has already reported challenges in working on big data analytics. This paper has considered a case study of generation of big data from the healthcare section in the form of Electronic Health Record (EHR). Owing to adoption of cloud-based services, various healthcare units are also adopting it in hosting their application. This process leads to generation of abundant amount of medical big data that are quite difficult to be stored as to be assessed also. At present, there are various research work carried out towards medical big data e.g. [7–10]; however, such practical constraints are less found to be adopted. Therefore, this manuscript presents a novel and quite simplistic idea

R. Silhavy et al. (Eds.): CoMeSySo 2018, AISC 860, pp. 12–21, 2019.
https://doi.org/10.1007/978-3-030-00184-1_2

about dealing with this problem of complex big data in order to obtain faster response time. The paper discusses its solution with respect to an experimental model. The organization of the complete paper is as follows: Sect. 2 discusses about the existing research work followed by problem identification in Sect. 3. Section 4 discusses about proposed methodology followed by elaborated discussion of algorithm implementation in Sect. 5. Comparative analysis of accomplished result is discussed under Sect. 6 followed by conclusion in Sect. 7.

2 Related Work

This section discusses about the existing research work being carried out toward big data analytics techniques adopted for medical and health care. The previous work of Chandru and Seetharam et al. [11] discussed review on the medical data analysis for effective detection of diseases. The works carried out by Chen et al. [8] have designed an advance Diabetes model composed of sensing layer, sharing layer and diagnosis layer. The objective of presented model is to reduce cost and provide a sustainable personalized treatment. Shafqat et al. [12] have carried surveyed on big data analytics based healthcare systems in order to provide better understanding to construct cost-effective and smart healthcare system. Chen et al. [13] and Zhong et al. [14] have used machine learning technique over big data analytics to predict diseases based on structured and unstructured medical data and to preserve patient data privacy. Chrimes and Zamani [15] have suggested using HBase approach in bid data analytics to achieve secured patient record for the clinical services. Garattini et al. [16] have discusses various key factor for big data analytics to control infectious diseases. Magarino et al. [17] have developed smart bed system based on sensor and IoT to recognize the sleeping action. Sarkar et al. [18] has worked to construct secure distributive model to protect patient medical data record. Tawalbeh et al. [19] have discussed the role of cloud computing and big data technique in networked healthcare system. Zhang et al. [20] have tried to bring attention of researchers in machine learning, biomedical and mobile computing to display the progressive system and application for the smart health system. Wu et al. [21] have carried deep studied on the Electronic health record system and Omic data and similarly, Viceconti et al. [22] have disclosed some major problem to analyze the performance of big data analytics for healthcare monitoring and treatment system. Zhang et al. [23] have proposed a smart and intelligent healthcare model based on big data and cloud computing technology in order to provide better treatment environment and better application interface to patient so that they can take effective and better advantages of smart healthcare system. Srinivasan and Arunasalam [24] have presented a framework to analyze the health insurance record by using big data analytics to detect fraud and waste claims in order to reduce healthcare costs. In the same way [25–27], and [28] have discussed the potential, capability and challenges for integrating big data analytics to networked and smart healthcare systems for effective treatment.

3 Problem Description

After reviewing the existing approaches in prior section, it was found that existing system significantly suffers from practical implementation of data structurization. Majority of the experiment calls for using dataset, which is less likely maps with big data, and applying analytical approach on it. However, the problem associated with unstructured data retains as it is. Usage of various tools and approaches used in existing system are already reported by many researchers to have flaws. Therefore, there is no off-the mainstream work to highlight the significant contribution towards medical big data approach. Another essential problem found is lack of consideration of response time or time complexity which is yet an open-end problem in medical big data. Therefore, the problems associated with data structurization are yet found computationally challenging and hence is addressed in the present research work.

4 Proposed Methodology

The prime purpose of the proposed system is to offer a simplistic solution towards addressing the problem of variety in big data. The implementation of the proposed system is carried out considering the case study of Electronic Health Record (EHR) which is highly unstructured while the proposed system targets to not only perform template-based data structurization but also should be able to perform analytical operation for extraction knowledge associated with disease.

According to Fig. 1, the proposed system introduces a novel concept of virtual template that is designed using matrix-based principle. This mechanism allows all the incoming data to formulate in the form of data stream where each data streams consists of chunks of information. The proposed system obtained semi-structured data by further obtaining the core fields and values within the data chunk of each individual data streams. Further, a syntactical parsing operation is applied to the structured data in order to obtain the knowledge. The next section discusses about the algorithm implementation of proposed system.

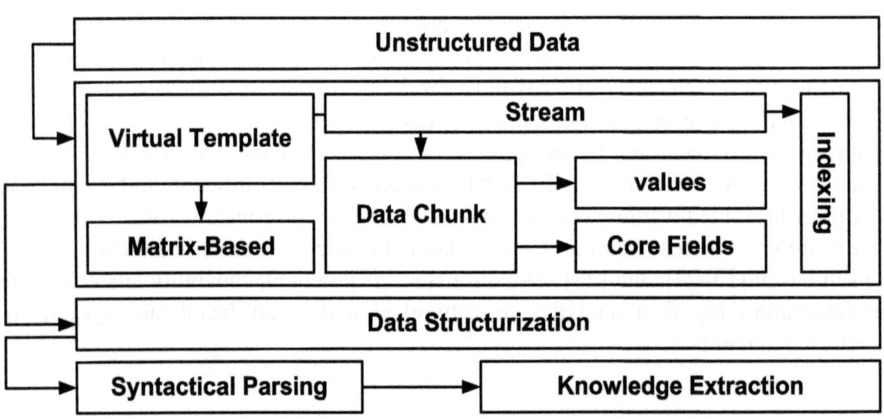

Fig. 1. Flow of proposed system

5 Algorithm Implementation

The primary aim of the proposed algorithm is to perform a proper structurization towards the unstructured EHR data. A practical case study is considered for this algorithm implementation. Consider a chain of hospital is running on different location globally where a localized server is positioned in each hospital in order to maintain daily records of EHR. It is very obviously that such EHR records will be highly unstructured and massive in size that is further hosted on cloud where the computation of the algorithm is carried out. A pictorial diagram is presented to understand the possible flow of proposed algorithm implementation (Fig. 2).

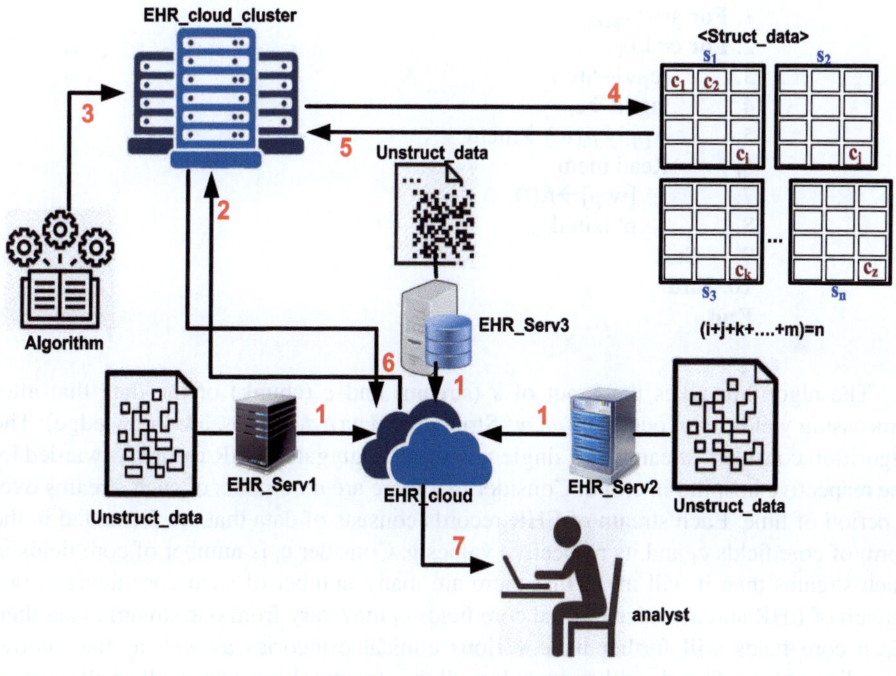

Fig. 2. Flow of algorithm implementation

According to Fig. 2, the study consider that there are three sample location of chains of hospital and is assumed to save their EHR records on the basis of certain templates in EHR server. The *first* process is that all the aggregated EHR records from all three sample hospitals are required to be forwarded to cloud environment. It is also assumed that these EHR records are highly unstructured and not feasible to be stored back in any structured storage model. The *second* process is to forward a request to the EHR cloud cluster in order to consider the aggregated data which will be further subjected to processing. The *third* process prompts the implementation of the algorithm for performing structurization and analytical operation. The *fourth* process results in

EHR cloud cluster to execute the algorithm in order to develop a virtual memory system where the structured data is constructed. The proposed system performs the construction of the data in terms of streams for assisting on live processing of data unlike the existing system that performs offline processing of big data. The *fifth* process results in generation of structured data along analytical outcome. The *sixth* process results in storage of structured data in cloud storage units while the extracted knowledge is forwarded to the analyst for exploring future possibilities of the data. The algorithm implemented over the *third* process is illustrated below:

Algorithm forPerforming Structured Mining
Input: *s* (stream), *c* (chunk)
Output: *mem* (Structured data), d_{know}(disease knowledge)
Start
1. **For** s=1:s_{max}
2. **For** c=1:c_i
3. $[c_f\ v]=f_1(c_i)$
4. tag id→c_i
5. Apply f_2(id)→mem
6. Read mem
7. [w p]→f_3(t)
8. obtain d_{know}
9. **End**
10. **End**
End

The algorithm takes the input of *s* (stream) and *c* (chunk) of the data that after processing yields to an output of *mem* (Structured data), d_{know}(disease knowledge). The algorithm considers streams as a single unit of an aggregated EHR record forwarded by the respective hospital in cloud. Consider that there are *n* numbers of such streams over a period of time. Each stream of EHR records consists of data that are formatted in the form of core fields c_f and its respective values *v*. Consider c_i is number of core fields in each streams than it will mean that there are many number of such core fields in one stream of EHR and also size of total core fields c_f may vary from one stream to another. Each core fields will further have various clinical categories as well as their corresponding values. The algorithm considers all the streams (Line-1) as well as all the core fields within them (Line-2). A simple function $f_1(x)$ is applied on the core fields to discrete extracts all the *corresponding core fieldc*$_i$ and its *corresponding valuev* (Line-3). The process is continued for all the core fields in all the streams. In order to facilitate faster query processing, the algorithm performs indexing each core fields with an identity *id* (Line-4). A function $f_2(x)$ is constructed that is responsible for managing such id for better accessibility of the metadata (Line-5). This process further stores all the ids in a virtual memory *mem* (Line-5). This results in formation of semi-structured data. The defined identities of each core fields are then used for accessing all the streams in order to read the semi-structured data thereby forming a virtually structured data that is further stored in cloud storage unit (Line-6). The next part of the algorithm implements a function $f_3(x)$ that applies syntactical parsing to generate word w and position of the word p (Line-7). This process further complements to form a structured data as well as knowledge of disease too (Line-8).

6 Results Discussion

One of the challenging phases of the analysis of proposed system is related to the input file, which is required to be in the form of big data. A data could be called as big data if (i) its size is ever increasing especially in the form of streams and not any offline file system and (ii) it should be associated with all the 5V problem (or atleast 3V to some extent). Unfortunately, there is no such database system that offers such forms of data. There are some claimed dataset that are publicly available e.g. [29–31]; however, they lacks above mentioned characteristics. Hence, construction of big data for carrying out the investigation was a quite challenging aspect in proposed system. This section will discuss about the methodology used for constructing input file, construction of test-bed, and result observation.

(i) **Constructing Input File:**

After reviewing the publically available dataset related to healthcare sector [32, 33], it was found that such form of individual file of one data varies in the range of Kilobytes to Megabytes. There are various formats of such available data viz. JSON format, comma separated value (.csv) format, XLS format, XML format etc. These datasets are highly scattered and there is lack of spontaneity in the year of collection of data. For analyzing the proposed system, what is required is a stream of data that could be anticipated to form a shape of big data. Owing to absence of any such data, a synthetic dataset is designed considering the practical scenario of EHR record management in mind. After reviewing the work of Butt [34], it was realized that an EHR record has various forms of data within it. Hence, the synthetic data is constructed considering existing big dataset and EHR records in mind. In order to retain better form of simplicity, the proposed system designs synthetic streams of data that considers primary information of patient along with response generated by the physician. It is intentionally designed in such a way that this unstructured information should be used for extracting knowledge about the criticality of the disease that the patient is suffering from.

(ii) **Construction of Test-Bed:**

Owing to adoption of experimental approach, a test-bed is designed using a system with simple core i3 processor and 4 GB RAM. The scripting of the proposed system is carried out in MATLAB as proposed system demands various forms of transformation-based operation where matrix-based approach will be the most precise way to represent the set of big data. In order to perform an effective analysis, the complete dataset is split up in different numbers where size of one datasets exceeds twice the size of the others. The experiment is carried out by giving an input to the smallest size of streamed data and observed the data structurization time and normalized data structurization time. The experiment was also carried out with different processor and different set of data. An interesting part of the proposed test-bed is that it is now completely independent of any forms of medical dataset. This will mean that proposed test bed is now capable of not only performing data structurization but also performs a very simplistic extraction of knowledge that is directly connected to disease.

(iii) **Result Observation:**

The analysis is carried out by computing time for data structurization for each individual unit of data streams. The normalized time is obtained by subtracting the prior recorded time with recent record of data structurization algorithm. Basically, these performance factors are used for assessing the response time as usually response time is the primary assessment factor to check the algorithm complexity.

The outcome shown in Figs. 3 and 4 highlights that overall time to perform data structurization is within 60–70 s. A closer look into both the trend will show that significant increase in time consumption for data structurization only shoots after 6th or 7th experimental trials, whereas in each experimental trials the data is incremented in double sized. It basically shows that proposed algorithm is capable of offering approximately 50–60% of consistency in offering minimal response time, which also indirectly tells that it is also computationally efficient (Fig. 5).

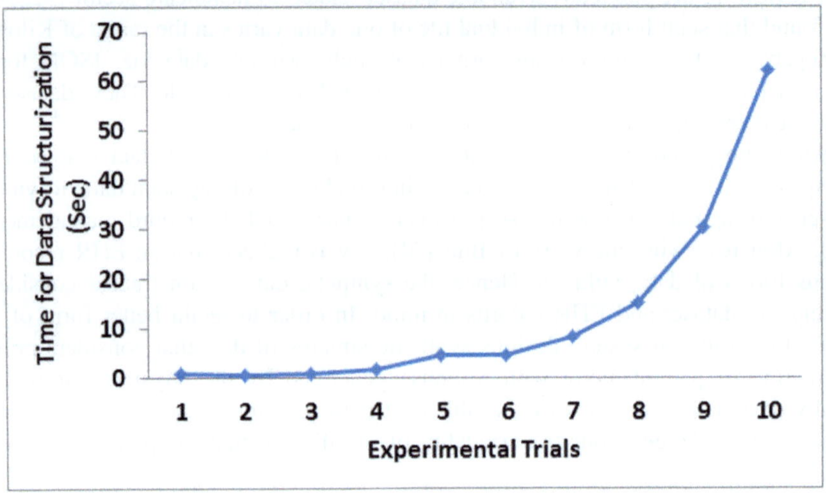

Fig. 3. Analysis of time for data structurization

When the analysis is carried out with respect to overall computation time, it was found that proposed system offers much reduced data structurization time and slightly increases in knowledge extraction time. The different between the trends are approximately less than 5–8% and this outcome could be claimed to be in tolerable limits.

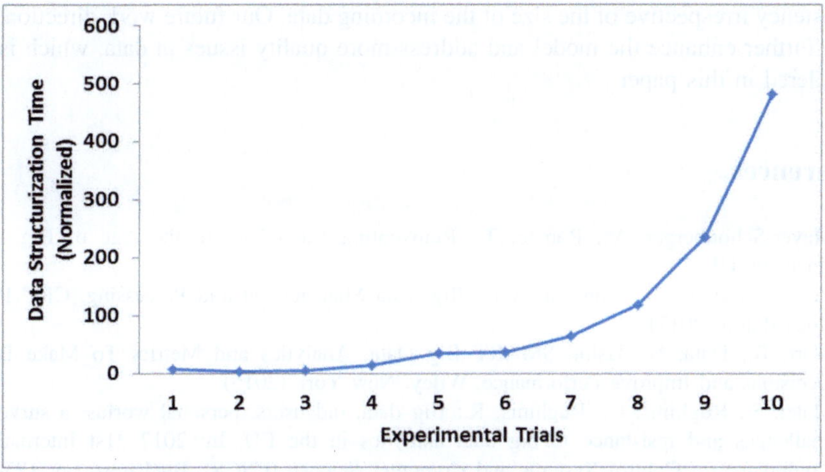

Fig. 4. Analysis of time for data structurization (Normalized)

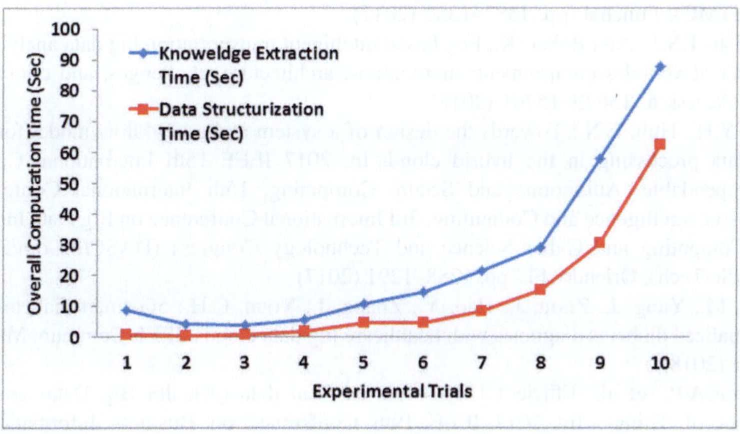

Fig. 5. Analysis of overall computational time

7 Conclusion

Big data is just in its nascent stage and still the technologies are evolving. The proposed paper is a novel attempt to solve an open-end problem associated with data structurization taking the case study of medical big data. The significant research contributions of the proposed system are as follows: (i) the proposed system is completely independent of any tools e.g. Hadoop/MapReduce those are frequently used in big data analysis and is also reported to have flaws in it, (ii) the complete model offers both structurization of data as well as knowledge extraction together without using any complex mechanism, unlike existing approaches, (iii) the potential beneficial factor of proposed system is its response time which is successful in maintaining increasingly consistency irrespective of the size of the incoming data. Our future work direction will be to further enhance the model and address more quality issues in data, which is not considered in this paper.

References

1. Mayer-Schonberger, V., Ramge, T.: Reinventing Capitalism in the Age of Big Data, Hachette, UK (2018)
2. Li, K.-C., Jiang, H., Zomaya, A.Y.: Big Data Management and Processing. CRC Press, Boca Raton (2017)
3. Marr, B., Data, B.: Using SMART Big Data, Analytics and Metrics To Make Better Decisions and Improve Performance. Wiley, New York (2015)
4. Matei, S., Rughinis, C., Rughinis, R.: Big data, old users, personal worlds: a survey of challenges and resistance to big data analytics in the EU. In: 2017 21st International Conference on Control Systems and Computer Science (CSCS), Bucharest, pp. 175–181 (2017)
5. Biliri, E., et al.: Big data analytics in public safety and personal security: challenges and potential. In: 2017 International Conference on Engineering, Technology and Innovation (ICE/ITMC), Funchal, pp. 1382–1386 (2017)
6. Darwish, T.S.J., Abu Bakar, K.: Fog based intelligent transportation big data analytics in the internet of vehicles environment: motivations, architecture, challenges, and critical issues. IEEE Access **6**, 15679–15701 (2018)
7. Kim, Y.H., Huh, E.N.: Towards the design of a system and a workflow model for medical big data processing in the hybrid cloud. In: 2017 IEEE 15th International Conference on Dependable, Autonomic and Secure Computing, 15th International Conference on Pervasive Intelligence and Computing, 3rd International Conference on Big Data Intelligence and Computing and Cyber Science and Technology Congress (DASC/PiCom/DataCom/CyberSciTech), Orlando, FL, pp. 1288–1291 (2017)
8. Chen, M., Yang, J., Zhou, J., Hao, Y., Zhang, J., Youn, C.H.: 5G-smart diabetes: toward personalized diabetes diagnosis with healthcare big data clouds. IEEE Commun. Mag. **56**(4), 16–23 (2018)
9. Plageras,A.P., et al.: Efficient Large-scale medical data (eHealth Big Data) analytics in Internet of Things. In: 2017 IEEE 19th Conference on Business Informatics (CBI), Thessaloniki, pp. 21–27 (2017)
10. Setiawan, A.W., Utami, N., Mengko, T.R., Indrayanto, A.: Implementation of electronic medical record in community health center towards medical big data analytics application. In: 2014 International Conference on Electrical Engineering and Computer Science (ICEECS), Kuta, pp. 30–35 (2014)
11. Chandru, A.S., Seetharam, K.: Reviewing the techniques of disease detection and classification from the challenging medical data. Int. J. Comput. Appl. **125**(6) (20150
12. Shafqat, S., Kishwer, S., Rasool, R.U., Qadir, J., Amjad, T., Ahmad, H.F.: Big data analytics enhanced healthcare systems: a review. J. Supercomput. 1–46 (2018)
13. Chen, M., Hao, Y., Hwang, K., Wang, L., Wang, L.: Disease prediction by machine learning over big data from healthcare communities. IEEE Access **5**, 8869–8879 (2017). https://doi.org/10.1109/access.2017.2694446
14. Zhong, H., Xiao, J.: Enhancing health risk prediction with deep learning on big data and revised fusion node paradigm. Sci. Program. **2017**, 18 (2017)

15. Chrimes, D., Zamani, H.: Using distributed data over HBase in big data analytics platform for clinical services. In: Computational and Mathematical Methods in Medicine (2017)
16. Garattini, C., Raffle, J., Aisyah, D.N., Sartain, F., Kozlakidis, Z.: Big data analytics, infectious diseases and associated ethical impacts. Philos. Technol., 1–17 (2017)
17. García-Magariño, I., Lacuesta, R., Lloret, J.: Agent-based simulation of smart beds with Internet-of-Things for exploring big data analytics. IEEE Access **6**, 366–379 (2018)
18. Sarkar, B.K.: Big data for secure healthcare system: a conceptual design. Complex Intell. Syst. **3**(2), 133–151 (2017)
19. Tawalbeh, L.A., Mehmood, R., Benkhlifa, E., Song, H.: Mobile cloud computing model and big data analysis for healthcare applications. IEEE Access **4**, 6171–6180 (2016)
20. Zhang, Y., Zhang, L., Oki, E., Chawla, N.V., Kos, A.: IEEE access special section editorial: big data analytics for smart and connected health. IEEE Access **4**, 9906–9909 (2016)
21. Wu, P.Y., Cheng, C.W., Kaddi, C.D., Venugopalan, J., Hoffman, R., Wang, M.D.: Omic and electronic health record big data analytics for precision medicine. IEEE Trans. Biomed. Eng. **64**(2), 263–273 (2017)
22. Viceconti, M., Hunter, P., Hose, R.: Big data, big knowledge: big data for personalized healthcare. IEEE J. Biomed. Health Inform. **19**(4), 1209–1215 (2015)
23. Zhang, Y., Qiu, M., Tsai, C.W., Hassan, M.M., Alamri, A.: Health-CPS: healthcare cyber-physical system assisted by cloud and big data. IEEE Syst. J. **11**(1), 88–95 (2017)
24. Srinivasan, U., Arunasalam, B.: Leveraging big data analytics to reduce healthcare costs. IT Prof. **15**(6), 21–28 (2013)
25. Andreu-Perez, J., Poon, C.C.Y., Merrifield, R.D., Wong, S.T.C., Yang, G.Z.: Big data for health. IEEE J. Biomed. Health Inform. **19**(4), 1193–1208 (2015)
26. Das, N., Das, L., Rautaray, S.S., Pandey, M.: Big data analytics for medical applications. Int. J. Mod. Educ. Comput. Sci. **10**(2), 35 (2018)
27. Wang, Yi., Kung, L., Byrd,T.A.: Big data analytics: understanding its capabilities and potential benefits for healthcare organizations. Technol. Forecasting Soc. Change **126**, 3–13 (2018)
28. Prasad, S.T., Sangavi, S., Deepa, A., Sairabanu, F., Ragasudha, R.: Diabetic data analysis in big data with predictive method. In: 2017 International Conference on Algorithms, Methodology, Models and Applications in Emerging Technologies (ICAMMAET), Chennai, pp. 1–4 (2017)
29. Census. http://www.census.gov/data.html. Accessed 06 June 2018
30. Physionet. http://www.physionet.org/physiobank/database/, Accessed 06 June 2018
31. Opendata. https://registry.opendata.aws/. Accessed 06 June 2018
32. DataGov. https://data.gov.uk/dataset/f259a7b7-6c97-4632-93d0-38b93afb324f/nhs-continuing-healthcare-activity. Accessed 06 June 2018
33. DataGovHealthcare. https://data.gov.uk/search?q=healthcare. Accessed 06 June 2018
34. Butt, N., Shan, J.: CyberCare: a novel electronic health record management system. In: 2016 IEEE First International Conference on Connected Health: Applications, Systems and Engineering Technologies (CHASE), Washington, DC, pp. 326–327 (2016)

Indirect Optimal Approach Applied to H1N1 Spread Through Moroccan Regions

Amine Bouaine[1,2,3](✉) and Mostafa Rachik[3]

[1] Institut Supérieur du Génie Appliqué, IGA Casabalanca, Casabalanca, Morocco
amine.bouaine@gmail.com

[2] Department of Industrial Engineering, École Nationale Supérieure de Mines de Rabat, Rabat, Morocco

[3] Department of Mathematics and Computer Science, Faculty of Sciences Ben M'sik, Hassan II University Casablanca, Casablanca, Morocco

Abstract. This study concerns modeling the H1N1 spread through the twelve official regions of Morocco using discrete spatiotemporal epidemic model including many subsystems which describe thoroughly the local characteristics inside each region and take into account the neighborhood impact. Each region has a known neighborhood deducted from the geographic map, the effect of this neighborhood is included in the model as a second process of transmission. We endeavor to reduce infectious individuals of the target area, inaccessible region, through control of certain accessible regions. Hence, we propose indirect optimal control approach in order to reduce infected individuals in the target region with the minimum cost. The optimality system is a two-point boundary value problem, with separated boundary conditions. It's resolved by an iterative method related to Forward-Backward Sweep Method (FBSM). The effectiveness of the proposed approach is examined via numerous numerical illustrations.

Keywords: H1N1 spread · Diffusive models
Discrete spatiotemporal epidemic models · Optimal control

1 Introduction

Despite the heightened and growth improvement of medicine in many areas and at many levels as the development of antibiotics and vaccines, infectious diseases still contribute significantly to deaths worldwide and have caused several epidemics, leaving behind them not only millions of dead people but also several socioeconomic consequences. The prevalence of disease in a population is

R. Silhavy et al. (Eds.): CoMeSySo 2018, AISC 860, pp. 22–33, 2019.
https://doi.org/10.1007/978-3-030-00184-1_3

often described by an SIR model where the population is subdivided into three different classes: susceptible, infected and recovered [1,2].

Optimal control theory has successful applications in many disciplines, namely, economics, environment, management, engineering, etc. [3–5]. In this context, control of many contagious diseases is investigated using optimal control theory such as tuberculosis [6], cancer [7] and HIV [8,9].

In this study, we attempt to present adequate modeling for H1N1 spread through Moroccan regions using a new approach as explained in the rest of this paper. We develop discrete spatiotemporal H1N1 model for many reasons. First, the statistics data concerning H1N1 infection are collected in discrete-time. Second, we may avoid some mathematical complexities like adequate topology, regularity of solutions, and the choice of state space. Third, to simulate continuous models, we are in need to discretize the concerned model, while discretization may not reflect exactly the continuous one.

Control of all susceptible or infected regions is difficult or impossible, especially in the case of developing countries due to the shortage of materials and human resources. In addition to this, medicines can't be provided and the medical team can't have access to all infected areas for several reasons such as: logistic obstacles (1) the deteriorated state of infrastructures or lack of transportation; medical equipment (2) concerning some epidemics, vaccines and medical equipment are too expensive especially for a huge population in low-income regions; unsafety (3) access to certain insecure regions is a major concern; climatic and geographical conditions (4) for example some mountainous regions are extremely cold and snowy in the winter, whereas the weather is too hot in the summer.

To cope with this problematic, we propose distributed H1N1 infection model including many local subsystems which describe thoroughly the local characteristics of H1N1 prevalence inside each region and different interactions relying each region to its neighbors. We aim to reduce infected people in the target region where the control isn't possible, in other words, the control isn't permitted in all regions but is only limited to certain infectious regions due to the reasons above. One the one hand, restricted control means the control is limited to regions that belong to the controlled ones, on the other hand, the control is indirect, we are interested in acting positively on the target region by applying optimal control strategies in the controlled regions. The main objective is not only modeling H1N1 infection and prediction of its evolution but is also fighting the spread of H1N1 prevalence through development appropriate economical and medical policies.

The advanced regionalization in Morocco started in September 2015 and provided for the division of the country into 12 regions [10], see Fig. 1. This regional configuration was made according to criteria based on the principles of efficiency, consistency, proportionality and balance, but also accessibility and proximity. In Fig. 1, the new regional re-partition of Morocco is presented, in addition to this the names of regions and the number of their inhabitants [10] are shown in Table 1. Unfortunately, the infrastructure in a deplorable state and the roads are usually blocked when it snows, the region also suffers from a lack

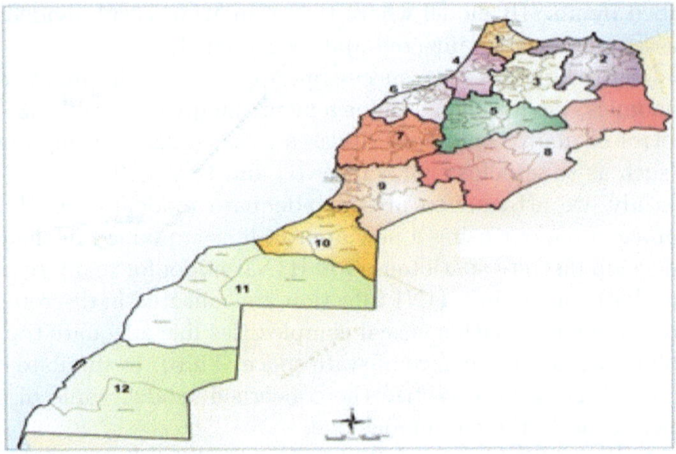

Fig. 1. Repartition of Moroccan regions

of hospitals and medical facilities. For these reasons control of this region is still difficult and too expensive.

The target region Beni Mellal-Khenifra has five regions in its neighborhood as shown in Fig. 1. The neighbors of this region are as follows: region (3) Fez-Meknes; region (4) Rabat-Sale-Kenitra; region (6) Casablanca-Settat; region (7) Marrakech-Safi; region (8) Draa-Tafilat. We take into account the effective contact between individuals from the target region (5) and individuals from other regions of its neighborhood and we think about controlling the accessible regions of the neighborhood in order to reduce the number of infected people in the concerned region (5). In fact, we are interested in controlling the regions (4) and (6) in such a way to act on the region (5). The region (4) and region (6) respectively contain the political capital Rabat and the economic capital Casablanca of the kingdom; these latter are financially rich, hospitals are well equipped and have well-trained medical teams. The climate is very favorable throughout the year, the infrastructures are good compared to those in region (5).

This paper is organized as follows. In the next section, we present a discrete H1N1 model taking into consideration the impact of the neighborhood regions, and we incorporate controls in the subsystems describing H1N1 evolution inside the accessible regions. A new approach called indirect optimal control under restricted control formulated, developed and analyzed in Sect. 3. In Sect. 4, the models are simulated, compared, commented and interpreted to assess the effectiveness of the optimal controls. To end, achievements are summarized and future work is explored in Sect. 6.

2 Modeling H1N1 Spread Through Moroccan Regions

We consider a spatiotemporal SIR epidemic model for modeling H1N1 spread. The population in each region is divided into three disease-state compartments: susceptible individuals (S_k^r), people present in region (r) who can catch the disease at time k, infectious (infected) individuals (I_k^r), people who have the disease in region (r) at time k and can transmit the disease, recovered individuals (R_k^r), people present in region (r) who have recovered from the disease at time k. We assume that an individual can be infected only through contacts with infectious individuals and that immunity is permanent. There are 12 region (r) considered in this research, $r \in [1, 12]$. For example, if $r = 7$ the corresponding region (7) is Marrakech-Safi, see Fig. 1 and Table 1.

Table 1. Initial population of each region

Index	The initial population	Related region (r)	Values (millions)
1	N_0^1	Tangiers-Tetouan	2.83
2	N_0^2	Oriental and Rif	2.434
3	N_0^3	Fez-Meknes	4.22
4	N_0^4	Rabat-Sale-Kenitra	4.272
5	N_0^5	Beni Mellal-Khenifra	2.611
6	N_0^6	Casablanca-Settat	6.862
7	N_0^7	Marrakech-Safi	4.108
8	N_0^8	Draa-Tafilat	1.392
9	N_0^9	Souss-Massa	2.475
10	N_0^{10}	Guelmim-Oued Noun	0.429
11	N_0^{11}	Laayoune-Saguia al Hamra	0.364
12	N_0^{12}	Ed Dakhla-Oued ed Dahab	0.152

N_k^r is the population of region (r) at time k. Not to mention that, N_0^r is the initial population and equal to the number of inhabitants of region (r). The initial population of each region (r) is presented in Table 1. We consider two processes of disease transmission:

Inside the Same Region (First Process): disease transmission is assumed to occur between individuals of the same region, then β_{rr} stands for effective contact rate between susceptible (S_k^r) and infected (I_k^r) of the same region at any time k, where $r \in [1, 12]$.

With the Neighbor Regions (Second Process): disease transmission is assumed to occur between susceptible of region (r) and infected of its neighbor region (j), then β_{rj} stands for effective contact rate between susceptible (S_k^r) of the region (r) and infected (I_k^j) of the region (j) at any time k, where $r \in [1, 12]$ and $j \in v(r)$.

Table 2. List of neighbors and β_{rj}

Region (r)	$v(r)$	β_{rj}
1	2, 3, 4	$\beta_{12}, \beta_{13}, \beta_{14}$
2	1, 3, 8	$\beta_{21}, \beta_{23}, \beta_{28}$
3	1, 2, 4, 5, 8	$\beta_{31}, \beta_{32}, \beta_{34}, \beta_{35}, \beta_{38}$
4	1, 3, 5, 6	$\beta_{41}, \beta_{43}, \beta_{45}, \beta_{46}$
5	3, 4, 6, 7, 8	$\beta_{53}, \beta_{54}, \beta_{56}, \beta_{57}, \beta_{58}$
6	4, 5, 7	$\beta_{64}, \beta_{65}, \beta_{67}$
7	5, 6, 8, 9	$\beta_{75}, \beta_{76}, \beta_{78}, \beta_{79}$
8	2, 3, 5, 7, 9	$\beta_{82}, \beta_{83}, \beta_{85}, \beta_{87}, \beta_{89}$
9	7, 8, 10	$\beta_{97}, \beta_{98}, \beta_{910}$
10	9, 11	$\beta_{109}, \beta_{1011}$
11	10, 12	$\beta_{1110}, \beta_{1112}$
12	11	β_{1211}

$v(r)$ stands for the neighbors of region (r), see Table 2. As an example, for $r = 8$, $v(8) = \{2, 3, 5, 7, 9\}$ that means the neighbors of region (8) are regions 2, 3, 5, 7 and 9.

The Table 2 is based on the geographic border of each region, see Fig. 1. The country is bounded by the Atlantic Ocean from the west side and the Mediterranean Sea on the north side. The land borders are closed on the east side, while the south region (12) is the least populated and the least dense. Further, the contact with the southern country is neglected.

In morocco, annual rate of population growth has reached 0.0125, that signifies the daily rate of population growth is just 0.000034. In real models, we take into account generally days as unit of time. For this reason daily rate of population growth is neglected. Therefore, birth rate and death rate would not be included in the concerned model. Furthermore, the population of each region will be considered constant.

The following parameters will be also introduced

γ is the recovery rate;
d is the disease induced death rate;
T is the terminal time.

The spatiotemporal epidemic model is described as follows

$\forall r \in [1; 12]\ \forall k \in [0; T-1]$.

$$\begin{cases} S_{k+1}^r &= S_k^r - \beta_{rr} S_k^r I_k^r / N_k^r - \sum_{j \in v(r)} \beta_{rj} S_k^r I_k^j / N_k^r - \pi_D(r) u_k^r S_k^r \\ I_{k+1}^r &= (1 - \gamma - d) I_k^r + \beta_{rr} S_k^r I_k^r / N_k^r + \sum_{j \in v(r)} \beta_{rj} S_k^r I_k^j / N_k^r \\ R_{k+1}^r &= R_k^r + \gamma I_k^r + \pi_D(r) u_k^r S_k^r \end{cases} \tag{1}$$

where $r \in [1;12]$, the first process described above concerning H1N1 transmission is given by $\beta_{rr} S_k^r I_k^r / N_k^r$, whereas the second process of H1N1 transmission is given by $\sum_{j \in v(r)} \beta_{rj} S_k^r I_k^j / N_k^r$. On the one hand, the control isn't possible in the target region $G = \{5\}$ for many reasons explained above, on the other hand the control is recommended in Rabat-Sale-Kenitra the region (4) and Casablanca-Settat region (6) thanks to favorable conditions in term of abundant medical equipment, human resources well formed, adequate climatic and geographical conditions, quality of infrastructures, appropriate financial circumstances and so on. Then, we define the controlled regions $D = \{4; 6\}$, where the control is possible and we introduce the functions

$$\pi_D : \begin{array}{ll} [1;12] & \longrightarrow \{0,1\} \\ r & \longrightarrow \pi_D(r) \end{array} \tag{2}$$

If $r \in D$ then $\pi_D(r) = 1$, the control will appear in the model above, else $\pi_D(r) = 0$ the control will disappear. By and large, the control appears only in the subsystems which imitate H1N1 spread inside regions (4) or (6).

$$\pi_G : \begin{array}{ll} [1;12] & \longrightarrow \{0,1\} \\ r & \longrightarrow \pi_G(r) \end{array} \tag{3}$$

If $r \in G$ then $\pi_G(r) = 1$ else $\pi_G(r) = 0$.

u_k^r is the percentage of susceptible individuals being vaccinated in the region (r) at time k, we note that if $r \in D$ then $u_k^r S_k^r$ will move from the susceptible compartment to the recovered compartment at time k.

3 Indirect Optimal Approach Under Restricted Control

First, restricted control means the control is limited to region belonging to the controlled regions D, in this case just susceptible S_k^4 and S_k^6 can benefit from the control. Second, the control is indirect because we are interested in applying controls in the regions (4) and (6) in order to act on the target region (5).

3.1 Indirect Optimal Control Approach

Our goal is to reduce infected individuals I_k^5 present in the region (5) during the time steps $k = 0$ to T and also to minimize the costs of controls applied in the region (4) and (6). To simplify, we assume that the cost of administering the controls are quadratic. Then, the objective functional is presented as follows

$$J^5(U) = -\sum_{k=0}^{T} A_k^5 I_k^5 - \sum_{k=0}^{T-1} \sum_{j=1}^{12} \pi_D(j) B_k^j \left[u_k^j \right]^2 \tag{4}$$

where A_k^5 and B_k^j are the cost coefficients and depend respectively on the relative importance of I_k^5 and u_k^j. These positive constants A_k^5 and B_k^j represent desired weight related to the benefit and the cost. We are looking for optimal control U^* which verifies

$$J^5(U^*) = \max_{U \in U_{ad}} J^5(U) \tag{5}$$

where U_{ad} is the set of admissible controls defined by

$$U_{ad} = \left\{ U = \begin{bmatrix} u_0^4 \, u_1^4 \, ... \, u_{T-1}^4 \\ u_0^6 \, u_1^6 \, ... \, u_{T-1}^6 \end{bmatrix} ; 0 \leq u_{min}^j \leq u_k^j \leq u_{max}^j \leq 1; \; k \in [0; T-1] \, ; \; j \in D \right\} \tag{6}$$

For the optimal control given by (4) along with the state equations (1), there exists a control $U^* \in U_{ad}$ such that

$$J^5(U^*) = \max_{U \in U_{ad}} J^5(U)$$

Proof. There are a finite number of time steps and the coefficients of the state equations (1) are bounded then $(S_0^r, S_1^r, .., S_T^r)$, $(I_0^r, I_1^r, .., I_T^r)$, and $(R_0^r, R_1^r, .., R_T^r)$ are uniformly bounded for all $U \in U_{ad}$. As a result of that $J^5(U)$ is uniformly bounded for all U in the control set U_{ad}. Based on [[11], Theorem 1], we deduce that $\sup_{U \in U_{ad}} J^5(U)$ is achieved and the sufficient conditions are satisfied for the existence of an optimal control.

3.2 Characterization of the optimal control

We apply the Pontryagin's maximum principle in discrete time, the system (1) will be converted into a problem of minimizing the Hamiltonian, at each step k, defined by

$$H_k^r = -A_k^5 I_k^5 - \sum_{j=1}^{12} \pi_D(j) B_k^j \left[u_k^j\right]^2 + \sum_{j=1}^{3} \theta_{j,k+1}^r f_{j,k+1}^r \tag{7}$$

where $f_{j,k+1}^r$ is the right side of the difference equation concerning the region (r) of the jth state variable at time step $k+1$.

θ_j is the adjoint function given by

$$\theta_j = \begin{bmatrix} \theta_j^1 \\ . \\ . \\ \theta_j^r \\ . \\ . \\ \theta_j^{12} \end{bmatrix} = \begin{bmatrix} \theta_{j,1}^1 \, ... \, \theta_{j,T}^1 \\ . \\ . \\ \theta_{j,1}^r \, ... \, \theta_{j,T}^r \\ . \\ . \\ \theta_{j,1}^{12} \, ... \, \theta_{j,T}^{12} \end{bmatrix} \tag{8}$$

where $j \in \{1, 2, 3\}$.

Given an optimal control $U^* \in U_{ad}$ and solutions S_k^{r*}, I_k^{r*}, and R_k^{r*}, there exist adjoint functions θ_1, θ_2 and θ_3 which satisfy

$$\begin{cases} \theta_{1,k}^r = \left(\sum_{j\in v(r)} \beta_{rj} I_k^j / N_k^r + \beta_{rr} I_k^r / N_k^r\right)\left(\theta_{2,k+1}^r - \theta_{1,k+1}^r\right) \\ \qquad + \theta_{1,k+1}^r \left(1 - \pi_D(r) u_k^r\right) + \theta_{3,k+1}^r \pi_D(r) u_k^r \\ \theta_{2,k}^r = -\pi_G(r) A_k^r + \beta_{rr} S_k^r / N_k^r \left(\theta_{2,k+1}^r - \theta_{1,k+1}^r\right) + \theta_{2,k+1}^r (1 - \lambda - d) + \gamma \theta_{3,k+1}^r \\ \theta_{3,k}^r = \theta_{3,k+1}^r \end{cases} \tag{9}$$

where $r \in [1;12]$ and $k \in [0; T-1]$, with the transversality conditions at terminal time T

if $r = 5$ then $\theta_{1,T}^5 = 0$, $\theta_{2,T}^5 = -A_T^5$ *and* $\theta_{3,T}^5 = 0$.
if $r \in [1;12] \setminus \{5\}$ then $\theta_{1,T}^r = 0$, $\theta_{2,T}^r = 0$ *and* $\theta_{3,T}^r = 0$.

In addition to this, the optimal control is given by

$$U^* = \begin{bmatrix} u_0^{4*} \; u_1^{4*} \; ... \; u_{T-1}^{4*} \\ u_0^{6*} \; u_1^{6*} \; ... \; u_{T-1}^{6*} \end{bmatrix} \tag{10}$$

further

$$u_k^{j*} = min\left[u_{max}^j, max\left(u_{min}^j, S_k^j\left(\theta_{1,k+3}^j - \theta_{1,k+1}^j\right)/2B_k^j\right)\right] \tag{11}$$

where $j \in D$ and $k \in [0; T-1]$

Proof. The corresponding hamiltonian is given by

$$\begin{aligned} H_k^r = & -A_k^5 I_k^5 - \textstyle\sum_{j=1}^{12} \pi_D(j) B_k^j \left[u_k^j\right]^2 \\ & + \theta_{1,k+1}^r \left(S_k^r - \beta_{rr} S_k^r I_k^r / N_k^r - \textstyle\sum_{j\in v(r)} \beta_{rj} S_k^r I_k^j / N_k^r - \pi_D(r) u_k^r S_k^r\right) \\ & + \theta_{2,k+1}^r \left((1-\gamma-d) I_k^r + \beta_{rr} S_k^r I_k^r / N_k^r + \textstyle\sum_{j\in v(r)} \beta_{rj} S_k^r I_k^j / N_k^r\right) \\ & + \theta_{3,k+1}^r \left(R_k^r + \gamma I_k^r + \pi_D(r) u_k^r S_k^r\right) \end{aligned} \tag{12}$$

According to the Pontryagin's maximum principle in discrete time, see [12], the equations for the adjoint variables, are

$$\theta_{1,k}^r = \frac{\partial H_k^r}{\partial S_k^r};\ \theta_{2,k}^r = \frac{\partial H_k^r}{\partial I_k^r};\ \theta_{3,k}^r = \frac{\partial H_k^r}{\partial R_k^r} \tag{13}$$

The optimal control u_k^{j*} is obtained by

$$\frac{\partial H_k^r}{\partial u_k^j} = 0 \ for \ k = 0, 1.., T-1 \ and \ j = 4, 6. \tag{14}$$

Subject to the lower and upper bounds for u_k^j, the characterization becomes

$$u_k^{j*} = min\left[u_{max}^j, max\left(u_{min}^j, S_k^j\left(\theta_{3,k+1}^j - \theta_{1,k+1}^j\right)/2B_k^j\right)\right] \tag{15}$$

4 Numerical Illustrations: Application to Target Beni Mellal-Khenifra Region

In this section, we implement improved numerical method for solving this problem. This system consists of the state system, adjoint variables, initial and final time conditions, and the controls characterization. There are initial conditions for the state variables and terminal conditions for the adjoints. That is, the optimality system is a two-point boundary value problem, with separated boundary conditions at time step $k = 0$ and $k = T$. So, we make use of Forward-Backward Sweep Method (FBSM) to solve the optimality system by an iterative method with forward solving of the state system followed by backward solving of the adjoint system. We start with an initial guess for the controls at the first iteration and before the next iteration, we update the controls by using the values of state and co-state variables obtained at the previous steps.

Our study covers the 12 regions of Morocco, the actual population of each region is considered as initial population, each region has a known neighborhood deducted from the map, the effect of this neighborhood is reflected in the expression of the second process.

At the initial state, the disease appears in all regions and the infected are not yet recovered. Unfortunately, population-based epidemiological data are extremely limited in Morocco. Hence, we are forced to make some initial assumptions to simulate the concerned model. The rest of parameters and desired weight factors are assumed as follows: $\beta_{rr} = 0.0287$, $\beta_{rj} = 0.1095$, $d = 0.0063$, $\gamma = 0.1$, $u^4_{min} = 0.2$, $u^4_{max} = 0.8$, $u^6_{min} = 0.3$, $u^6_{max} = 0.7$, $B^j_k = 10,000$, and $A^5_k = 500$. Note that initial infected and initial recovered are shown in Table 3, whereas initial susceptible are derived from $S^r_0 = N^r_0 - I^r_0$.

Table 3. Initial state

Region(r)	Initial Infected I^r_0	Initial Population N^r_0	Initial Recovered R^r_0
r = 1	20,500	2.83×10^6	0
r = 2	17,000	2.434×10^6	0
r = 3	5,000	4.272×10^6	0
r = 4	30,000	4.272×10^6	0
r = 5	50,000	2.611×10^6	0
r = 6	28,000	6.862×10^6	0
r = 7	23,300	4.108×10^6	0
r = 8	15,600	1.392×10^6	0
r = 9	34,200	2.475×10^6	0
r = 10	10,000	0.429×10^6	0
r = 11	5,800	0.364×10^6	0
r = 12	4,600	0.152×10^6	0

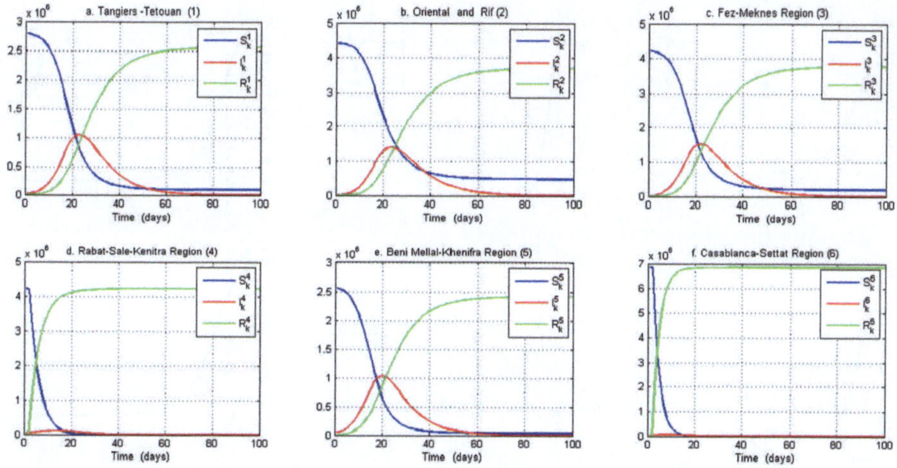

Fig. 2. Evolution of the state variables in the six first regions (1)–(6) in case of indirect control.

Figure 2 shows the evolution of susceptible, infected and recovered of each region (1) to (6). On the whole, the infectious individuals grow rapidly until the top and decrease exponentially until disappearance after duration of treatment. We can observe the positive impact of indirect optimal control on the controlled regions $D = \{4; 6\}$, the infected of Rabat-Sale-Kenitra region (4) and Casablanca-Settat region (6) respectively don't exceed 300,000 and 200,000 individuals.

Duration of the appearance of infectious disease is 70 days in regions Tangiers-Tetouan (1), Oriental and Rif (2), and Fez-Meknes (3), 65 days in the target region Beni Mellal-Khenifra (5), 30 days in the controlled region Rabat-Sale-Kenitra (4), and 20 days in the controlled region Casablanca-Settat (6).

In Fig. 3, we can observe the evolution of susceptible, infected and recovered of each region (7) to (12). Similarly to the first six regions, the infectious individuals increase at the first until the top and decrease until disappearance after duration of treatment. Duration of the appearance of infectious disease is 70 days in regions Marrakech-Safi (7), Souss-Massa (9), Laayoune-Saguia al Hamra (11), and Ed Dakhla-Oued ed Dahab (12), 55 days in region Draa-Tafilat (8) and region Guelmim-Oued Noun (10).

The system evolves in space and time, so for a clear presentation and an overview of the evolution of the system over its coordinates on which it depends we use three-dimensional simulations. Figure 4 gives us the opportunity to observe evolution of susceptible and infected over all Moroccan regions and from the initial time till the terminal time T. We remark that the susceptible individuals stabilizes in a fixed value after 40 days, the infected individuals tend towards zero after 70 days in the worst case. In addition to this, in Fig. 4(b) we can observe two gaps for $r = 4$ and $r = 6$, which confirm the powerful effect of the controls on the accessible regions.

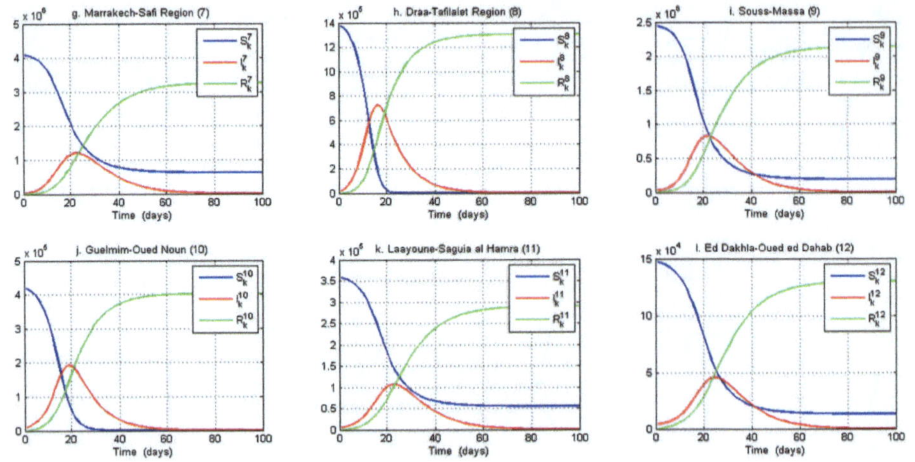

Fig. 3. Evolution of the state variables in the six last regions (7)–(12) in case of indirect control.

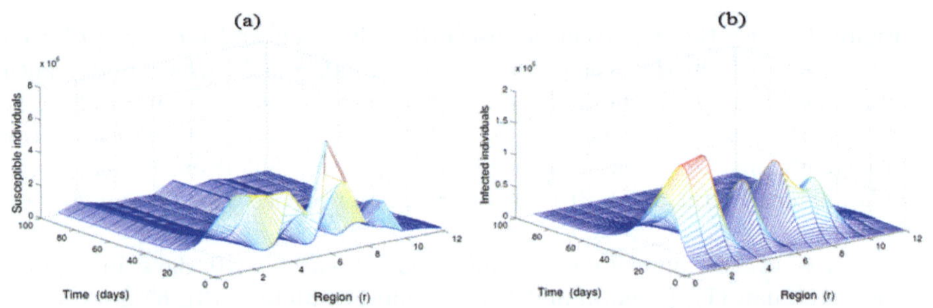

Fig. 4. (a) Spatiotemporal evolution of susceptible individuals. (b) Spatiotemporal evolution of infected individuals

References

1. Chen, Y., Zou, S., Yang, J.: Global analysis of an SIR epidemic model with infection age and saturated incidence. Nonlinear Anal. RWA **30**, 16–31 (2016)
2. Takeuchi, Y., Ma, W., Beretta, E.: Global asymptotic properties of a delay SIR epidemic model with finite incubation times. Nonlinear Anal. RWA **42**, 931–947 (2000)
3. Bouaine, A., Rachik, M.: Adjustment submersible cages for optimal fish farming: improved spatiotemporal Lotka-Volterra model for n species in partitioned area. Int. J. Dyn. Control (2017). https://doi.org/10.1007/s40435-017-0329-3
4. Elouarrak, H., Bouaine, A., Rachik, M., Mestari, M.: Trajectory planning for a four-wheel robot using decomposition-coordination principle. In: 2015 Third World Conference on Complex Systems (WCCS) (2016). https://doi.org/10.1109/ICoCS.2015.7483265

5. Bouaine, A., Rachik, M.: Advanced approach for observability of distributed systems using internal pointwise sensor. In: Advances in Intelligent Systems and Computing, vol. 662, pp. 298–309. Springer, Cham (2018)
6. Silva, C.J., Torres, D.F.M.: Optimal control for a tuberculosis model with reinfection and post-exposure interventions. Math. Biosci. **244**(2), 154–164 (2013)
7. Ledzewicz, U., Schättler, H.: Antiangiogenic therapy in cancer treatment as an optimal control problem. SIAM J. Control Optim. **46**(3), 1052–1079 (2007)
8. Yang, Y., Xiao, Y., Wu, J.: Pulse HIV vaccination: feasibility for virus eradication and optimal vaccination schedule. Bull. Math. Biol. **75**(5), 725–751 (2013)
9. Bouaine, A., Rachik, M.: Optimal control of a discrete HIV model with two classes of antiretroviral therapy. Int. J. Sci. Res. **5**(2), 1670–1677 (2016)
10. Commission consultative de la régionalisation, Rapport sur la régionalisation avancée, Royaume du Maroc (2015). http://www.regionalisationavancee.ma/PDF/Rapport/Fr/L3_Dev_Eco_So.pdf
11. Dabbs, K.: Optimal control in discrete pest control models. University of Tennessee Honors. Thesis Projects (2010)
12. Pontryagin, L.S., Boltyanskii, V.G., Gamkrelidze, R.V., Mishchenko, E.F.: The Mathematical Theory of Optimal Processes. Wiley, New York (1962)
13. Bouaine, A., Rachik, M.: Modeling the impact of immigration and climatic conditions on the epidemic spreading based on cellular automata approach. Ecol. Inform. **46C**, 36–44 (2018). https://doi.org/10.1016/j.ecoinf.2018.05.004

Masking Operator Based Genetic Algorithm Optimization for Edge Detection in Images

G. Kavitha(✉)

Hindustan Institute of Technology and Science,
Chennai 603 103, Tamil Nadu, India
kavithateam@gmail.com

Abstract. In the paper proposed, a genetic algorithm optimization on masking operator for edge detection is implemented to give an optimized working model. Genetic algorithm helps in finding the operator values by optimizing the parameters to reach the target solution. Berkeley Segmentation Database images and their ground truths are used for experimentation and qualitative evaluation of the proposed model. In addition to other methods available, Genetic Algorithm optimized detection of edge, play a key role in understanding the features of images. Genetic Algorithm optimized Edge detection helps in matching the pattern, and recognizing an object in an accurate way.

Keywords: Edge detection · BSD (Berkeley Segmentation Database) Ground truth · Masking operator · Genetic Algorithm

1 Introduction

Detecting the edges in an image is a challenging task. The basic task in image processing is edge detection. The main objective of the paper is to propose and obtain a perfect masking operator to detect the edges in a better way using Genetic Algorithm approach. There are many techniques proposed and implemented in the literature [1–13] for detecting the edge of an image with accuracy.

2 Methodology

In Image processing, mask is a small matrix or a kernel or a convolution matrix. It is used in edge detection for fine tuning the edges. This is achieved by doing a convolution between an image and a kernel. The images were taken from Berkeley Segmentation Database (BSD). The Original BSD image was fed as an input image and the Ground Truth image was fed as the Expected Ideal Output. These two pair of images were fed into the Genetic Algorithm for getting the desired output. The methodology used for edge detection in the proposed algorithm is masking operator based Genetic Algorithm optimization for edge detection of images. The main objective function of the Genetic Algorithm was to minimize the Mean Square Error. The constraints of the Algorithm was the sum of all elements in masking operator is zero. The nine elements

R. Silhavy et al. (Eds.): CoMeSySo 2018, AISC 860, pp. 34–43, 2019.
https://doi.org/10.1007/978-3-030-00184-1_4

in the operator constitute the genes. Following are the steps for the optimized GA algorithm.

Step 1: Load the input query image and the ideal expected output image
Step 2: Apply GA algorithm for a 3*3 operator mask. The masking operators used are Sobel, Prewitt and Laplacian operator
Step 3: Perform edge detection on the input query image using above masking operators
Step 4: Compare the result obtained with ideal expected output using GA fitness function and update the mask.
Step 5: Repeat Step 3 till stopping criteria
Step 6: Display the result

3 Results and Discussion

The proposed approach has been simulated in MATLAB 2018(a). The image database includes insect and human image. The proposed methodology has been tested on various types of images with different masking operators. The output images obtained gives better results after optimizing using Genetic Algorithm for all the input images considered.

Figures 1 and 2 gives the Sobel operator output, GA optimized Sobel operator output and the 3D axes view for Insect image. Figures 3 and 4 gives the Prewitt operator output, GA optimized Prewitt operator output and the 3D axes view for Insect image. Figures 5 and 6 gives the Laplacian operator output, GA optimized Laplacian operator output and the 3D axes view for Insect image.

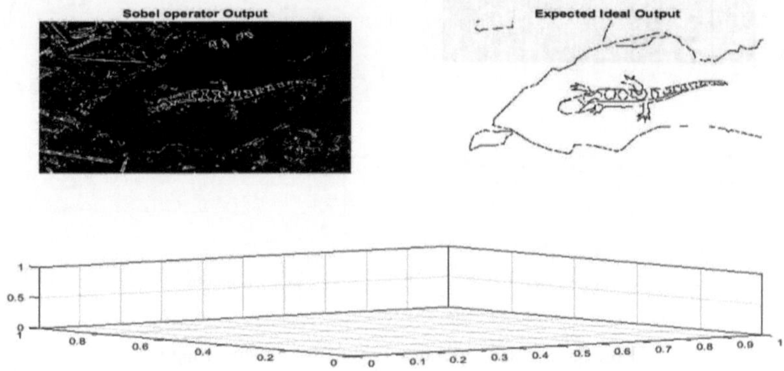

Fig. 1. Sobel operator output for Insect image

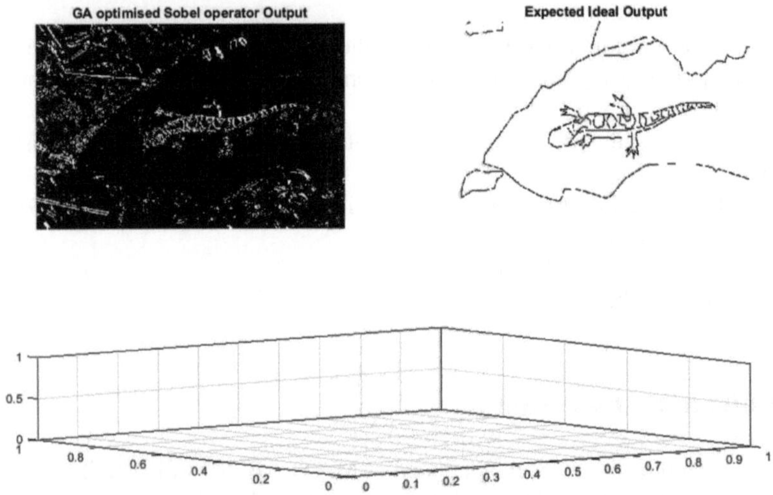

Fig. 2. GA optimized Sobel operator output for Insect image

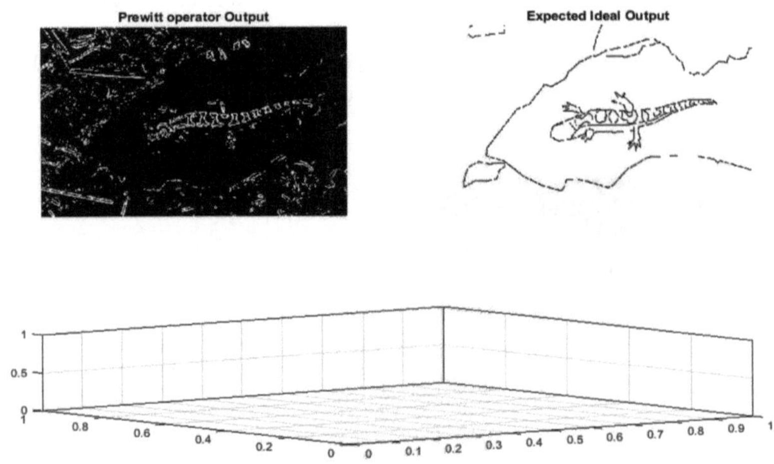

Fig. 3. Prewitt operator output for Insect image

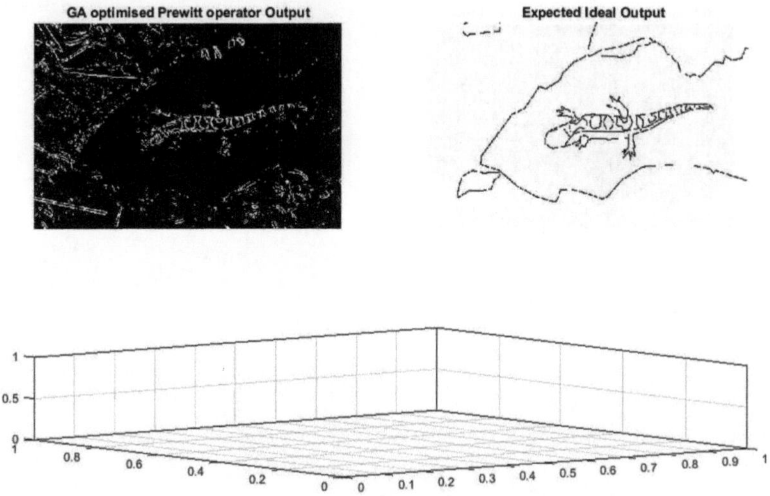

Fig. 4. GA optimized Prewitt operator output for Insect image

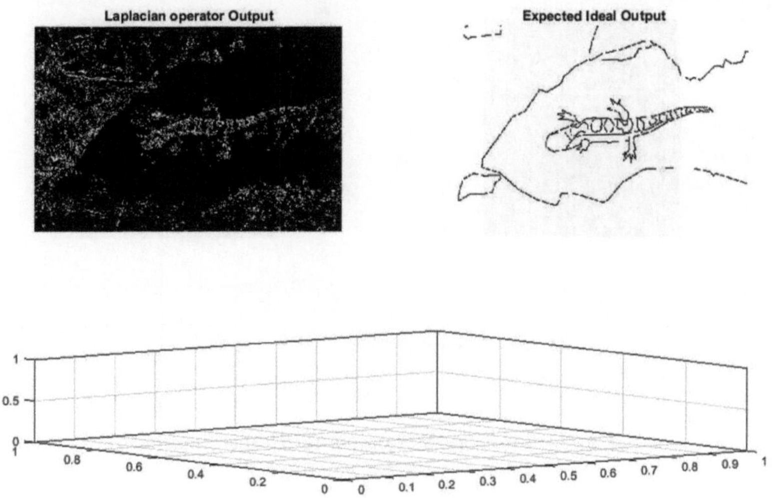

Fig. 5. Lapalcian operator output for Insect image

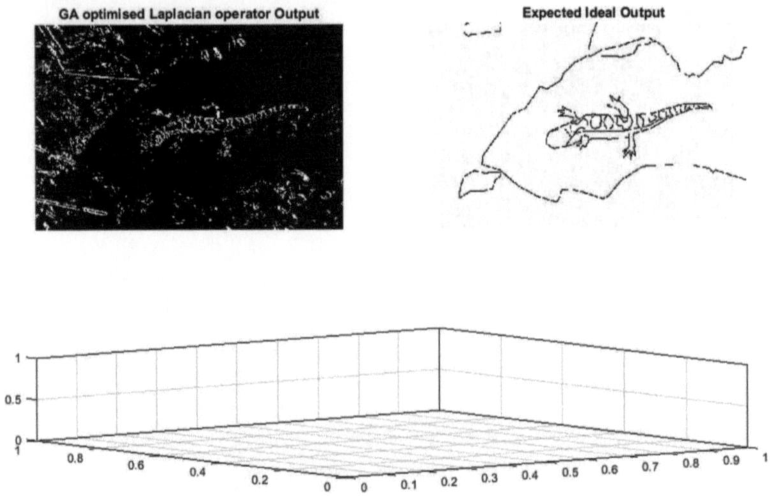

Fig. 6. GA optimized Laplacian operator output for Insect image

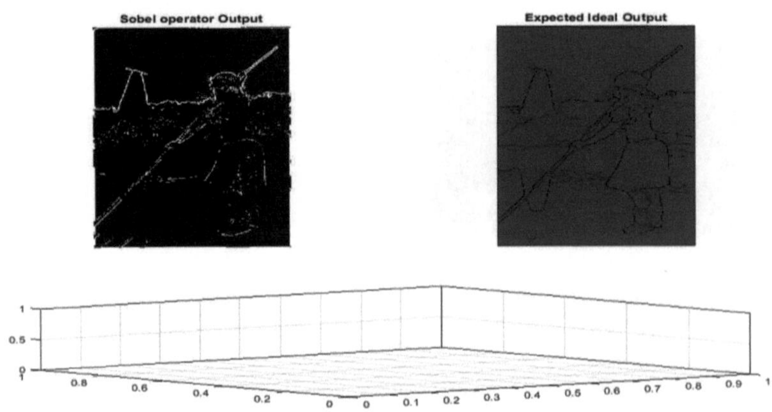

Fig. 7. Sobel operator output for Human image

Figures 7 and 8 gives the Sobel operator output, GA optimized Sobel operator output and the 3D axes view for Human image. Figures 9 and 10 gives the Prewitt operator output, GA optimized Prewitt operator output and the 3D axes view for Human image. Figures 11 and 12 gives the Laplacian operator output, GA optimized Laplacian operator output and the 3D axes view for Human image.

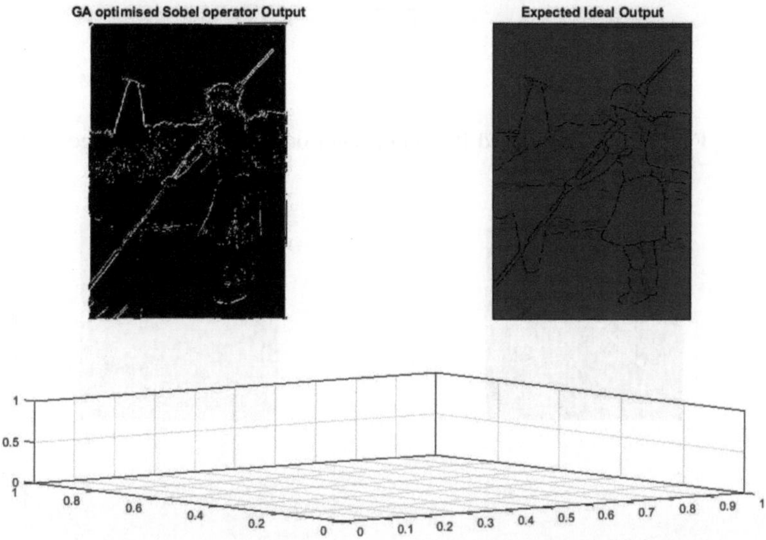

Fig. 8. GA optimized Sobel operator output for Human image

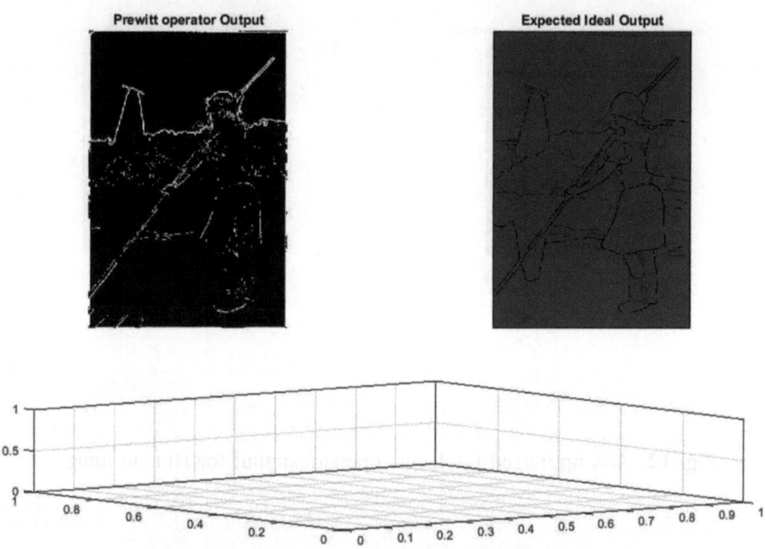

Fig. 9. Prewitt operator output for Human image

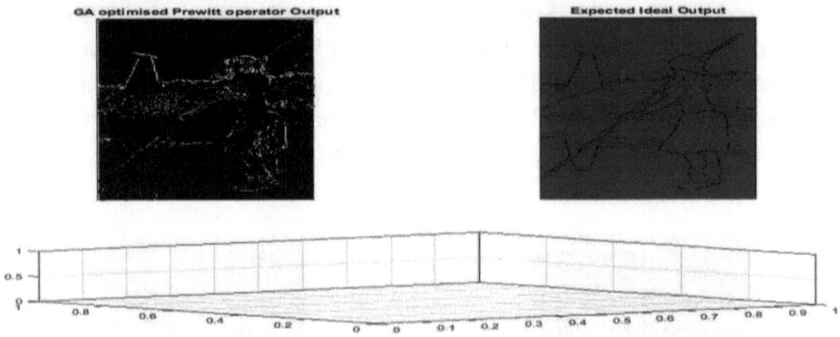

Fig. 10. GA optimized Prewitt operator output for Human image

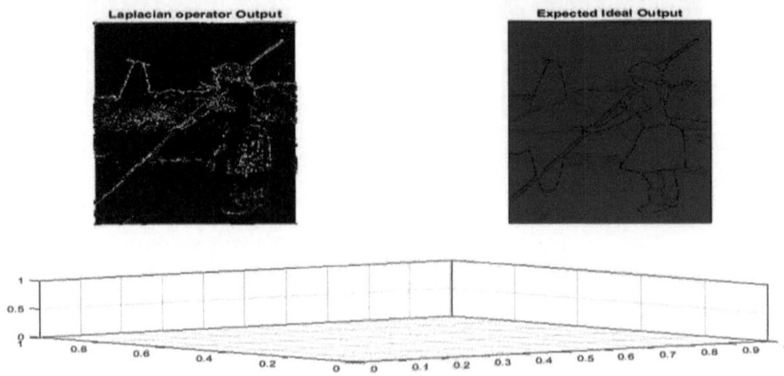

Fig. 11. Laplacian operator output for Human image

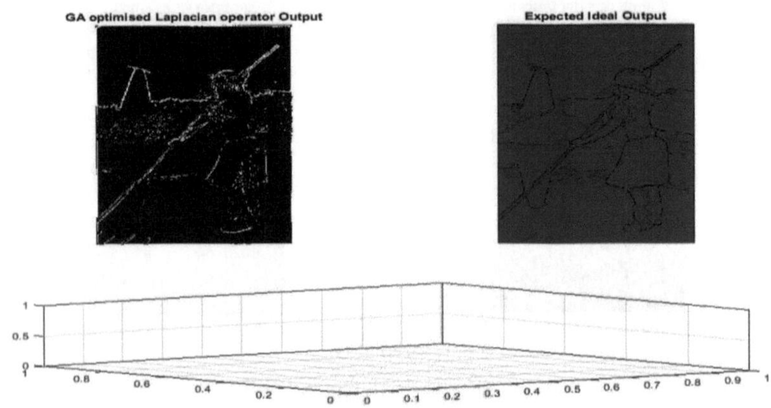

Fig. 12. GA optimized Laplacian operator output for Human image

The Expected mean-square error is calculated for the three different images considered. The peak signal to Noise Ratio denoted by PSNR is also calculated for the input images. For higher PSNR, the quality of the image would be better. $PSNR = 10.\log_{10}(\frac{MAXP_I^2}{MSEr})$, where MAXPI is the maximum pixel value possible in the image.

Table 1 shows GA optimized operator output for Insect image using Sobel, Prewitt and Laplacian operator.

Table 1. GA optimized operator output for Insect image using Sobel, Prewitt and Laplacian operator

Generation	Func-count	GA Optimized best f(x)			Max constraint		
		Sobel	Prewitt	Laplacian	Sobel	Prewitt	Laplacian
1	2400	0.933414	0.926568	0.937332	3.438	0.01846	5.763
2	4600	0.924845	0.928958	0.92842	0.03175	0.02439	0.05382
3	6800	0.927235	0.928796	0.92842	0.002841	0.001146	0.008267
4	9000	0.927235	0.928802	0.92842	0.002841	0.008606	0.008267
5	11200	0.927235	0.927293	0.92842	0.002841	0.01073	0.008267
6	13400	0.927235	0.927293	0.92842	0.002841	0.0008502	0.008267
7	15600	0.927235	–	0.928524	0.002841	–	0.005448
8	17800	0.927235	–	0.928524	0.002841	–	0.005448
9	20000	0.927235	–	0.928524	0.002841	–	0.005448
10	22200	0.927235	–	0.928524	0.002841	–	0.005448

Table 2 shows the calculated values of EMSE and PSNR output value Before and After GA Optimization for Insect image using Sobel, Prewitt and Laplacian operator

Table 2. EMSE and PSNR output value Before and After GA Optimization for Insect image using Sobel, Prewitt and Laplacian operator

Calculated value	Output before GA Optimization for Insect image			Output after GA Optimization for Insect image		
	Sobel	Prewitt	Laplacian	Sobel	Prewitt	Laplacian
EMSE	0.9261	0.9263	0.9241	0.9272	0.9273	0.9285
PSNR	0.7676	0.7658	0.7894	0.7555	0.7549	0.7416

The optimization terminated when the maximum number of generations exceeded. Table 3 shows GA optimized operator output for Human image using Sobel, Prewitt and Laplacian operator.

Table 3. GA optimized operator output for Human image using Sobel, Prewitt and Laplacian operator

Generation	Func-count	GA optimized best f(x)			Max constraint		
		Sobel	Prewitt	Laplacian	Sobel	Prewitt	Laplacian
1	2400	0.297392	0.298864	0.28797	0.2665	0.01122	1.477
2	4600	0.30054	0.292119	0.299229	0.0005967	0.01731	0.005554
3	6800	–	0.291812	0.299229	–	0.005822	0.005554
4	9000	–	0.291809	0.299229	–	0.002523	0.005554
5	11200	–	0.291809	0.29924	–	0.002523	1.775e-05
6	13400	–	0.291791	–	–	0.008283	–
7	15600	–	0.291809	–	–	0.002523	–
8	17800	–	0.291809	–	–	0.002523	–
9	20000	–	0.291809	–	–	0.002523	–
10	22200	–	0.291809	–	–	0.002523	–

Table 4 shows the calculated values of EMSE and PSNR output value Before and After GA Optimization for Human image using Sobel, Prewitt and Laplacian operator.

Table 4. EMSE and PSNR output value Before and After GA Optimization for Human image using Sobel, Prewitt and Laplacian operator

Calculated value	Output before GA Optimization for Flower image			Output after GA Optimization for Flower image		
	Sobel	Prewitt	Laplacian	Sobel	Prewitt	Laplacian
EMSE	0.2999	0.2997	0.3023	0.3005	0.2918	0.2992
PSNR	0.6020	0.5971	0.6823	0.6242	0.3294	0.5809

4 Conclusion

The proposed masking operators using Genetic Algorithm technique is better for all kind of images compared to other classical techniques. By optimizing the operators like Sobel, Prewitt and Laplacian using Genetic Algorithm, a better edge detection is obtained. The outputs were simulated before optimization and after optimization. Optimization of different masking operators have been tried to view the accuracy of edge detection. Future direction is to detect the object in the given input image to classify the image further. The Algorithm can also be further fine-tuned by optimizing the threshold parameters and detecting the edge in image. Furthermore many optimized masking operators can be developed in future for better accuracy.

References

1. Om, P.V.: An optimal fuzzy system for edge detection in color images using bacterial foraging algorithm. IEEE Trans. Fuzzy Syst. **25**(1), 114–127 (2017)
2. Gonzalez, R.C., Woods, R.E.: Digital Image Processing, 3rd edn., pp. 692–794. Prentice Hall, Englewood Cliffs (2013)
3. Gaur, S.B.C., Vajpai, J.: Comparison of edge detection techniques for segmenting car license plates. Spec. Issue Int. J. Comput. Appl. Electron. Inf. Commun. Eng. ICEICE (5), 8–12 (2011).(0975–8887)
4. Basu, M.: Gaussian based edge detection methods-a survey. IEEE Trans. System Man Cybern. **32**(3), 252–260 (2002)
5. Canny, J.: A computational approach to edge detection. IEEE Trans. Pattern Anal. Mach. Intell. **8**(6), 679–687 (1986)
6. Patel, J., et al.: Fuzzy inference based edge detection system using Sobel and Laplacian of Gaussian operators. In: ICWET 2011, pp. 694–697. ACM (2011). ISBN 978-1-4503-0449-8
7. Setayesh, M., et al.: Detection of continuous, smooth and thin edges in noisy images using constrained particle swarm optimization. In: GECCO 2011, Dublin, Ireland, pp. 45–52. ACM (2011). ISBN 978-1-4503-0557-0
8. Alshennawy, A.A., Aly, A.A.: Edge detection in digital images using fuzzy logic technique. World Acad. Sci. Eng. Technol. **51**, 178–186 (2009)
9. Mendhurwar, K., et al.: Edge-detection in noisy images using independent component analysis. ISRN Signal Processing, 9 p. (2011)
10. Wang, S., et al.: Evaluating edge detection through boundary detection. EURASIP J. Appl. Signal Process., Article ID 76278, pp. 1–15 (2006)
11. Bhadauria, H.S., Dewal, M.L.: Comparison of edge detection techniques on noisy abnormal Lung CT image before and after using morphological filter. Int. J. Adv. Eng. Appl. **1**, 272–275 (2010)
12. Maini, R., Aggarwal, H.: Study and comparison of various image edge detection techniques. Int. J. Image Process. (IJIP) **3**, 1–12 (2010)
13. Panda, C.S., Patnaik, S.: Filtering corrupted image and edge detection in restored grayscale image using derivative filters. Int. J. Image Process. (IJIP) **3**, 105–119 (2010)

A Novel Design and Implementation of Imaging Chip Using AXI Protocol for MPSOC on FPGA

H. R. Archana(✉) and K. S. Vasundara Patel

Department of Electronics and Communication Engineering,
BMS College of Engineering, Bengaluru, India
hraarchana@gmail.com

Abstract. The medical images are the main source for analysis of diseases. These images are having visual quality (VQ) issues like contrast, glare, noise etc, which restricts the proper analysis of disease. In order to overcome these image related issues, various techniques are presented in last decade. Digital image enhancement (IE) is the technique which improves the VQ of the image. The IE converts the desired image as image with particular VQ for an application. This paper introduces a novel approach of IE using Field Programmable Gate Array (FPGA) by implementing the different algorithms (brightness control, contrast adjustment, thresholding, negative transformation, filtering) for brain imaging. The proposed IE technique implements the Multi-Processor System on Chip (MPSoC) over FPGA. The MPSoC is distinct computer architecture while FPGA is hardware where the image processing algorithms can be implemented properly. The proposed IE methodology assures the low cost system with high accuracy. The proposed technique is incorporated with Advanced Microcontroller Bus Architecture (AMBA) which provides the interface for communication among master and slave module.

Keywords: Field Programmable Gate Array (FPGA) · Image enhancement Medical image · Video quality · Multi-Processor System on Chip (MPSoC)

1 Introduction

The Image Enhancement (IE) is a technique of digital image processing having wide range of application in medical and non-medical areas [1]. The principle objective of image enhancement is manipulating image attributes to make it more appropriate for an observer to see. This can be implemented adapting algorithms developed on FPGA in very large scale integration (VLSI) [2]. The MPSoC is a class of VLSI and is a system on chip (SOC) integrated with multiple programmable processors as its system components. Widely having its applications in the field of digital signal processing, communication, networking and multimedia MPSoC incorporates all the required components for their operation. MPSoC is a distinct and an essential branch of multiprocessors. Complexity of the application to be designed would demand "high-performance computing" and hence it fulfills the demand of being application specific, as it is divided into homogenous and heterogeneous multiprocessors. The advantage of

R. Silhavy et al. (Eds.): CoMeSySo 2018, AISC 860, pp. 44–57, 2019.
https://doi.org/10.1007/978-3-030-00184-1_5

real time application performing ability, functioning at a low power, handling multiple tasks at the same instant of time, when combined with the reconfigurable hardware technologies (FPGAs) MPSoC posses the capacity of building a custom hardware system [2]. A homogenous MPSoC system is easier to implement due the presence of same type of nodes in the processor having lower power and performance throughput. In case of heterogeneous MPSoC systems, inbuilt with different processing nodes and hardware components, provides higher speed of computing as a function of programming flexibility. Application Specific Instruction-set Processor (ASIP) based MPSoC system is proposed to balance both higher performance parameter and programming flexibility. ASIP is capable to perform many image processing algorithms by the use of same data path. When compared to conventional RISC processor, ASIP works at 16 times higher rate and performs several ID filtering processing in 8cycle/pixel along with advantage of achieving FullHD (1920 × 1080) application [3]. Interconnections in MPSoC are established by Network – on- chip (NOC) to ensure high efficiency and scalability. The development of an low cost MPSoC platform and its application is implemented using an algorithm to enhance a video based on Super Resolution (SR) [4]. A technique to synthesize a heterogeneous architecture, having many processors, to speed up a specific application. The methodology to implement the proposed heterogeneous system synthesis is in the accordance with to a design flow commercially built on the platform of Xtensa^{TM} from Tensilica Inc. This is the first tool integrating extensible processors for the synthesis of customized MPSoC [5]. Design space exploration (DES) is a method required to assess the different design alternatives for utilizing the immense hardware resources available in MPSoC efficiently. A framework is developed for the faster simulation and evaluation of the performance of the MPSoC which reduces the time to market and within the approach of Transaction level Modeling (TLM), a unique way of defining the Programmer's View (PVT) level is introduced by two complementary modeling sublevels [6]. The paper is organized with following sections which includes Sect. 2 briefs the background of the image processing concept. The Sect. 3 discusses the survey of various researches towards image enhancement techniques using hardware design language. Section 4 illustrates the problem statement. The Sect. 5 explains the design and implementation of the proposed system. The next Sect. 6 discusses the analysis of the obtained results. Finally Sect. 7 concludes paper contribution with conclusion.

2 Background

The Image Enhancement (IE) is process which composed of different image processing techniques and can be utilized to enhance the visual appearance of the image i.e., image quality [7]. The IE is a process by which the image quality is improved and it helps to extract the spatial features of the image. The images captured from the mobile phones will have poor quality. Hence, these images needs contrast enhancement which can be done by using IE algorithms. The significance of the IE is that it helps to enhance the information interpretability or perception exists in the image for human viewers. This generates the output image which looks better than originally captured image. The enhanced image will have improved pixel intensity than original image [8, 9]. The IE is a significant technique in medical image processing, microscopic imaging, remote sensing, hyper spectral image processing, speech recognition, high definition television (HDTV), industrial XRAY image processing and other image processing applications. Some of the IE techniques need interactive procedures to get the desired output image and hence it does not meant for routine application [10]. The following Fig. 1 represents an example of IE.

(a) (b)

Fig. 1. (a) Original image, (b) Enhanced image

The categorization of IE approaches can be done as Spatial Domain (SD) and Frequency Domain (FD) methods. The SD is meant to image plane itself and mechanism in and are based on direct image pixels manipulation. The FD techniques works on the basis of modifying the Fourier transform of an image. Thus transformation process can be indicated by:

$$g_{(x,y)} = T[f_{(x,y)}] \tag{1}$$

In the above Eq. (1), f(x, y) indicates the input image and the processed image is indicated with g(x, y). The term "T" is an operator on 'f' defined over some neighborhood of (x, y). These neighborhoods are defined at a point (x, y) to utilize a square or rectangular sub image area centered at (x, y). This sub image center is moved from pixel to pixel starting, consider at the top of left corner. Then T is subjected at every '(x, y)' to get the output (g) at same location. The process utilizes only the pixels in the image area spanned by the neighborhood [11].

The IE methods consists of the Wavelet Transform to enhance the medical image, which can be utilized over high frequency (HF) sub images and can be decomposed by applying Haar WT [12]. Then, the noise in the HF sub-bands can be minimized by implementing the soft-thresholding. The final enhanced image can be obtained through inverse WT. The histogram equalization (HE) can also used for the medical image enhancement which improves the image contrast making changes in the distribution of gray levels [13]. The drawback of HE is that it doesn't offer significant results as it damages the image and its border areas. Thus, to overcome these drawbacks there were various methods are presented which generalizes the HE to improve its performance. An combined method of WT and HE offers significant results for medical images by making improvement in edges and is been proven good in real time applications. The method called "Gamma correction (GC)" optimizes the every pixels gamma values through co-occurrence matrix homogeneity of the original image. The GC method improves the dynamic range and improves the image [14, 15].

In order sharp the medical image, Morphological filters (MF) can be used which locates the edges through gradient-based operators. To locate the edges in an image using morphology gradient can be used which exhibit performance ability like Canny and Sobel [16–19] edge detectors.

The main applications of IE are Satellite imaging, Aerial imaging, Medical imaging, Digital camera application and also in remote sensing. The IE are techniques widely utilized in forensics, Astrophotography and in finger print matching etc. The techniques like color brightening, contrast enhancement and sharpening provide the clear image output. The medical imaging implements the above mechanism to reduce the noise and details of sharpening and provides the visually improved image. All these techniques help to review the atomic region in MRI, ultrasound images, and fingerprint recognition, analysis of the security videos and for investigation of crime scene that helps to find the culprits and protect the victims [20, 21].

3 Existing Researches

This section discusses the existing researches of image enhancement using various approaches.

The work of Salcic and Sivaswamy [22] introduced a computationally efficient image enhancement approach of high speed by utilizing a low FPGA system. This enhancement approach uses high boost filter and histogram equalization algorithms and FPGA approach helps to attain better implementation of hardware. The proposed approach achieves the high performance and offers low cost solution.

Author Jung et al. [23] presented real time implementation of IE pre-processor for complementary metal-oxide-semiconductor (CMOS) image sensor. In this, spatially adaptive IE algorithm is utilized to compensate the physical limitations of CMOS sensor. The proposed system was implemented over FPGA chip and obtained the improved quality of image.

Similarly, author Long et al. [24] presented adaptive IE algorithm and operated with noise reduction algorithm. The proposed method uses parallel and pipelined structure to reduce the processing time of image by designing hardware over FPGA at frequency of 90 MHz.

The work towards image edge detection was presented in Guo et al. [25] by using FPGA. Author described a parallel processing sobel edge detection algorithm for IE by using FPGA chip. The proposed algorithm offers efficient and quick edge detection.

The work presented by Sree and Rao [26] illustrates the implementation of IE algorithms such as contrast adjustment, brightness control and histogram equalization over FPGA to attain high performance. This system offers highly efficient IE over FPGA by using MATLAB based Simulink.

Author AlAli et al. [27] described the FPGA based hardware design by using various image processing and filtering algorithms. These algorithms are implemented with Xilinx Spartan-6 board and found significant outcomes as per area and power is concerned.

The work introduced by Akkala et al. [28] discussed the ultrasound backend system to perform the IE by using FPGA. The outcomes suggested that image resolution, hardware resolution are improved. The work of Yen et al. [29] expressed the algorithm of contrast enhancement mechanism and hardware implementation for real time applications. This work uses the histogram equalization algorithm with MATLAB based simulation. In real time applications this work outcomes with significant results for many surveillance system.

The direct implementation of verilog hardware description language is introduced for medical IE in Chiuchisan [30]. In this, the latest series of filters are designed at hardware level to perform the edge detection, contrast enhancement, brightness adjustment and sharpen operation. The proposed system achieves the greater potential of IE.

The work of Singh et al. [31] expressed the real time implementation of the IE using FPGA. In this, various detection algorithms are implemented to improve the image quality. This system was applied for electro optical surveillance system under naval and land platforms. The outcomes suggest the improvement in the image quality.

Nelakuditi et al. [32] introduced a real time implementation of the IE applications. The proposed study uses the hardware based co-simulation algorithm over FPGA. This study implements the edge detection, negative image and image thresholding algorithm by using MATLAB based Simulation. The proposed study outcomes with low cost, less complex system. The work of Nityananad et al. [33] presented an review over the IE techniques based on the histogram equalization. Author has discussed different IE techniques such as numerical methods, signal processing and statistics. This work studied the Histogram Equalization (HE) and discussed their advantages, limitations and applications.

Another survey work on IE techniques is done by Sharumathi and Priyadarshini [34], in which found that transformation mechanisms like wavelet, curve let and contour let are useful to takeout the speckle noises of acoustic images. Also, the author presented a framework to understand various noises exist in the acoustic image which can be removed by filtering mechanisms.

An comparative analysis of different IE techniques is discussed in Jarande et al. [35]. Author proposed a new interpolation based IE technique having HF sub-bands obtained by lifting WT and low resolution input image. The implementation of stationary WT helps to enhance the image edges by introducing an intermediate stage. The lifting WT was used to perform the decomposition of the input image to different HF sub-bands. Later, these sub bands and input image are interpolated in parallel. These sub bands are modified by using obtained sub-bands from stationary WT. Finally, these sub bands and estimated LR image are reconstructed by inverse lifting WT to obtain new image. The quantitative and visual images of the proposed technique indicate the significance of proposed method over conventional and state-of-the-art methods.

The work of Alareqi et al. [36] introduced the design and implementation of FPGA for biomedical applications. Author explained various IE mechanisms like inverting image operation, brightness control, contrast stretching and segmentation. An comparative analysis was performed among these mechanisms to know the significant one. From the outcomes found that the brightness control mechanism is efficient than other mechanisms. The work addressed in Su and Jung [37] introduced the noise (two-step) suppression based IE (low light) method. Author has performed noise aware contrast enhancement by adapting the noise level function (NLF). The NLF was utilized to get the noise aware histogram that restricts the noise amplification. Also, author implemented the noise aware histogram for contrast enhancement. The outcomes of the proposed method present the contrast enhancement of low light images while minimizing noise amplification.

4 Problem Statement

Due to the growth and demand of cost effective image system operating at a minimal amount of power like video cam recorder, PC camera and digital camera algorithm implied to it are a necessity. Algorithms applied for enhancement are critical in terms of constraints and hence to maintain the visual appearance of the image in signal processing, these methodologies are utilized [38]. Serial processors being deployed in the processing of real-time multimedia image manipulation is problematic with the size of an image being so large having a higher resolution. To make the image processing system work efficiently providing high performance parameter in a short span of time is a demanding task. The implementation of image processing algorithms with respect to hardware architecture is difficult in real time scenario [39]. Factors namely chip area, computational speed constraints; architecture maintained at low power should be effectively optimized. The design modeling should be carried out by hardware based languages VHDL, Verilog or Xilinx system generator supporting DSP [40].

5 Proposed System

The presented methodology for the research of MPSoC consists of three modules namely, master module design using image enhancement algorithm, advanced extensible interface (AXI) interconnect design and slave module design using dual port memory. The architecture of the proposed system is given in Fig. 2. The master module has the image processing block which receives the information from the input data/image block, following which the image is operated via the image enhancement algorithms. The implemented algorithms are classified as:

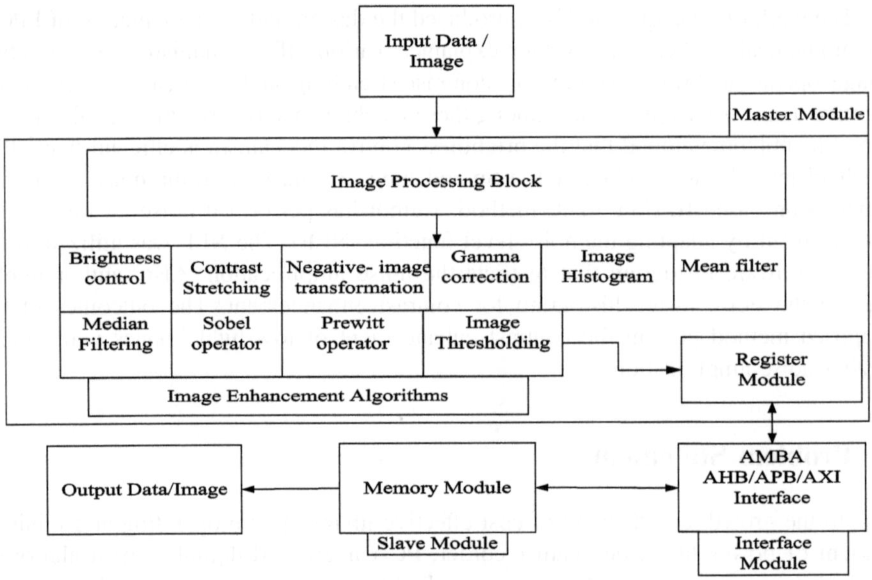

Fig. 2. Proposed system architecture

5.1 Brightness Control

By adding/subtracting the constant value to gray level of each pixel individually, this technique is effectively involved in the image enhancement algorithms to increase or decrease the brightness control of the digital image captured.

5.2 Contrast Stretching

The value of the intensity range is stretched to improve the quality of the image by making use of the possible existing values. This method gives the image with better quality.

5.3 Negative Image Transformation

For applications in areas like medical image display which applies the negatives of digital-images, this algorithm is implemented. The grey level in the range of [0, N − 1] is processed to obtain the negative of an input image.

5.4 Gamma Correction or Power Law Transformation

The requirement of displaying the image at different intensity and clarity is met by the gamma correction algorithm or the power law transformation.

5.5 Image Histogram

This algorithm is used to enhance the contrast of the image. The equalization would stretch the uniformly distributed histogram from gray levels at darker ends. Different filtering techniques with noise reduction—below mentioned are ways of filtering by simple sliding window spatial filtering technique.

5.6 Mean Filter

This filtering technique replaces the centre pixel value from the window along the mean value of all the pixels present in the window. Allows the reduction of the noise in the images and is held responsible to decrease the variation in the intensity from one pixel to the next pixel that smoothens the image.

5.7 Median Filtering

This performs the same function as that of mean filter with the advantage of preserving the valuable data of the image. This filtering technique has adaptive nature and in hardware architecture we are incorporating the noise reduction, adaptive computation module to filter the images.

5.8 Edge Detection Techniques

These are incorporated in the computer vision and image processing to enable detection of the edges using operators namely sobel and prewitt. They calculate the approximate image function gradient.

5.9 Sobel Operator

These work on the basis of convolution performed on the original pixels of the image with filter values in vertical and horizontal directions.

5.10 Prewitt Operator and Other Techniques

This is mostly used for higher frequency changes in the image produced by gradient approximation.

5.11 Image Thresholding

It is categorized as an image segmentation technique, which allows the creation of binary images in gray level.

The register module holds the new image after it is operated by different algorithms. The interfacing module consisting of AMBA AHB/APB/AXI Interface is the combination of advanced microcontroller bus architecture, advanced extensible interface, advanced peripheral bus and advanced high performance bus which provides an interface medium between the slave (memory) and master module.

The Fig. 3 represents the AMBA AXI-4 protocol. The interfacing module consisting of AMBA AHB/APB/AXI Interface is the combination of advanced micro-controller bus architecture, advanced extensible interface, advanced peripheral bus and advanced high performance bus. It supports burst length up to 256 bits, requirements of updated write and read operations and data on component interoperability. It has 16 masters and 16 slaves interfacing ability. The slave is connected to interconnect via master interface whereas the master is connected to interconnect via slave interface. To perform the function of read and write, the master module validates the VALID signals and is fed to the interface unit. The initiation is continued with concatenating the input of [awcache, awprot, awlock, awburst, awid, awaddr]. The arbiter and decoder are present in the interconnect block. The access of bus priority is assigned by arbiter as two master's access simultaneously. The role of the decoder is to decode addresses forwarded by the master allowing the control to be given to one of the slaves out of 16. The interface decoder is a centralized digital block. Slave module consists of general read/write storage element (buffer) which stores the address of read/write and data. The remaining read addresses to be sent are stored in pending address list and same goes for the write addresses. The test bench of AXI4 initiates the read or writes transaction access and the read/write buffers are obtained as output from it. Slave module of the proposed architecture stores the intermediate interface results and displays them. It has the memory module that can be single or dual port in nature.

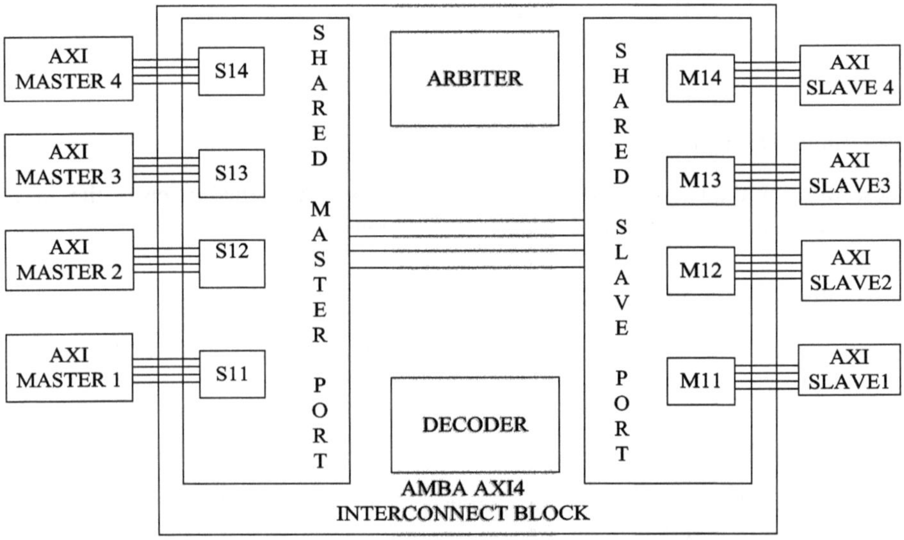

Fig. 3. Block diagram of AMBA AXI-4 protocol

6 Results and Analysis

The outcomes of the proposed system are discussed below. The proposed system is designed by using MATLAB based image processing technique and implemented over Xilinx based FPGA board. The following Fig. 4 represents the input image of brain MRI and outcome of brightness control image is given in Fig. 5 (Figs. 6, 7, 8, 9, 10, 11, 12, 13, 14 and 15).

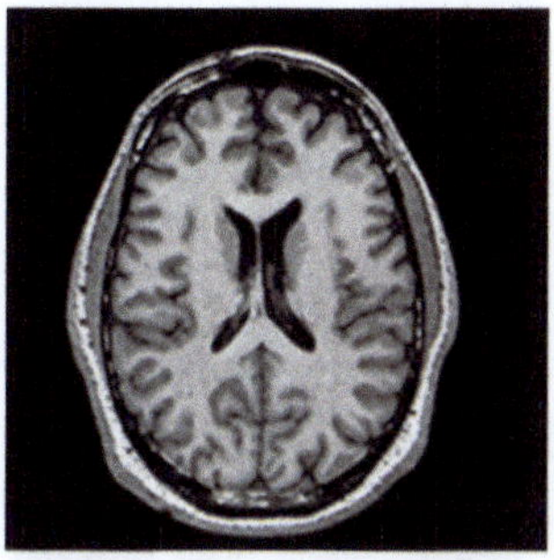

Fig. 4. Input image

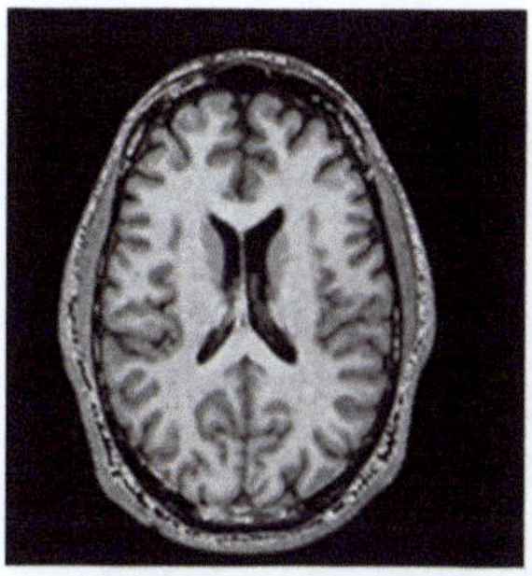

Fig. 5. Brightness control

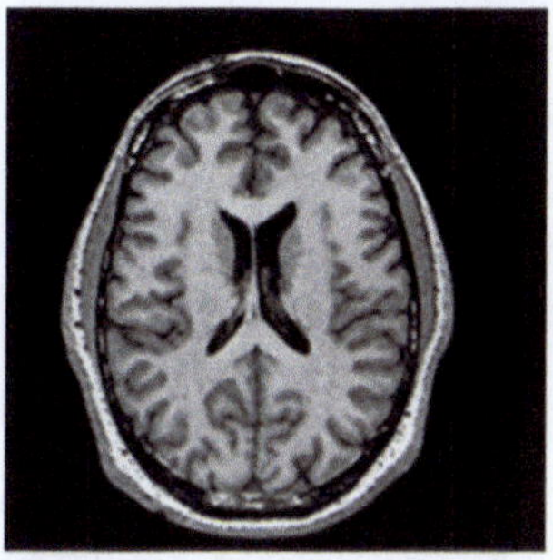

Fig. 6. Contrast adjusted image

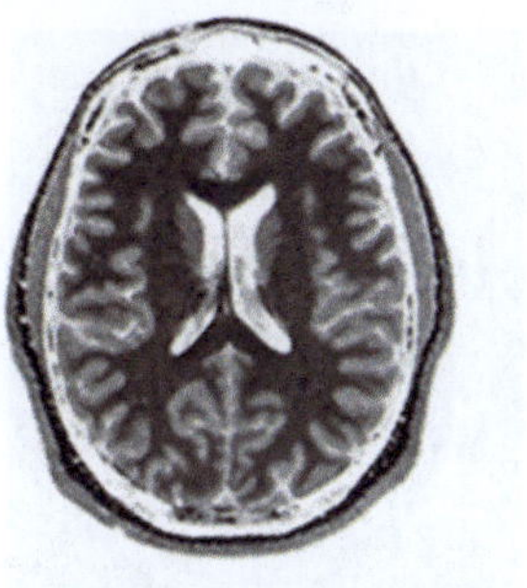

Fig. 7. Negative transformed image

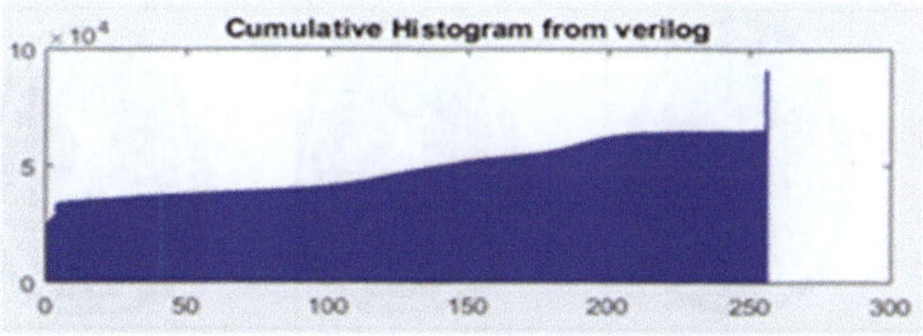

Fig. 8. Histogram from Verilog

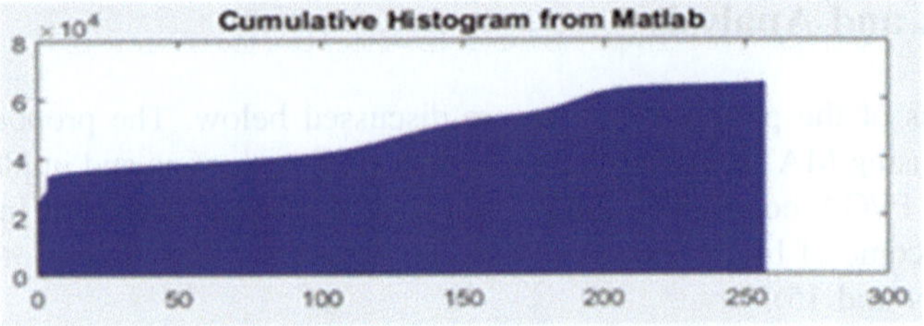

Fig. 9. Histogram from MATLAB

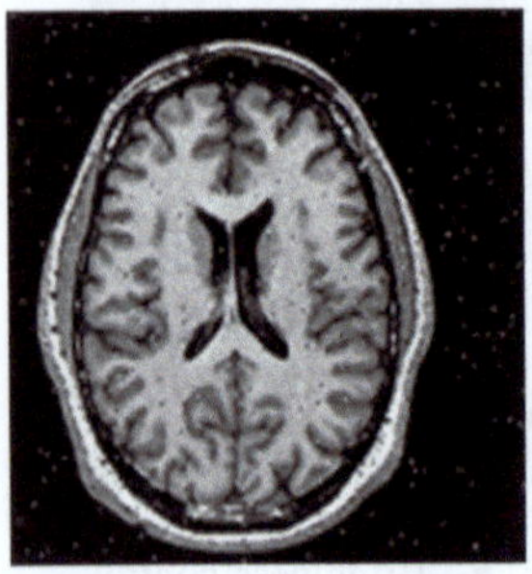

Fig. 10. Noisy image

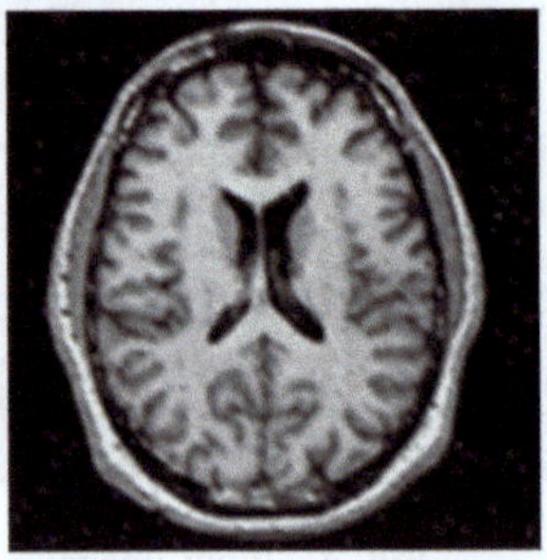

Fig. 11. Mean filtered image

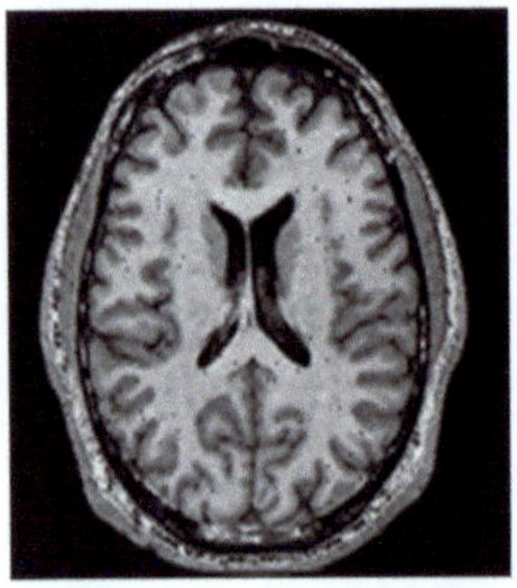

Fig. 12. Median Filtered image

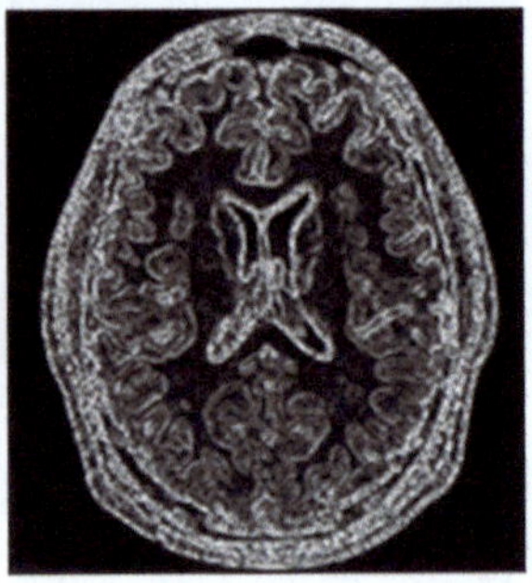

Fig. 13. Sobel edge detected Image

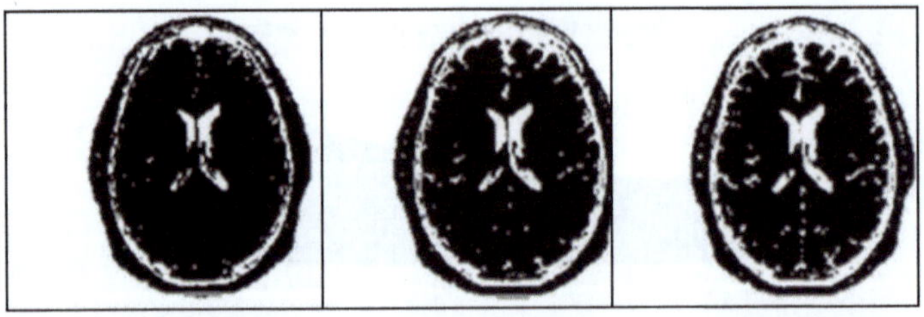

Fig. 14. Image thresholding with threshold value = 20, 50, 70

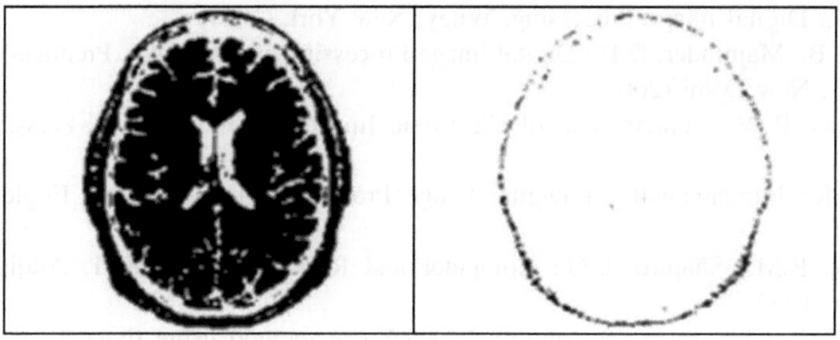

Fig. 15. Image thresholding with threshold value = 90 and 240

7 Conclusion

This paper provides a novel framework to increase the efficiency with which the image for an application is enhanced. The inclusion of the ten algorithms namely brightness control, image construction, edge detection via sobel, prewitt technique, mean and median filtering, thresholding of an image, contrast stretching work identically to improve the quality of the image. The interface established using the AMBA protocol performs the function of bus to transmit the image. The above tabulated results work in the favor of intensifying the image quality in medical field.

References

1. Daboczi, T., Bako, T.B.: Inverse filtering of optical images. IEEE Trans. Instrum. Meas. **50**(4), 991–999 (2001)
2. Stark, J.A.: Adaptive image contrast enhancement using generalizations of histogram equalization. IEEE Trans. Image Process. **9**(5), 889–896 (2000)
3. Menacer, M., Aroussi, A., Guendouz, C.: Adaptive contrast enhancement method for typical histogram configuration. Electron. Lett. **35**(15), 22 (1999)
4. Chen, T., Ren, H.: Application of partition-based median type filters for suppressing noise in images. IEEE Trans. Image Process. **10**(6), 829–836 (2001)
5. Spears, W.M.: Evolutionary Algorithm. Springer, Heidelberg (2000)
6. Goldberg, D.E.: Genetic Algorithm in Search, Optimization and Machine Learning. Addison Wesley, Boston (1989)
7. Schettini, R., Corchs, S.: Underwater image processing: state of the art of restoration and image enhancement methods. EURASIP J. Adv. Signal Process. **2010**(1), 746052 (2010)
8. Jarrett, D.P., Habets, E.A.P., Naylor, P.A.: Introduction. Theory and Applications of Spherical Microphone Array Processing, pp. 1–10. Springer, Cham (2017)
9. Oliyide, O.A., et al.: Image enhancement in digital image processing. U.S. Patent No. 5,978,518, 2 November 1999
10. Davidson, M., Kaufman, K., Mazor, I.: Method of image enhancement for the coherence probe microscope with applications to integrated circuit metrology. U.S. Patent No. 5,112,129, 12 May 1992
11. Krupinski, E.A., et al.: Differential use of image enhancement techniques by experienced and inexperienced observers. J. Digit. Imag. **18**(4), 311–315 (2005)

12. Pratt, W.: Digital Image Processing. Wiley, New York (2001)
13. Chanda, B., Majumder, D.D.: Digital Image Processing and Analysis. Prentice-Hall of India Pvt. Ltd., New Delhi (2002)
14. Jr. Weeks, R.W.: Fundamental of Electronic Image Processing. SPIE Press, Bellingham (1996)
15. Jain, A.K.: Fundamentals of Digital Image Processing. Prentice Hall, Englewood Cliffs (1989)
16. Haralick, R.M., Shapiro, L.G.: Computer and Robot Vision, vol. 1. Addison Wesley, Reading (1992)
17. Park, G.-H., Cho, H.-H.: A contrast enhancement method using dynamic range separate histogram equalization. IEEE Trans. Commun. Netw. Broadcast. **54**, 1981–1987 (2008)
18. Kim, Y.-T.: Contrast enhancement using brightness preservation bi-histogram equalization. IEEE Trans. Commun. Netw. Broadcast. **43**, 1–8 (1997)
19. Chen, S.-D., Ramli, A.R.: Contrast enhancement using recursive mean-separate histogram equalization for scalable brightness preservation. IEEE Trans. Commun. Netw. Broadcast. **49**, 1301–1309 (2003)
20. Kim's, M.C.: Recursively separate and weighted histogram equalization for brightness preservation and contrast enhancement. IEEE Trans. Commun. Netw. Broadcast. **54**, 1389–1397 (2008)
21. Reza, A.M.: Realization of the contrast limited adaptive histogram equalization (CLAHE) for real-time image enhancement. J. VLSI Signal Process. **38**(1), 35–44 (2004)
22. Salcic, Z., Sivaswamy, J.: IMECO: a reconfigurable FPGA-based image enhancement co-processor framework. In: TENCON 1997, IEEE Region 10 Annual Conference. Speech and Image Technologies for Computing and Telecommunications, Proceedings of IEEE, vol. 1. IEEE (1997)
23. Jung, Y.H., et al.: Design of real-time image enhancement preprocessor for CMOS image sensor. IEEE Trans. Consum. Electr. **46**(1), 68–75 (2000)
24. Long, J., Cui, X., Yu, D.: An adaptive edge enhancement algorithm and hardware implementation. In: 2010 IEEE International Conference of Electron Devices and Solid-State Circuits (EDSSC). IEEE (2010)
25. Guo, Z., Xu, W., Chai, Z.: Image edge detection based on FPGA. In: 2010 Ninth International Symposium on Distributed Computing and Applications to Business Engineering and Science (DCABES). IEEE (2010)
26. Sree, V.K., Rao, P.S.: Hardware implementation of enhancement of retinal fundus image using simulink. In: 2013 IEEE Asia Pacific Conference on Postgraduate Research in Microelectronics and Electronics (PrimeAsia). IEEE (2013)
27. AlAli, M.I., Mhaidat, K.M., Aljarrah, I.A.: Implementing image processing algorithms in FPGA hardware. In: 2013 IEEE Jordan Conference on Applied Electrical Engineering and Computing Technologies (AEECT). IEEE (2013)
28. Akkala, V., et al.: FPGA based ultrasound backend system with image enhancement technique. In: Biosignals and Biorobotics Conference (2014): Biosignals and Robotics for Better and Safer Living (BRC), 5th ISSNIP-IEEE. IEEE (2014)
29. Yen, J.-C., et al.: Modified contrast enhancement algorithm and its hardware design for real-time applications. In: 2014 International Symposium on Computer, Consumer and Control (IS3C). IEEE (2014)
30. Chiuchisan, I.: An approach to the Verilog-based system for medical image enhancement. In: E-Health and Bioengineering Conference (EHB) 2015. IEEE (2015)

31. Singh, M., et al.: FPGA based implementation of real-time image enhancement algorithms for Electro-Optical surveillance systems. In: 2015 12th International Conference on Electrical Engineering/Electronics, Computer, Telecommunications and Information Technology (ECTI-CON). IEEE (2015)
32. Nelakuditi, U.R., Naresh Babu, M., Narayana Bhagirath, T.: Efficient real time hardware co-simulation for image enhancement applications. In: 2015 2nd International Conference on Electronics and Communication Systems (ICECS). IEEE (2015)
33. Nithyananda, C.R., Ramachandra, A.C.: Review on histogram equalization based image enhancement techniques. In: International Conference on Electrical, Electronics, and Optimization Techniques (ICEEOT). IEEE (2016)
34. Sharumathi, K., Priyadharsini, R.: A survey on various image enhancement techniques for underwater acoustic images. In: International Conference on Electrical, Electronics, and Optimization Techniques (ICEEOT). IEEE (2016)
35. Jarande, S.S., Kadbe, P.K., Bhagat, A.W.: Comparative analysis of image enhancement techniques. In: International Conference on Electrical, Electronics, and Optimization Techniques (ICEEOT). IEEE (2016)
36. Alareqi, M., et al.: Design and FPGA implementation of Real-Time Hardware Co-Simulation for image enhancement in biomedical applications. In: 2017 International Conference on Wireless Technologies, Embedded and Intelligent Systems (WITS). IEEE (2017)
37. Su, H., Jung, C.: Low light image enhancement based on two-step noise suppression. In: 2017 IEEE International Conference on Acoustics, Speech and Signal Processing (ICASSP). IEEE (2017)
38. Shirodkar, S.U.: Image resolution enhancement using various wavelet transform. Int. J. Adv. Sci. Eng. Technol. **1**(3) (2014). ISSN 2321-9009
39. Piao, Y., Shin, I., Park, H.W.: Image resolution enhancement using inter-subb and correlation in wavelet domain. In: Proceedings of International Conference Image Process, vol. 1, pp. I-445–448 (2007)
40. Carey, W.K., Chuang, D.B., Hemami, S.S.: Regularity preserving image interpolation. IEEE Trans. Image Process. **8**(9), 1295–1297 (1999)

Simple Modeling of Mobile Foreground Detection Using Probabilistic Linear Estimation Approach

G. Madhu Chandra[1,2(✉)] and G. M. Sreeramareddy[1]

[1] Department of Electronics and Communication Engineering, CBIT, Kolar, India
madhu.guru1984@gmail.com, sreeramareddy90@gmail.com
[2] Visvesvaraya Technological University, Belagavi, India

Abstract. The problems associated with precise, simple, and instantaneous identification of mobile objects has been researched by various investigators. However, upon reviewing existing system shows that they are quite complex, time consuming, and could not assists in identifying objects from multi-feed events. Hence, the proposed system offers a novel framework that is capable of performing identification of multiple objects from multiple feeds at different scale of time. The implementation is carried out using analytical modeling approach that uses probability theory to perform discrete time modeling integrated with linear state space modeling. The proposed system also introduced a simple clustering process that uses depth map for facilitating superior identification of foreground. The study outcome was found to offer almost instantaneous identification of the multi video feeds.

Keywords: Object identification · Mobile object tracking
Video surveillance system · Probability theory · Motion blob

1 Introduction

With the wide range of styles and features, video surveillance systems are commonly using everywhere around the world. These digital surveillance systems can range from theft and vandalism deterrence to control traffic, weather monitoring and several purposes. In the present digital world, video surveillance system plays a significant role for security concerns like example: in banking sector, defense etc. [1, 2]. A digital video surveillance system is a type of surveillance system which able to capture the objects (i.e. image or videos) that can be compressed, save or transmit over communication networks [3]. There are several advantages associated with video surveillance systems. By comparing with analogous systems, digital video surveillance systems are simple and easy to install and exploit. Those systems contain superior quality (i.e. video clearness and sound effect) and compared with analog signals, digital signals has the capability to transmit the information in faster way. Additionally, compared to conventional analog system, digital video surveillance systems have the scalability feature as well as can provide higher viewing angles and more clarity for capturing the image or videos [4].

R. Silhavy et al. (Eds.): CoMeSySo 2018, AISC 860, pp. 58–68, 2019.
https://doi.org/10.1007/978-3-030-00184-1_6

In several applications, the digital video surveillance systems are widely utilizing for the purpose of capturing, monitoring and recording of human activities [5–7]. The most common applicatory areas are; in industrial field, intrusion and access control monitoring, energy management, visual employee escort and scout a head etc. There are several different areas where real monitoring of video surveillance systems are commonly utilizing in industrial processes. In some companies, electric power and temperature level must be kept constant so as to manage product integrity, product delivery, and additionally save lives. While, exploiting real monitoring video surveillance systems for intrusion and access control rescues not only man-power, it also prevent equipments as well as energy. Video-archiving and dispatching are the few other services which can be offered from your video surveillance monitoring company. If you require extra space for video storage, those companies can help you with that. These monitoring companies are responsible to store the videos on their servers and offer you with original or replica files if requested.

As mentioned above, digital video surveillance cameras have the better image clarity than analog systems. In today's digital world surveillance systems has the capability to control and save more videos. Applications can be downloaded to allow for access control, remote access, and building management systems. By remote accessing system, it is much easier for the companies to monitor real time video like as centralized visioning to monitor property. With the correct applications video feed can be captured any location via internet which eases the need to have on site monitors.

Practically, the video surveillance system divides the video file into segments which are categorized as temporal and spatial segmentation. The temporal video segmentation is constructed from time based, video shots in videos. While spatial segmentation is based on digital image segmentation process [8]. The temporal segmentation approach is the initial process towards automatic annotation of digital video streams, and its initial goal is to segment the video sequences into a group of video shots which are utilized as fundamental attributes for indexing. Hence, every video shot is then characterized by electing the key-frames and indexed by extraction of spatial & temporal features [9]. The process of video retrieval is depends upon similarity among feature vectors and already saved video features.

Whereas, digital image segmentation is typically, utilized for the partition or segmentation of image in bi-level or high-level into various regions which are related to given image or video file [10]. However, tracking and detecting of moving objects in real time is the challenging problems in video surveillance processing. Thus, many researchers designed different video processing tools to predict the location and mobility of the object in each video frames.

The paper presents a novel and very simple mechanism of video surveillance system that is capable of performing object identification. Section 2 discusses about the existing research work followed by problem identification in Sect. 3. Section 4 discusses about proposed methodology followed by elaborated discussion of algorithm implementation in Sect. 5. Comparative analysis of accomplished result is discussed under Sect. 6 followed by conclusion in Sect. 7.

2 Related Work

This section discusses about the existing research work being carried out toward techniques conceded for visual traffic surveillances systems. To appropriately recognize the different types of motor vehicle through images captured from the traffic surveillance visual systems, the work of Liu et al. [11] have presented a classification model which uses balancing scheme with neural network and deep learning techniques. The outcomes of this study display that the proposed system is able to classify the image with high accuracy. Javanbakhti et al. [12] have tried to facilitate efficient decision making task from video database by extracting spatial background data from a video content. The author has used a color & texture features in semantic region and discrete cosine transformation for salient region detection to measure the accuracy and computational complexity of their proposed technique. The Experimental result of this study demonstrates the presented approach can be applicable in various surveillance applications with achieving good accuracy and with least computational cost. The studies of Chen et al. [13] have concentrated on mobile visual surveillance network with moving cameras in order to improve human action analysis with quality-of-content. The author has presented a novel model which is considered as optimization problem which is solved by using joint source coding and rate allocation approach. Another work is carried out by the Yuan et al. [14] have addressed the issue of anomaly detection in self-directed vehicle that is the analyzing of the traffic view from the driver perspective. A spatial localization scheme is used to control the abnormality of motion orientation of mobile camera's and also a Bayesian model is used to provide robustness feature and to handle other kinds of abnormal events. The real world visual surveillance system suffers from various challenges such as intra-class target variation, poor recognition tracking results and limited image resolution. To tackle these types of situation the study of Wang et al. [15] have offered context-augmented recognition program that captures capture different environments that including prior level, image level and semantic level in respect to improve the performance of event detection. A sampling scheme is adopted by the Liu et al. [16] in which the author have presented a multi-localization approach to capture the event of crowed area with dense population. Filtering technique is also designed to minimize the computational time of event localization and the experimental outcomes of presented approach reveals efficient performance in terms of localization accuracy when compared to existing methods. The work carried out by the Lai et al. [17] have used hysteresis thresholding technique with motion compensation to solve the problem of mining mobile objects from the surveillance video. The efficiency of presented approach is calculated after experimental task which display presented approach outperform the related existing approach. Kong and Dai [18] have study and measure the performance of lossy compression on object detection and presents fluctuation reduced framework to mitigate the time-based fluctuation in background of video. The proposed system utilizes the combined multiple absolute frame difference to evaluate the amount of temporal fluctuation. The proposed approach shows effectiveness in object-detection with lower bit rate. Human walker detection framework is designed by Bilin et al. [19] for the video surveillance system. The implementation of proposed approach is based on the

FPGA device which consumes less energy when compared to existing techniques. Similarly, another work is carried out by the Cosar et al. [20] for the detection of a normal event from the video objects in surveillance system. [20] have used unsupervised method that include trajectory analysis with motion features. In the study of Xian et al. [21] have tried to evaluate the performance of detection system in the surveillance system with taking mixture of surveillance event datasets. The evaluation task uses the Fisher vector method and Random forest algorithm for each video clips. Chiranjeevi and Sengupta [22] have focused on the detection system when it having under background of dynamic uncertainties. The author have used the fuzzy logic sets. Full qualitative and quantitative experiments are taken in order to obtaining effectiveness of the presented approach in regards to mitigating unwanted dynamic situations in the background but the presented approach having some limitation that it efficient structure localization of video objects is a bit lower than existing AFABS technique. In the work of Bloisi et al. [23] have designed a modular framework for self oriented maritime-surveillance-systems. The main aim of this proposed system is to use camera as the place of main sensor that is totally different from the existing surveillance systems with the advantage of achieving global capture view with lower cost. Zhong et al. [24] have carried work toward improving the background modeling for the segmentation of foreground video objects analysis. The presented model is utilized with the pixel-based segmentation technique which results in object to be motionless that can be efficiently taken in the background. The presented work is tested in various situations and found to be applicable in various real-time surveillance systems. A modeling framework is designed in the work of Motiian et al. [25] which have tried to provide optimal online detection feature and recognition of human behavior through multiple cameras. The author have utilizes the proposed framework with the kernel methods to boost the recognition processes. The experimental performance of presented scheme suggests that it achieves effective process for analyzing the real-time human interaction form the multiple camera views. The increasing growth of monitoring system creates collection of huge multimedia data that to be analyzed whereas scalar sensor are incapable of producing big data so multimedia sensor is taken over the scalar sensor that produces a lot of data and for which it there is requirement of more manpower which became an serious problem so the work of Civilek et al. [26] have designed an architecture of WMSN which is integrated with sensor node for classification task without using lots human power. The study of Wahyono et al. [27] have addressed the issue of illegal parking of vehicles on the road and introduced the novel detection method which depend on the Gaussian model and temporal analysis. The experimental performance of presented scheme delivers that it can efficiently detect the illegally parked motor vehicles and can be considered as intelligent traffic control system. Bolvardi et al. [28] have investigated the issue related with dynamic clutter and multi-target detection in video based on passive bistatic radar. The author has presented Suppression technique with frequency based multiplexing channel to mitigate the problem of dynamic Clutter. Yuan et al. [29] have worked to improve detection and tracking system by presenting sequential exposure model. Another work is carried out by Wen et al. [30] which provide a learning scheme based on Gaussian process for controlling the speeding vehicle in through surveillance systems. The author have also constructed Fisher model and weighted sampling strategy to provide effective performance to their presented model. Experimental task is performed on real surveillance videos by which proposed method achieves better results compared with existing approaches.

3 Problem Description

At present, there are various studies towards video processing as well as video analytics, but majority of the studies implemented till date performs only one set of operations. It can be also said that existing techniques are more focused in addressing single or smaller number of research problems in one solution. This is one challenging issue as video analytics as there are multiple challenging traits that can pose serious problems in analysis e.g. cluttered scene, partial occlusion, poor illumination condition, dynamic movement of subjects, static subject and moving background, moving subject and static background, or moving subject and moving background, etc. Hence, a single video clip can itself have all these problems that need to be removed in order to analyze the video. At present, we don't have availability of any protocol suite that can solve such issues altogether. Apart from these, there is another serious issue-in present time, there are multiple camera to track single subject, hence associating the features information from different camera information and performing distributed operation (e.g. identification followed by tracking, and recognition) of one subject itself is a challenging problem.

4 Proposed Methodology

The proposed study is an extension of our prior work toward outlier detection from the video [31]. The prime purpose of this part of the study is to introduce a framework incorporated with multiple video analyzing operations within its scope. The computational capability of developing such video analytics framework is further accelerated using probabilistic theory and linear quadratic estimation. The indicative schema of the proposed study in this stage is shown in Fig. 1.

The proposed study considers developing multiple modules first for performing extraction of multiple information from the mobile subject. The indicative steps for the proposed stage of the study are as follows: The system performs a novel and simple segmentation process for enhancing the mask of the subject as well as for visualizing the foreground and background of the given scene. The system than performs the analysis of the subject surveillance using (i) surveilling subjects within a view of camera and (ii) surveilling subjects from multiple camera. The proposed system will use conventional clustering algorithms as well as use of 3D graphics for extracting the information pertaining to distance of the surface of the subject from the reference viewpoint. This mechanism will be mainly adopted to solve the problem of occlusion during segmentation process in video object detection. The system will then perform modeling of discrete time and linear state space model for the purpose of tracking the object. The proposed system than attempts to normalize the state as well as density of noise using parametric probability density function approach and will perform multi-camera surveillance by use of a novel probabilistic cum linear quadratic estimations.

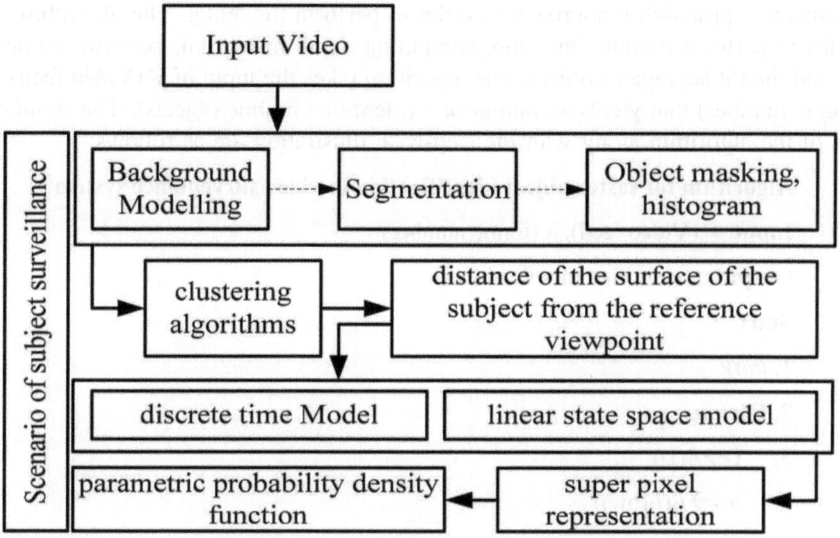

Fig. 1. Schema using probabilistic linear estimation

The identification of the subject is finally made on the basis on shape information as well as color emergence to be developed using super pixel representation. The next section discusses about the algorithm implementation.

5 Algorithm Implementation

This algorithm is meant for introducing a simple object detection mechanism from the video file using a discrete time model as well as a linear state space model. The complete algorithm design principle is like this-the algorithm captures the frames and performs segmentation process and then it is followed by eliminating all the reduced regions from the mask of an original image. A simple indexing mechanism is applied over all the connected regions that are present in the mask of the binarized image. The system will only retain those components of a mask that has increasing quantity of pixels. The algorithm also uses a cut-off value in order to check if the proportion between the pixels quantity over foreground as well as cumulative quantity of the targeted segment is found more than the cut-off level than that part of the object is termed as foreground object. Interestingly, this switching over is quite faster process and hence detection response time is quite almost instantaneous. Another significance of this algorithm design is that it uses a clustering process for offering distinction between occluded regions that is present in the mask in order to facilitate more degree of superior identification process. The significant advantage of this algorithm is that it emphasizes on the image depth which would be very unique for different forms of images. This phenomenon is utilized in proposed study in order to identify a distinct object exclusively for the objects suffering from occlusion. The algorithm also implements a probability approach in order to perform modeling. The algorithm contributes to perform motion detection, computing depth histogram, perform segmentation, and then it segregates object. The algorithm takes the input of *V* (Video feed) and *n* (frame number) that yields an output of z (identified mobile objects). The significant steps of the algorithm along with the step-wise illustration are as follows:

Algorithm for faster object identification in video surveillance system

Input: *V* (Video feed), *n* (frame number)

Output:*z* (identified mobile object)

Start

1. *initV*
2. **For**$i=n$
3. $I_d \rightarrow f_1(I_d)$
4. $bw \rightarrow f_3(f_2(bw))$
5. **For**$j=1:c$
6. $[x, y]=H(b, arg_{min}(b): arg_{max}(b))$
7. **End**
8 $I_{seg} \rightarrow I_d(bw)$
9. **If**$s \leq m$
10. $z \rightarrow [f(mean(x_{ts}, y_{ts}, I_d))]$
11. **Else**
12. $zs \rightarrow f4(b)$
13. **End**

End

The step-wise illustration of the algorithm is: - The algorithm takes the input of video feed (Line-1) as well as extracts the depth map of it. The algorithm than performs computation of frame range. For a given number of frames (Line-2), the algorithm applies two dimensional median filtering on image depth in the form of function $f_1(x)$ (Line-2). It than extract background *bw* with a criteria if the image depth is found within a range of minimum and maximum value of it. The algorithm then performs morphological operation using function $f_2(x)$ on the background image *bw* followed applying function $f_3(x)$ responsible for eliminating all the reduced size objects from the binary image (Line-4). For all the *c* number of binary levels of the *bw* image (Line-5), the algorithm obtains histogram of the min-max scope of the depth image (Line-6). The next part of the algorithm further reformulates the *bw* image of the image depth (Line-8) in order to obtain the finally segmented image. For all the index of binary image m (Line-9), the algorithm obtains the object and computes the standard deviation of it. This operation is further followed by applying a function $f(x)$ using the input arguments

(x_{ts} and y_{ts}) that matches with the foreground object (Line-10). The algorithm also consider mean of image depth and it finally applies threshold function $f_4(x)$ in order obtain the detected foreground object (Line-12). The next section outlines the result obtained by implementing this algorithm.

6 Results Discussion

The implementation of the proposed algorithm for performing identification of an object for a given vide o frame where the scripting of the proposed logic is carried out using MATLAB. For this purpose, a real time multiple event was video-graphed in digital camera with the size of each test clip being in the range of 5999 Megabytes to 9731 Megabytes. The dimension of the frame is retained uniformly same as 640 × 480 with a frame-rates of 30 frames per second, data bit rate of 7981 kilobytes per second and total bitrate of 7981 kilobytes per second. This section discusses the outcomes obtained for object detection.

A closer look into Fig. 2 shows the mechanism of processing the video feed into two different ways. The actual colored video feed is taken as an input which is then transformed to the depth frames in order to facilitate the proper identification of each and every moving object for the given video feed. The algorithm than extracts the

Fig. 2. Processing input video feed

binary version of an image (Fig. 3) followed by extraction of depth histogram (Fig. 4). Such information significantly assists in identifying an unique characteristics of every mobile object. Finally segmentation is carried out in order to identify every foreground

Fig. 3. Identified motion blob

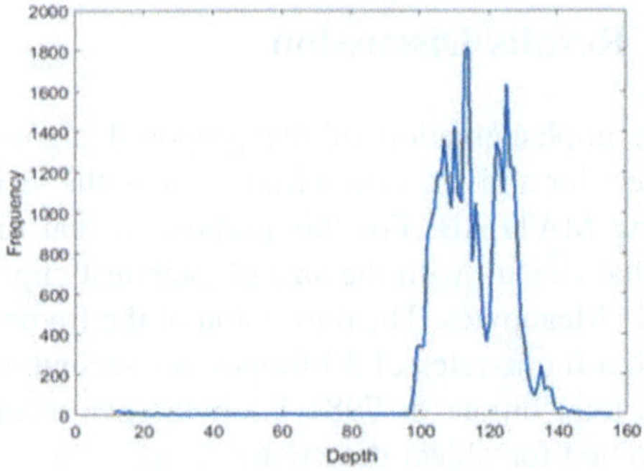

Fig. 4. Depth histogram

Fig. 5. Segmented depth

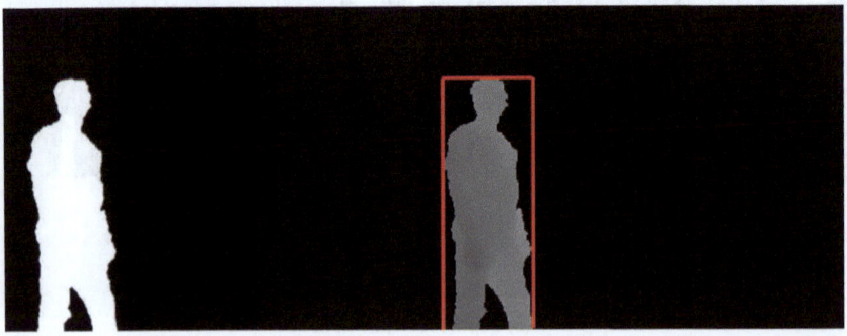

Fig. 6. Final object detection

mobile objects. Owing to adoption of clustering process, the algorithm could efficient perform faster identification of any region that could be either foreground or background (Figs. 5 and 6).

The accuracy level of the proposed system for all the test video feeds is found to be approximately 99.98% with maximum level of response time. The complete algorithm takes approximately 0.25549 s to perform the accurate detection for any form of mobile object. This faster computational response time is considered to be a potential benefit for any form of real-time application that demands instantaneous response. Therefore, the proposed system offers very simplified mechanism that lets mobile foreground object to be identified instantaneously without using any form of complex and computational expensive algorithms.

7 Conclusion

Usage of probabilistic model as well as linear quadratic estimation techniques, the proposed system can be ensured for higher accuracy with extremely less extent of false positives from the video captured from multi-events. The proposed system is capable of identifying different foreground mobile objects for any number of multi-events video feeds. The study outcome shows that proposed system offers good accuracy with faster computational response time.

References

1. Mandal, R., Choudhury, N.: Automatic video surveillance for theft detection in ATM machines: an enhanced approach. In: 2016 3rd International Conference on Computing for Sustainable Global Development (INDIACom), New Delhi, pp. 2821–2826 (2016)
2. Virendra, V.S., Ukunde, N.: Intelligent embedded video monitoring system for home surveillance. In: 2016 International Conference on Inventive Computation Technologies (ICICT), Coimbatore, pp. 1–4 (2016)
3. https://www.techopedia.com/definition/4703/digital-video-surveillance-system
4. http://centralizedvision.com/CVWP01/video-surveillance-system-monitoring/
5. Babiker, M., Khalifa, O.O., Htike, K.K., Hassan, A., Zaharadeen, M.: Automated daily human activity recognition for video surveillance using neural network. In: 2017 IEEE 4th International Conference on Smart Instrumentation, Measurement and Application (ICSIMA), Putrajaya, pp. 1–5 (2017)
6. Wang, W.-J., Chang, J.-W., Haung, S.-F., Wang, R.-J.: Human posture recognition based on images captured by the Kinect sensor. Int. J. Adv. Rob. Syst. **13**(2), 54 (2016)
7. Zhang, S., Wei, Z., Nie, J., Huang, L., Wang, S., Li, Z.: A review on human activity recognition using vision based method. J. Healthc. Eng. **2017** (2017)
8. Roshan, A., Zhang, Y.: Using mel-frequency audio features from footstep sound and spatial segmentation techniques to improve frame-based moving object detection. IET Comput. Vis. **12**(3), 341–349 (2018)
9. Kim, H., Lee, H.J.: A low-power surveillance video coding system with early background subtraction and adaptive frame memory compression. IEEE Trans. Consum. Electron. **63**(4), 359–367 (2017)

10. Mohanapriya, D., Mahesh, K.: A video target tracking using shadow suppression and feature extraction. In: 2017 International Conference on Information Communication and Embedded Systems (ICICES), Chennai, pp. 1–6 (2017)
11. Liu, W., et al.: An ensemble deep learning method for vehicle type classification on visual traffic surveillance sensors. IEEE Access **5**, 24417–24425 (2017)
12. Javanbakhti, S., Zinger, S.: Fast scene analysis for surveillance & video databases. IEEE Trans. Consum. Electron. **63**(3), 325–333 (2017)
13. Chen, X., et al.: A Quality-of-Content-based joint source and channel coding for human detections in a mobile surveillance cloud. IEEE Trans. Circuits Syst. Video Technol. **27**(1), 19–31 (2017)
14. Yuan, Y., Wang, D., Wang, Q.: Anomaly detection in traffic scenes via spatial-aware motion reconstruction. IEEE Trans. Intell. Transp. Syst. **18**(5), 1198–1209 (2017)
15. Wang, X., Ji, Q.: Hierarchical context modeling for video event recognition. IEEE Trans. Pattern Anal. Mach. Intell. **39**(9), 1770–1782 (2017)
16. Liu, C.-W., et al.: Accelerating vanishing point-based line sampling scheme for real-time people localization. IEEE Trans. Circuits Syst. Video Technol. **27**(3), 409–420 (2017)
17. Yeh, C.-H., et al.: Three-pronged compensation and hysteresis thresholding for moving object detection in real-time video surveillance. IEEE Trans. Ind. Electron. **64**(6), 4945–4955 (2017)
18. Kong, L., Dai, R.: Object-detection-based video compression for wireless surveillance systems. IEEE MultiMedia (2017)
19. Bilal, M., et al.: A low-complexity pedestrian detection framework for smart video surveillance systems. IEEE Trans. Circuits Syst. Video Technol. **27**(10), 2260–2273 (2017)
20. Coşar, S., et al.: Toward abnormal trajectory and event detection in video surveillance. IEEE Trans. Circuits Syst. Video Technol. **27**(3), 683–695 (2017)
21. Xian, Y., et al.: Evaluation of low-level features for real-world surveillance event detection. IEEE Trans. Circuits Syst. Video Technol. **27**(3), 624–634 (2017)
22. Chiranjeevi, P., Sengupta, S.: Interval-valued model level fuzzy aggregation-based background subtraction. IEEE Trans. Cybern. **47**(9), 2544–2555 (2017)
23. Bloisi, D.D., et al.: Enhancing automatic maritime surveillance systems with visual information. IEEE Trans. Intell. Transp. Syst. **18**(4), 824–833 (2017)
24. Zhong, Z., et al.: An adaptive background modeling method for foreground segmentation. IEEE Trans. Intell. Transp. Syst. **18**(5), 1109–1121 (2017)
25. Motiian, S., et al.: Online human interaction detection and recognition with multiple cameras. IEEE Trans. Circuits Syst. Video Technol. **27**(3), 649–663 (2017)
26. Civelek, M., Yazici, A.: Automated moving object classification in wireless multimedia sensor networks. IEEE Sens. J. **17**(4), 1116–1131 (2017)
27. Jo, K.-H.: Cumulative dual foreground differences for illegally parked vehicles detection. IEEE Trans. Ind. Inf. **13**(5), 2464–2473 (2017)
28. Bolvardi, H., Derakhtian, M., Sheikhi, A.: Dynamic clutter suppression and multitarget detection in a DVB-T-based passive radar. IEEE Trans. Aerosp. Electron. Syst. **53**(4), 1812–1825 (2017)
29. Yuan, Y., Yuwei, L., Wang, Q.: Tracking as a whole: Multi-target tracking by modeling group behavior with sequential detection. IEEE Trans. Intell. Transp. Syst. **18**(12), 3339–3349 (2017)
30. Wen, J., et al.: Directional Gaussian model for automatic speeding event detection. IEEE Trans. Inf. Forensics Secur. **12**(10), 2292–2307 (2017)
31. Madhu Chandra, G., Sreerama Reddy, G.M.: Analytical framework for identification of outliers for unscripted video. Int. J. Comput. Sci. Inf. Secur. Comput. Intell. Sig. Image Process. **14** (2016)

Significant Simulation Parameters for RESTART/LRE Method in Teletraffic Systems of Ultra Broadband Convergence Networks

Elena Ivanova[1](✉), Teodor Iliev[1], Grigor Mihaylov[1], Ventsislav Keseev[1], and Ivaylo Stoyanov[2]

[1] Department of Telecommunication, University of Ruse, 7017 Ruse, Bulgaria
{epivanova, tiliev, gmihaylov, vkeseev}@uni-ruse.bg
[2] Department of Electrical Power Engineering, University of Ruse, Ruse, Bulgaria
stoyanov@uni-ruse.bg

Abstract. The rare events such as blocking, overflows and losses in broadband convergence networks, especially in systems with very low probability, are one of the major estimation problems for ordinary simulation methods. Speed-up simulation is the most convenient method for estimation of rare events. The main purpose of the research is to determine and compare the significant parameters for speed-up simulation of Ultra Broadband Convergence Networks. The simulations opportunities represent a chance to reduce the digital divide and to ensure the Ultra Broadband Convergence Networks with minimum cost of time and financials. The simulation of teletraffic models allows the modeling of resource management issues, admission of system users with different parameters.

Keywords: RESTART · LRE · Rare event
Ultra Broadband Convergence Networks · Broad-band handover
Teletraffic models

1 Introduction

The last years the telecommunication sector constantly evolve. The new vision of the network is about unlocking the every one of potential markets and profitable revenue growth by engaging the new technologies in the smart home, small and big business, all smart device and multi-user infrastructure of broadband network. Estimating parameters as the probability of a rare event has applications in reliability, telecommunications, insurance, and several other areas. For complicated models, this could be calculated done in principle by the Monte Carlo simulation, but when the event of interest is rare [1], straightforward simulation would in most cases require an excessive number of runs for the rare event to happen frequently enough so that the estimator is meaningful.

R. Silhavy et al. (Eds.): CoMeSySo 2018, AISC 860, pp. 69–77, 2019.
https://doi.org/10.1007/978-3-030-00184-1_7

Simulations represent a good opportunity to reduce the digital divide and to ensure the new ultra broadband networks. Many universities and researchers are currently involved in simulation modeling to provide better results. The implementation of simulation program is made with .NET. The teletraffic models with SSM, geometrical and Pareto arrival distributions, which are versatile, possessing a pseudo-memoryless property which makes the solution of many GE-type queuing systems and networks analytically tractable. The new communication systems are interested in the modeling and performance evaluation of single communication links, but in the modeling of more devices in the whole network. Typically, the performance of the whole network containing many links working in parallel and interfering with each other is of primary interest [2].

This type of simulation allows the modeling of resource management issues, admission of system users with different priority, interaction and network loading by several types of traffic sources, etc. The aim of such modeling is network capacity evaluation and probabilities of failure. Combination of RESTART and LRE (Limited Relative Error) is described by Carmelita Görg [3]. The LRE measures the complementary distribution function of the queue occupancy and performs the Run Time Control (RTC) of the simulation. The LRE performs the Run Time Control with two conditions: the Large Sample Conditions and the Relative Error Condition [4, 5]. The algorithm is described in our previous work [6, 7].

In order to be able to compare the project measures with known results, the part of the project is study different types of models of networks of queues. Evaluation of blocking probability for different teletraffic systems is simulated [8, 9]. RESTART/LRE is implemented the in Ptolemy, a powerful, object-oriented simulator in .NET [10]. The multifactor regression analysis is based on initial simulation results, which are manipulated with program Statistica v10 [9]. The purpose of the analysis is to explore the relationship among the variables of algorithm RETART/LRE and parameters of the teletraffic model for ultra broadband convergence networks and defining the significant simulation parameters for those systems.

2 Trends of Ultra Broadband Convergence Networks

The new vision of the network is about unlocking the every one of potential markets and profitable revenue growth by engaging the new technologies in the smart home, small and big business, all smart device and multi-user infrastructure of broadband network [11–13]. Market and technologies promise that:

- Ultra Fats Access technology bring high bandwidth without the need to re-wire;
- Internet of Things promise automation and intelligent remote managements;
- NFV insert the concept of low-cost networked compute and storage;
- SDN enable these functions and user control to create responsive experience;
- Rapid development of software oriented solutions via open source;
- The higher bandwidth is appropriate with new generation of 5G.

Software-defined networking (SDN), network functions virtualization (NFV), and network virtualization (NV) are all complementary approaches [14]. They each offer a new way to design deploy and manage the network and its services: SDN separates the network's control (brains) and forwarding (muscle) planes and provides a centralized view of the distributed network for more efficient orchestration and automation of network services; and NFV focuses on optimizing the network services themselves [15, 16].

NFV decouples the network functions, such as DNS, caching, etc., from proprietary hardware appliances, so they can run in software to accelerate service innovation and provisioning, particularly within service provider environments. NFV ensures the network can integrate with and support the demands of virtualized architectures, particularly those with multi-tenancy requirements.

The definition of the basic requirements and characteristics of this type of system sets the traffic parameters, which have to be described with heavy distributions. Determining the traffic systems and their simulation parameters, we can determine the significant parameters for reduction of the blocking events. And this cannot happen without generating a huge amount of traffic without speed-up simulation. On the other hand, the definition of significant parameters could greatly facilitate the introduction of these systems.

3 Teletraffic Models for Ultra Broadband Convergence Networks

The heavy traffic models must be simulated, because the various expectations must be fulfilled on very high level [7, 17, 18]. The type of investigated queueing systems is with different arrival distributions such as Pareto, Geo, Poisson, Erlang and etc. The service rates are exponential. A basic reference model for rare event simulation is the single server queuing system G/M/1/B - FIFO with a finite buffer size B and system with self-similar arrival traffic.

As is seen there exists a significant difference in the queueing performance between traditional models of teletraffic such as Poisson processes and Markovian processes, and those exhibiting self-similar behavior. More specifically, while tails of queue length distributions in traditional models of teletraffic decrease exponentially, those of self-similar traffic models decrease much slower [4, 5]. In practice, if self-similarity is not accounted for at the modelling stage, serious overestimations of performance predictions may result, and thus we establish quality requirements that are impossible to guarantee in a realistic network scenario. Thus, we investigate the potential impacts of traffic characteristics, including the effects of self-similar behavior on queuing and network performance, protocol analysis, and network congestion controls.

The time series of self-similar queuing processes exhibit burstiness over a wide range of time scales. For investigation of the influence of self-similarity in queuing processes onto the buffer overflow probability is used a sequential steady-state simulation of the SSM/M/1/B queuing system. As is seen there exists a significant difference in the queueing performance between traditional models of teletraffic such as Poisson processes and Markovian processes, and those exhibiting self-similar behavior. More specifically, while tails of queue length distributions in traditional models of teletraffic decrease exponentially, those of self-similar traffic models decrease much slower. In practice, if self-similarity is not accounted for at the modeling stage, serious overestimations of performance predictions may result, and thus we establish QoS requirements that are impossible to guarantee in a realistic network scenario [19]. Thus, we investigate the potential impacts of traffic characteristics, including the effects of self-similar behavior on queuing and network performance, protocol analysis, and network congestion controls.

The time series of self-similar queuing processes exhibit burstiness over a wide range of time scales. For investigation of the influence of self-similarity in queuing processes onto the buffer overflow probability is used a sequential steady-state simulation of the SSM/M/1/B queuing system (i.e. queuing systems with the finite buffer capacity).

For difference from M/M/1/B queuing system, the arrival rate λ_j into SSM/M/1/B queuing system is not a constant value. It depends on the sequential number of time-series i, the total number of observation n and the Hurst parameter H, which determine the rate of self-similarity [20].

The flow balance equations are shown with (Eq. 1)

$$\begin{array}{ll} \lambda_j = \lambda(i,n,H); & j = 1,2,\ldots,B \\ \lambda_j = 0; & j \geq B+1 \\ \mu_j = \mu; & j = 1,2,\ldots,B+1 \end{array} \tag{1}$$

This system is stable whether the throughput:

$$\rho = \frac{\lambda(i,n,H)}{\mu} < 1 \tag{2}$$

Let consider two separated cases: $\rho = 1$ and $\rho \neq 1$. For $j = 1,2,\ldots,B$ the distribution of number of customers in system is $P_j = \rho^j P_0$, which is determined according to (Eq. 3).

$$P_j = \frac{\rho^j(1-\rho)}{1-\rho^B}, \rho \neq 1, \quad P_j = \frac{1}{1+B}, \rho = 1 \tag{3}$$

Hence, the rate at which the customers are lost (blocked) is λP_B. The queuing process is described with the steady-state simulation algorithm. The self-similar traffic was generated and was received the sequence of arrivals. The fixed length of self-similar traffic is extracted via fixing the number of observations. As the service process is Markovian, the sequence of departures has exponential distribution, generated with inverse transform generator. The next step is calculation of buffer size. If the service size is greater than the size of arrivals, than the buffer size $B = 0$ (e.g. it is impossible to have negative buffer size). In cases when the buffer is overflow L, e.g. $B > L$ the traffic is lost, that's why is making an assumption that $B = L$ [14, 15, 21].

The blocking probability is determined by arrival sequence, independently by the rule of service FCFS (*First Come First Serve*), LCFS (*Last Come First Serve*) or HOL (*Head of Line*) static users.

4 Simulation Results of Teletraffic Models for Ultra Broadband Convergence Networks

Telecommunication tools for simulations are various, such as OPNET, NS2, other MATLAB, java implementations. The reasons of implementing a new tool are many, but we create new simulation module in teletraffic aspect. The program for Simulation of Rare events in teletraffic systems is made with the .NET Framework. The program product .NET is used because a large amount of generated data has to be processed efficiently. Standard simulators generate a pseudo random stream of information. The generated information that is needed for this study should be modelled by changing the specified parameters so that the RESTART method is applied. Simulations represent a good opportunity to reduce the digital divide and to ensure the new trend of SDN networks. The implementation of simulation program is made with .NET [22].

The simulation of the three main types of queuing systems is made of the AMD Athlon(tm) II X2 250 Processor 3.00 GHz and 3 GB RAM, running 64-bits operation system is Windows 10 Professional.

First simulations are made with M/M/1/N queuing systems with the .Net tools, because it was necessary to compare analytical and simulation results, to prove and validate the new results. The algorithm and results are described before [8].

After validation of results, the next step of determine significant parameters is calculating of correlational matrix for the teletraffic systems with general non Markovian distribution of arrival rate.

Correlational matrix for system G/M/1/N is calculated, after the important proof that the parameters L, n and RE are without linear dependence. The values of these parameters don't have such a linear correlation, because the correlation index is not bigger then 0.7 (Table 1).

Table 1. Correlational matrix proving nonlinear dependence between parameters.

	N	L	RE	λ	μ	FIFO	B
N	1	−0.533	−0.045	−0.407	−0.276	0.001	−0.149
L	−0.045	1	0.265	0.066	0.258	0	−0.407
RE	−0.045	0.265	1	0.139	−0.171	0.001	0.044
λ	−0.407	0.066	0.139	1	−0.553	0	−0.083
μ	−0.276	0.258	−0.171	-0.553	1	0	0.276
FIFO	0.001	0	0.001	0	0	1	−0.209
B	−0.149	−0.407	0.044	−0.083	0.276	−0.209	1

The next step of validation of significant parameters of the RESTART/LRE algorithm for such a system is multifunctional regression analysis. The analysis defines which parameters will have significant influence of final results.

The simulation of the SSM/M/1/N, Erlang/M/1/N and G/M/1/N queuing systems is made for different number of observations n. The increasing of arrival rate leads to insignificant increasing of blocking events. We choose arrival and service rate, to receive appropriate result. First simulations were made with different disciplines of service. The amount of rare events in system with disciplines of service: LIFO and FIFO are with insignificant difference [23].

The priorities in the modeled system G/M/1/N are dynamic; the service rates of G are with values 0.2, 0.3, 0.4, 0.44, 0.48 and 0.5. The number of examples are 10 000, 30 000, 50 000 and 100 000. The G service rate is with values from 0.2 to 0.5. The most interesting for teletraffic system with such parameters, is that the values of arrival rate must be with low intensity.

Significant parameters in the systems are intensities of arrival and service rates: λ and μ. If those parameters are bigger then 0.5, then the rare events are not already rare.

Please note that the first paragraph of a section or subsection is not indented. The first paragraphs that follows a table, figure, equation etc. does not have an indent, either.

The priorities in the modelled system are dynamic; the service rates of G are with values 0.2, 0.3, 0.4, 0.44, 0.48 and 0.5. The number of examples are 10 000, 30 000, 50 000 and 100 000. The G service rate is with values from 0.2 to 0.5. Table 2 shows simulation results with N = 60 places for service in the system, the relative error is 10%, the received blocking events. The results of teletraffic system with N = 60 and N = 90 are described in Table 2.

Table 2. Simulation results with $N = 60$ and $N = 90$ places for service in the system, the relative error is 10%, the received blocking events

L	n	N	Arrival rate λ	P_{BO}
2	10 000			1.6E−07
3	10 000	60	0.3	2.6E−07
3	50 000			1.9E−07
3	10 000			3.6E−9
3	30 000			4.6E−10
3	50 000	90	0.3	1.6E−10
3	100 000			1.6E−10
3	100 000	90	0.2	1.7E−12
2	10 000	60	0.3	3.43E−12

The simulation experiments show that the blocking events in the system increase with the increasing of *n* and arrival rate of the system (see Fig. 1).

The need of very low level of blocking events in new generations networks is very important part of the process of standardization. The potential users of the simulation product and software could be working groups in the area of the ultra-broadband convergence networks.

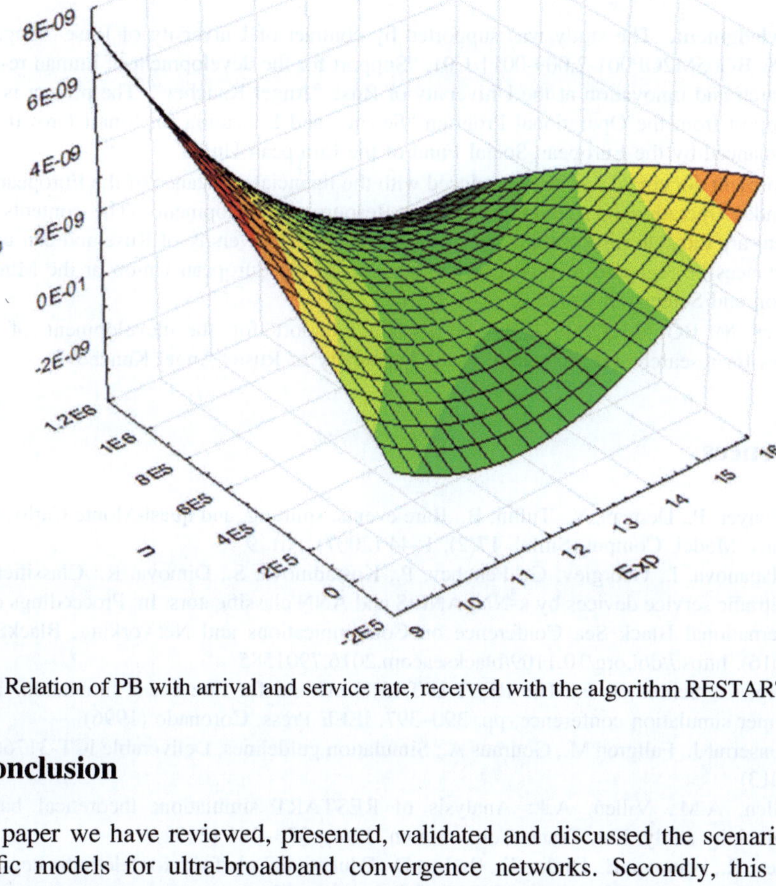

Fig. 1. Relation of PB with arrival and service rate, received with the algorithm RESTART/LRE

5 Conclusion

In this paper we have reviewed, presented, validated and discussed the scenarios for teletraffic models for ultra-broadband convergence networks. Secondly, this paper presents an implementation of teletraffic models developed in .NET for estimation of blocking events.

The simulation results are received for queuing system with non Markovian arrivals and Markovian service rate is implemented with geometrical distribution. The service is organized with FIFO and a finite buffer size N.

Significant parameters for estimating simulation results of RESTART/LRE algorithm for teletraffic system Geo/M/1/N are:

- The parameters of the thresholds L;
- Intensity λ of arrival non Markovian rate.

The future work is the study for the blocking probability with RESTART method for with c servers and finite capacity *N.* In this direction will focus on developing a multi-serving system. Introducing a heavy arrival and service distribution combined with a multi-service system will make simulation to unexpected results.

Acknowledgment. The study was supported by contract of University of Ruse “Angel Kanchev”, № BG05M2OP001-2.009-0011-C01, “Support for the development of human re-sources for research and innovation at the University of Ruse “Angel Kanchev”. The project is funded with support from the Operational Program “Science and Education for Smart Growth 2014 - 2020” financed by the European Social Fund of the European Union.

The present document has been produced with the financial assistance of the European Social Fund under Operational Programme “Human Resources Development”. The contents of this document are the sole responsibility of “Angel Kanchev” University of Ruse and can under no circumstances be regarded as reflecting the position of the European Union or the Ministry of Education and Science of Republic of Bulgaria:

Project № BG05M2OP001-2.009-0011-C01 “Support for the development of human resources for research and innovation at the University of Ruse “Angel Kanchev”.

References

1. L’Ecuyer, P., Demers, V., Tuffin, B.: Rare events, splitting, and quasi-Monte Carlo. J. ACM Trans. Model. Comput. Simul. **17**(2), 1–44 (2007). Art. 9
2. Balabanova, I., Georgiev, G., Penchev, P., Kostadinova, S., Dimova, R.: Classification of teletraffic service devices by κ-NN, ANFIS and ANN classificators. In: Proceedings of IEEE International Black Sea Conference on Communications and Networking, BlackSeaCom (2016). https://doi.org/10.1109/blackseacom.2016.7901585
3. Görg, C., Schreiber, F.: The RESTART/LRE method for rare event simulation. In: 28th Winter simulation conference, pp. 390–397. IEEE Press, Coronado (1996)
4. Monserrat J., Fallgren M., Gouraus A.: Simulation guidelines, Deliverable ICT-317669 D6.1 (2013)
5. Villen, A.M., Villen, A.J.: Analysis of RESTART simulation: theoretical basis and Sensitivity study. Eur. Trans. Telecommun. **13**(4), 373–386 (2002)
6. Raev, R., Ivanova, E., Dudin, E., Radev, D.: Educational platform for teletraffic engineering. In: Proceedings of Papers EDULEARN, Barcelona (2014)
7. Radev, D.: Teletraffic Engineering, University Publishing Center Ruse (2012). (in Bulgarian)
8. Ivanova, E.: Modelling of traffic in broadband convergence networks, Ph.D. thesis, University of Ruse Press, Ruse (2013)
9. Mitkov, A.: Theory of Experiments. University of Ruse Press, Ruse (2010)
10. https://www.microsoft.com/net/. Last Accessed 2018/106/10
11. Fallgren, M., Timus, B., Popovski, P., Braun V.: Deliverable D1.1 “Scenarios, Requirements and KPIs for 5G mobile and wireless system”, ICT-317669 (2013)
12. Stalling, W.: Foundation of Modern Networking, 1st edn. Pearson, Indianapolis (2016)
13. Pabst, R., Walke, B.H., Schultz, D.C., Herhold, P., Yanikomeroglu, H., Mukherjee, S., Viswanathan, H., Lott, M., Zirwas, W., Dohler, M., Aghvami, H., Falconer, D.D., Fettweis, G.P.: Relay based deployment concepts for wireless and mobile broadband radio. IEEE Commun. Mag. **42**, 80–89 (2004)
14. Garvels, M.: The splitting method in rare event simulation. Ph.D. thesis, University of Twente (2000)

15. Radev, D.: Modelling of Rare Events in Broadband Digital Networks. Kolbis, Sofia (2006). (in Bulgarian)
16. Cavendish, D., Murakami, K., Yun, S.-H., Matsuda, O., Nishihara, M.: TeraNets ultra-broadband communication networks in the terahertz band. IEEE Wirel. Commun. **21**, 130–135 (2007)
17. Boer. P.: Analysis and efficient simulation of queuing models of telecommunication systems. Ph.D. thesis, University of Twente (2000)
18. Bucklew, J.A.: Introduction to Rare Event Simulation. Springer Series in Statistics. Springer, New York (2004)
19. GPP TS 29.199-17 Open Service Access (OSA); Parlay X Web Services; part 17: Application driven Quality of Service (QoS), v7.0.2 (2007)
20. Ivanova, E., Raev, R., Radev, D.: Simulation of rare events in teletraffic systems with single queue. In: Proceedings of Papers of Conference ICEST, Veliko Tarnovo (2012)
21. Cerou, F., Guyader, A.: Adaptive multilevel splitting for rare event analysis. Research report, RR-5710, INRIA (2000)
22. Sells C.: Windows Forms Programming in C#, 1st edn. Addison-Wesley Professional, Boston (2003)
23. Bueno, D., Srinivasan, R., Nicola, V., van Etten, W., Tattje, H.: Adaptive importance sampling for performance evaluation and parameter optimization of communication systems. J. IEEE Trans. Commun. **48**(4), 557–565 (2000)

The Implementation of an SPC Chart to Improve the Forecasting Accuracy of the ARIMA Models

Karin Kandananond(✉)

Valaya Alongkorn Rajabhat University, Prathumthani 13180, Thailand
Karin@vru.ac.th

Abstract. The capability of forecasting techniques is based on the historical data and the autocorrelation is one of the structures used to construct a predicting model. However, a special cause, which cannot be explained by the model, always randomly occurs in the process and can significantly downgrade the performance of the forecasting model. Therefore, the statistical process control (SPC) technique widely used in the quality control area is applied to detect the outlier so that it will be removed and not be a part of the historical observation. However, since the time series data is mostly autocorrelated, the basic assumption of the SPC methodology is violated. In this research, the Box-Jenkins' autoregressive integrated moving average (ARIMA) models were used to filter the autocorrelation so there is only the pure residual to be monitored by the SPC. To illustrate the implementation, the actual data from the Singapore mercantile exchange, the daily price of the black pepper and the West Texas immediate crude oil future contracts, were used to represent two categories of autocorrelated data, stationary and non-stationary. As the first step, the historical data were modeled by the ARIMA models. Afterwards, the residuals from the models were monitored by the X-MR chart and the outliers were specified and removed. The study shows that the forecasting errors for both stationary and non-stationary cases are significantly improved after the outliers were systematically removed.

Keywords: Autocorrelation
Autoregressive integrated moving average (ARIMA) · Non-stationary
Root mean squared error (RMSE) · Stationary · X-MR chart

1 Introduction

The accuracy of any forecasting methods depends on the historical data and the observations regarding most business activities, e.g., stock prices, commodity prices and product demands, are autocorrelated. As a result, the correlation structure is used to construct a forecasting model and one of the most popular model is the Box-Jenkins' autoregressive integrated moving average (ARIMA) model. However, if there are outliers in the process, this scenario will downgrade the performance of any models to forecast the time series. Most forecasting techniques are designed to cope with only the natural variability (noise) which is, most of the time, small, and inherent in the process.

R. Silhavy et al. (Eds.): CoMeSySo 2018, AISC 860, pp. 78–85, 2019.
https://doi.org/10.1007/978-3-030-00184-1_8

On the other hand, the outlier resulted from an assignable cause is not a part of the actual process and its size is considered large when compared with the natural variability.

2 Review

An outlier in the time series can significantly affect the capability of any forecasting models and there are many authors trying to carry out different techniques to detect the outliers. The basic tool for identifying the outlier is the box-and-whisker plot but obviously it is not robust to the autocorrelation. Murihead [1] utilized the likelihood ratio to distinguish types of outliers, observation outliers, (wrong measurement or recording error) and innovation outliers (a perturbation of the system), in the time series process while Abraham and Yatawara [2] deployed a score test for the detection. Atkinson et al. [3] introduced the technique of considering the effect of the deletion on the residual sum of squares to detect a random shock or an outlier. A machine learning method, neural network, was utilized by Noorossana et al. [4] to detect an unusual shock in the autocorrelated processes simulated by following the ARIMA (1, 0, 0) or AR (1) model. Kakouros et al. [5] created a system to monitor a demand forecasting process and the outliers in the process were identified by measuring the forecast error variability. Besides focusing on the entire set of time series to distinguish the outliers, the adaptive algorithm for detecting the outliers was developed by Toshniwal and Shachi [6]. For the application of SPC, Jiang et al. [7] had pioneered the utilization of control charts to identify the critical change in the business activity process.

According to the literature, there are some evidences that the SPC method is utilized to monitor the business processes. Another advantage of this technique is its simplicity compared to other methods which are too complex and might not be suitable for use in the real life situation. Moreover, the SPC technique has been popularly used for a long time in the manufacturing area so it is one of the most important quality tools which cannot be ignored. For this reason, the SPC might be utilized as a tool to detect the outliers in the business processes in order to improve the predicting quality of any forecasting models. However, one of the challenges for using the SPC is its underlying assumption for not properly working in the autocorrelated scenario. As a result, to fix this flaw, the ARIMA method is used to filter the data so the SPC can efficiently detect the outliers without any violations of the underlying assumptions.

3 Proposed Methods

The SPC method is an efficient technique to detect the outliers. Anyway, it only works in the situation that the observations are independent. Moreover, most business processes are not independent and serially autocorrelated so the assumptions of SPC are violated. For this reason, the ARIMA model is utilized to filter the autocorrelation structure of the data and the selected model is the ARIMA models as shown in (1).

$$(1-\phi_1 B-\ldots-\phi_p B^p)(1-B)^d y_t=\xi+(1+\theta_1 B+\ldots+\theta_q B^q)\varepsilon_t \quad (1)$$

where

ϕ is the autoregressive (AR) coefficient
θ is the moving average (MA) coefficient
B is the backshift operator
ξ is the constant
ε_t is the residual at time t.

The values of ϕ and θ are between −1 and 1. After the autocorrelation is filtered, the residual from the filtering process is calculated and it is monitored by an individual control chart, X-MR chart, which is popular for its sensitivity to a small shift. The X-MR chart has the following control limits.

For X chart;

$$\begin{aligned} \text{Upper Control Limit (UCL)} &= \bar{x}+3*\overline{MR}/1.128 \\ \text{Center line (CL)} &= \bar{x} \\ \text{Lower Control Limit (LCL)} &= \bar{x}-3*\overline{MR}/1.128 \end{aligned} \quad (2)$$

For MR chart;

$$\begin{aligned} \text{Upper Control Limit (UCL)} &= 3.267*\overline{MR} \\ \text{Center line (CL)} &= \overline{MR} \\ \text{Lower Control Limit (LCL)} &= 0 \end{aligned} \quad (3)$$

Under the monitoring of an X-MR chart, the outliers are detected and flagged. Afterwards, they are removed from the process or not considered as a part of the historical data. Based on a new set of the data (without the outliers), a new forecasting model is repeatedly calculated. The root mean squared error (RMSE), as shown in (4),

$$RMSE=\sqrt{\frac{1}{n}\sum_{t=0}^{n}(Y_t-T)^2}, \quad (4)$$

is computed before and after the outliers are removed. The RMSE is used as the performance index of a forecasting model and the comparison is done in order to show that the new forecasting method has improved the forecasting errors. In conclusion, the above research procedures are depicted in Fig. 1. In order to validate the proposed research process, two sets of autocorrelated time series data are randomly selected to illustrate the application of SPC to the data. The data used is the daily price of future contracts on an agricultural commodity, black pepper futures (symbol: SMPEPPER), and the crude oil, West Texas intermediate crude oil futures (symbol: SMWTI), traded through Singapore mercantile exchange. It is interesting to note that the first set of data (SMPEPPER) is stationary while the second one (SMWTI) is non-stationary.

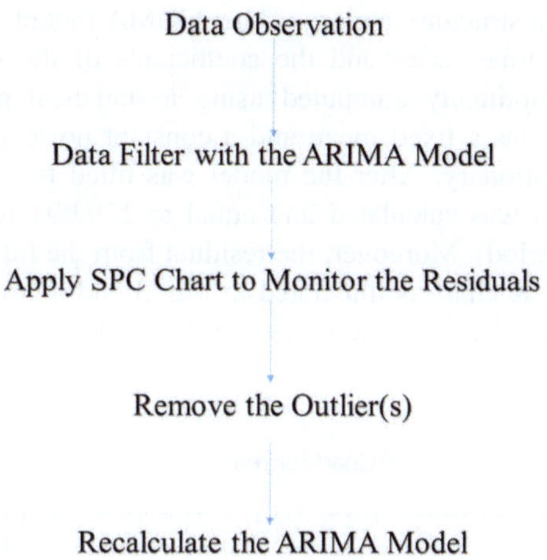

Fig. 1. Flow chart depicting the research procedures.

4 Empirical Results

The analysis are categorized into two cases due to the stationarity nature of the data, stationary and non-stationary.

4.1 Case 1: Black Pepper Futures (SMPEPPER)

Case 1 is used to represent the stationary behavior of the data. The daily price of black pepper futures (unit: $US per metric ton) from April 18, 2012 until August 21, 2012 (90 periods) is shown as a time series plot in Fig. 2.

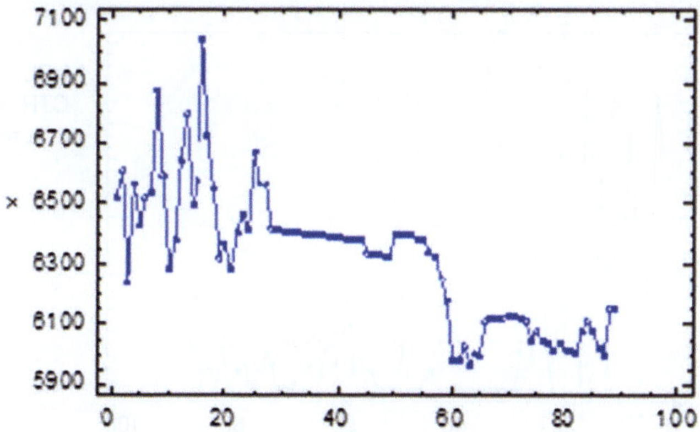

Fig. 2. Daily price of black pepper futures.

To explain the data structure, a class of the ARIMA model, ARIMA (1, 1, 1), was applied to the above time series and the coefficients of the model ($\phi = 0.480724$, $\theta = 0.997237$) were optimally computed using a statistical package, StatGraphics. Since this set of data has a fixed mean and a constant noise variance as well as the statistical test, it is stationary. After the model was fitted to the data, the root mean squared error (RMSE) was calculated and equal to 129.091 (estimation period) and 50.6145 (validation period). Moreover, the residual from the fitting was calculated and monitored by the X-MR chart as illustrated in Fig. 3 and 4. According to the control charts, they signify that there are the outliers at t = 3, 4, 5, 8, 9, 10, 14, 16, 17.

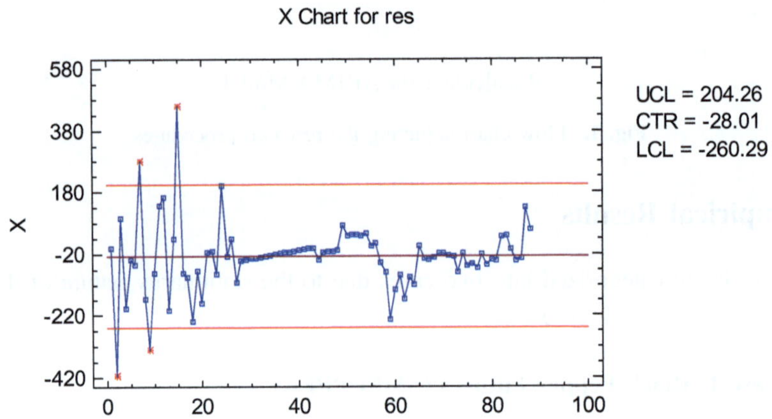

Fig. 3. X chart to monitor the residual.

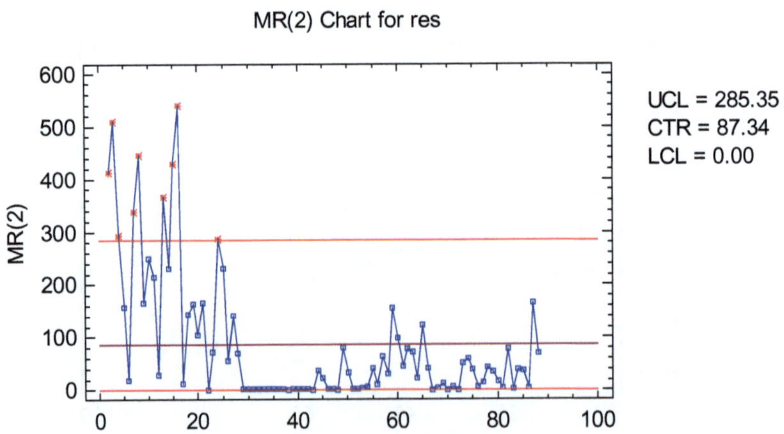

Fig. 4. MR chart to monitor the residual.

As a result, the outliers as specified by the charts were removed and the ARIMA model was computed. The best fitted model is the ARIMA (0, 1, 0) with the RMSE = 86.3483 (estimation period) and 45.1929 (validation period). According to the results, the RMSE of the forecasting is significantly reduced after the SPC is utilized to detect the outliers and they are excluded from being a basis for computing the forecasting model.

4.2 Case 2: West Texas Immediate Futures (SMWTI)

On the other hand, the non-stationary behavior of the data is demonstrated in case 2. The set of data used is the daily price of the West Texas immediate futures (unit: $US per barrel) and the time series plot of this data from April 2, 2012 until August 17, 2012 (99 periods) is shown in Fig. 5.

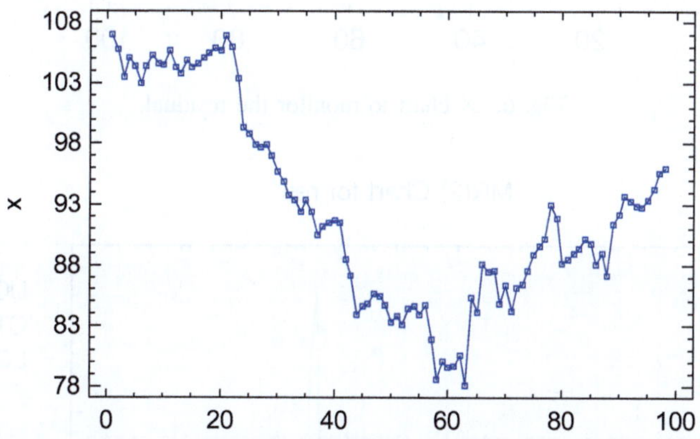

Fig. 5. Daily price of West Texas immediate futures.

The ARIMA (0, 1, 2) model with the coefficient $\theta_1 = 0.0778325$ and $\theta_2 = -0.344185$ was used to fit the time series. This set of data is obviously non-stationary because it does not have a fixed mean or contains a trend. For the forecasting error, the RMSE is equal to 1.61088 (estimation period) and 1.64559 (validation period) respectively. Similar to the first case, the X-MR chart was utilized to monitor the residual from the model (Figs. 6 and 7). Due to the chart, there are outliers among the observations at time t = 59, 64, 65 which might be the result of special causes. Therefore, they are removed from being a basis for calculating a new forecasting model.

After the removal, the ARIMA (0, 1, 0) is the best fitted model for the data with the RMSE = 1.51679 (estimation period) and 1.53535 (validation period). This result shows that the forecasting errors are significantly improved after the X-MR chart is utilized as the tool to monitor and detect the outliers embedded in the historical data. In conclusion, after implementing the proposed methods to the example sets of data, the results regarding forecasting errors are illustrated in Table 1.

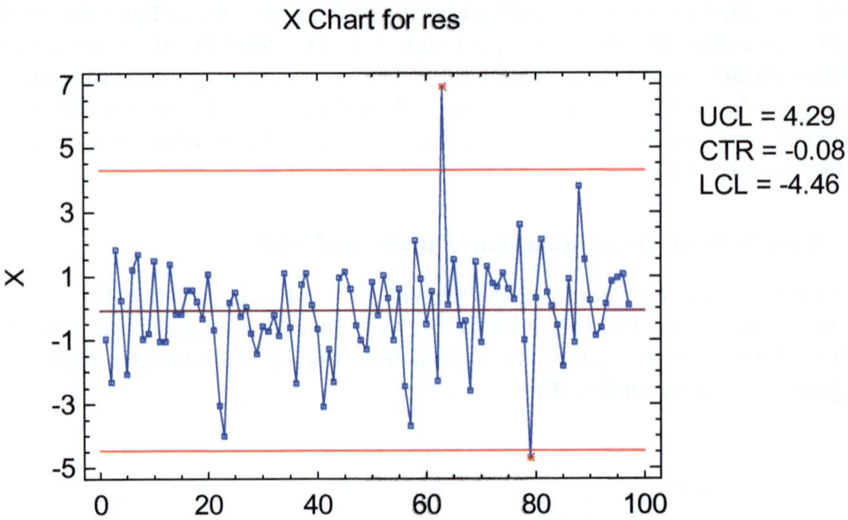

Fig. 6. X chart to monitor the residual.

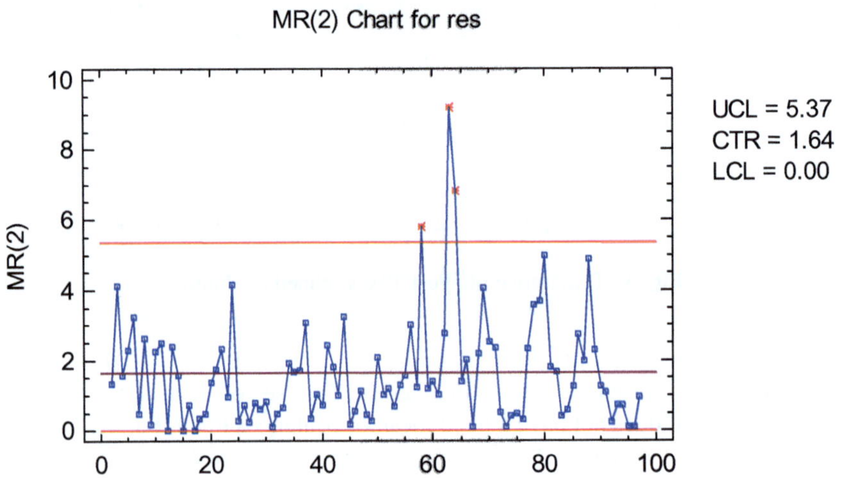

Fig. 7. MR chart to monitor the residual.

Table 1. Forecasting errors before and after the implementation.

Index	Period	Forecasting errors (RMSE)	
		Before	After
SMPEPPER	Estimation	129.091	86.3483
	Validation	50.6145	45.1929
SMWTI	Estimation	1.61088	1.51679
	Validation	1.64559	1.53535

5 Conclusions

The modeling of any forecasting models is based on the historical data. However, when there is a special cause which does not naturally occur in the process, it will significantly affect the accuracy of forecasting models to fit the observation since the outliers are not a part of the system. As a result, the statistical process control (SPC) was utilized to detect the data points which are considered as the outliers and should not be included in the historical data. The tests were conducted on two autocorrelated sets of data, stationary and non-stationary ones, which are represented by the daily price of black pepper futures and the daily price of West Texas immediate futures. After the ARIMA model was applied to fit the data, the residual from the model was computed. The results point out that the accuracy of the models is significantly improved (the RMSE decreases) after the outliers are identified by the X-MR chart and removed from the consideration. Moreover, they also indicate that the proposed method is able to be efficiently applied for both stationary and non-stationary cases of data.

References

1. Murihead, C.R.: Distinguishing outlier types in time series. J. Roy. Stat. Soc. Ser. B (Methodol.) **48**, 39–47 (1986)
2. Abraham, B., Yatawara, N.: A score test for detection of time series outliers. J. Time Ser. Anal. **9**(2), 109–119 (1988)
3. Atkinson, A.C., Koopman, S.J., Shephard, N.: Detecting shocks: outliers and breaks in time series. J. Econ. **80**, 387–422 (1997)
4. Noorossana, R., Farrokhi, M., Saghaei, A.: Using neural networks to detect and classify out-of-control signals in autocorrelated processes. Qual. Reliab. Eng. Int. **19**(6), 493–504 (2003)
5. Kakouros, S., Kuettner, D., Sasser, M., Smith, N.: Monitoring a demand forecasting process. US Patent, 7,584,116 B2 (2009)
6. Toshniwal, D., Shachi, Y.: Adaptive outlier detection in streaming time series. In: 2011 International Conference on Asia Agriculture and Animal, pp. 186–190. IACSIT, Singapore (2011)
7. Jiang, W., Au, T., Tsui, K.-L.: A statistical process control approach to business activity monitoring. IIE Trans. **39**(3), 235–249 (2007)

An Approach for Reduce Vulnerabilities in Web Information Systems

Gleidson Sobreira Leite(✉) and Adriano Bessa Albuquerque

Universidade de Fortaleza, Fortaleza, CE, Brazil
gleidson.sleite@gmail.com, adrianoba@unifor.br

Abstract. The evolution of the Internet has brought a significant change in the evolution of software, resulting in an increasing presence of Information Systems in Web environments and, consequently, an increase in security vulnerabilities and threats. In this context, secure application development has become a crucial component for information systems in the market. Exploring the main vulnerabilities in the coding of software for Web environments and the need for awareness and guidance for developers of information systems, this paper offers a approach for the treatment of vulnerabilities in Web applications composed by the selection and mapping of categories of major vulnerabilities, risks and existing proactive controls to prevent or treat occurrences of exploitation of vulnerabilities by malicious agents.

Keywords: Secure coding · Information security · Information systems
Web applications vulnerabilities

1 Introduction

Over time, the evolution and influence of the Internet has impacted not only on technical, academic or business, but also on society as a whole, as we increasingly use online services to access information or work, allowing the Web to consolidate itself as one of the main way of communication, relationships and business [1].

Due the ease and convenience provided, many internal business systems are migrating to Web, where, over time, they evolved rapidly from simple web sites, intended to enable online browsing of information, to complex information systems.

However, this universe of digital content is subject to various forms of threats that affect the security of people and their information by exploiting vulnerabilities. Just as the evolution of technology has made a major breakthrough in access to information, how vulnerabilities (failures or weaknesses) are exploited by malicious agents have also evolved, as noted in the Symantec Corporation report [2], in which 78% of web applications have at least one vulnerability, and 15% of them are critical.

Although becoming increasingly essential to society, the Internet, according to OWASP [3], is an infrastructure inherently insecure and dependent on a large number of programs developed by people (software) that define the rules of functioning of computers. As well as all human activity, software development is subject to errors that can generate failures as security, leading to exponential impacts to companies or individuals.

R. Silhavy et al. (Eds.): CoMeSySo 2018, AISC 860, pp. 86–99, 2019.
https://doi.org/10.1007/978-3-030-00184-1_9

With the increase of available and transacted information in the Web, a greater reflection on vulnerabilities is necessary in order to avoid fraud, invasion of intruders, propagation of viruses or means that circumvent the principles of information security (Confidentiality, Availability, Authenticity, Integrity and Non-repudiation) [4].

Unfortunately, most developers do not usually establish a template or internal guide of safe coding practices and their security controls to support the selection and identification of critical vulnerabilities to develop and deploying applications in unstable environments such as Internet.

Consequently, it is common to search for the main vulnerabilities identified in ranks of recognized international organizations to guide the development of secure software modules, or implement treatments after the assessment of security companies. According to [5], sometimes the companies recognize the importance of guaranteeing the security of the information late and after effectiveness of attacks, losses or destruction of information.

That is, only after the application has been developed that secure modules are included, which increase the time and resources spent on the project as a whole, increasing the risk of threat materializations, as well as the possibility of not being able to address all identified vulnerabilities as it is very costly to undertake a number of changes, which can lead to high risk and cost increases for organizations and companies.

Motivated by this scenario, there is a growing need to seek the minimization of vulnerabilities in software coding.

In this context, this article proposes to make available a proposal for the treatment of vulnerabilities in Web information systems, consisting of the selection and categorization of the main vulnerabilities and controls identified in three selected international security agencies/institutions. Besides, it is presented the result of an experience composed by tests of modules of a real system of a Brazilian Public Institution.

This work also intends to be a contribution to the studies related to security in information systems, regarding the treatment of vulnerabilities with the intention to minimize the action of malicious agents.

This article is organized as follow: the theoretical foundation and work related to the study (Sect. 2); the methodology and procedures performed in the work (Sect. 3); the result of the selection of vulnerabilities and risks (Sect. 4) and selection of the proactive controls (Sect. 5); and finally, an experience with its results (Sect. 6), followed by the final considerations and opportunities of future works (Sect. 7).

2 Related Works

According to [6], currently there are several methodologies and practices in information security because of the difficulties of technology companies in updating their models and control structures due to the constant change of technologies.

In order to assist organizations with these difficulties, there are worldwide-recognized standards such as NBR ISO/IEC 27001 and NBR ISO/IEC 27002, which together [7] are considered the most complete standard of information security management. They guide the creation of secure systems based on controls defined by international standards and good practices. Besides, there is the NBR ISO/IEC 15408 (Common Criteria), which provides a set of criteria to ensure the safety of the application through safety specifications based on the environment's characteristics.

Even after the definition and implementation of security policies in a company based on the adoption of standards and good practices of security management, software vulnerabilities can become a gateway to attacks that may lead, according to [8], to unauthorized access to privileged data, destruction of critical data, or even destruction of an entire system.

According to [9], the initial focus of software development is fundamentally to meet functional requirements that meet customer needs, where security requirements are often dealt with late, during testing and validation of software, because of the lack of preparation of customers and developers to address the issue in advance. This can produce a high negative impact on the project and even on companies.

Therefore, it is essential to adequately complement the concepts of information security in the specification and coding of information systems, especially in open environments such as Web, in order to minimize the possibility of greater occurrences of system failures or attacks.

For achieve this goal, some international agencies and institutions have carried out studies and published work results, defining ranks of the main vulnerabilities, weaknesses (insufficient controls) and risks (possibilities of threats) to support de software industry.

2.1 OWASP Top 10 – Most Critical Web Application Security Risks

The Open Web Application Security Project group is a global, nonprofit organization created to enable the improvement of software development companies to develop trusted Web-centric applications, increasing the visibility of application security, helping individuals from whole world to make better decisions regarding software security risks [10].

The work products produced by OWASP are available to the entire international community and used as a reference by entities and organizations in the area of technology and security, such as the US Defense Information Systems Agency (DISA), the US Federal Trade Commission and PCI Council [11].

According to [12], OWASP Top 10 groups the highest risks of critical attacks exploited by vulnerabilities in Web applications. It has periodic updates based on research data and statistics on the major attacks around the world.

Table 1 lists the major security risks of applications made available in Web environments, according to OWASP study [13], ranked in order of increasing prevalence and criticality, in combination with consensus related to estimates of exploration, detection and impact.

Table 1. Owasp Top 10 - 2017

Rank	Name
A1	Injection
A2	Broken Authentication
A3	Sensitive Data Exposure
A4	XML External Entities (XXE)
A5	Broken Access Control
A6	Security Misconfiguration
A7	Cross-Site Scripting (XSS)
A8	Insecure Deserialization
A9	Using Components with Known Vulnerabilities
A10	Insufficient Logging & Monitoring

2.2 CWE/SANS Top 25 Most Dangerous Software Errors

The CWE (Common Weakness Enumeration) is a project sponsored by the US Department of Homeland Security's National Cyber Security Division to classify security errors into a formal list that serves as a common language for describing software security weaknesses in architecture, design, or coding, providing a standard for efforts to identify, mitigate and prevent such weaknesses [14].

A collaboration between the SANS/MITER institute and several experts from the United States and Europe has defined the "2011 CWE/SANS Top 25 Most Dangerous Software Errors", which is a result of the selection of 25 most critical errors in software coding list CWE. These errors can lead to serious vulnerabilities in information systems [15]. They are listed in Table 2.

Table 2. CWE/SANS Top 25 Most Dangerous Software Errors

Rank	Name	Rank	Name
1	Improper Neutralization of Special Elements used in an SQL Command ('SQL Injection')	14	Download of Code Without Integrity Check
2	Improper Neutralization of Special Elements used in an OS Command ('OS Command Injection')	15	Incorrect Authorization

(*continued*)

Table 2. (*continued*)

Rank	Name	Rank	Name
3	Buffer Copy without Checking Size of Input ('Classic Buffer Overflow')	16	Inclusion of Functionality from Untrusted Control Sphere
4	Improper Neutralization of Input During Web Page Generation ('Cross-site Scripting')	17	Incorrect Permission Assignment for Critical Resource
5	Missing Authentication for Critical Function	18	Use of Potentially Dangerous Function
6	Missing Authorization	19	Use of a Broken or Risky Cryptographic Algorithm
7	Use of Hard-coded Credentials	20	Incorrect Calculation of Buffer Size
8	Missing Encryption of Sensitive Data	21	Improper Restriction of Excessive Authentication Attempts
9	Unrestricted Upload of File with Dangerous Type	22	URL Redirection to Untrusted Site ('Open Redirect')
10	Reliance on Untrusted Inputs in a Security Decision	23	Uncontrolled Format String
11	Execution with Unnecessary Privileges	24	Integer Overflow or Wraparound
12	Cross-Site Request Forgery (CSRF)	25	Use of a One-Way Hash without a Salt
13	Improper Limitation of a Pathname to a Restricted Directory ('Path Traversal')		

2.3 Web Applications Security Consortium (WASC) – Threat Classification

The Web Application Security Consortium (WASC) is composed of an international group of experts, industry professionals and representatives of organizations that seek to produce best practices of security standards, facilitating the exchange and dissemination of ideas, techniques, articles, security guidelines, among other useful documentation used by various companies, educational institutions, governments and software developers worldwide [16].

The WASC Threat Classification is a project resulting from a cooperative effort to clarify and organize attacks and weaknesses that can jeopardize a Web information system, its data or its users. It is compound, according to Table 3, by 49 items arranged in alphabetical order, grouped by attacks and weaknesses, with their respective numbering in the priority classification rank [17]:

Table 3. WASC Threat Classification

(Nr. Rank) Attacks	(Nr. Rank) Attacks	(Nr. Rank) Weaknesses
03- Integer Overflows	29- LDAP Injection	01- Insufficient Authentication
05- Remote File Inclusion	30- Mail Command Injection	02- Insufficient Authorization
06- Format String	31- OS Commanding	04- Insufficient Transport Layer Protection
07- Buffer Overflow	32- Routing Detour	13- Information Leakage
08- Cross-Site Scripting	33- Path Traversal	14- Server Misconfiguration
09- Cross-Site Request Forgery	34- Predictable Resource Location	15- Application Misconfiguration
10- Denial of Service	35- SOAP Array Abuse	16- Directory Indexing
11- Brute Force	36- SSI Injection	17- Improper Filesystem Permissions
12- Content Spoofing	37- Session Fixation	20- Improper Input Handling
18- Credential/Session Prediction	38- URL Redirector Abuse	21- Insufficient Anti-automation
19- SQL Injection	39- XPath Injection	22- Improper Output Handling
23- XML Injection	41- XML Attribute Blowup	40- Insufficient Process Validation
24- HTTP Request Splitting	42- Abuse of Functionality	47- Insufficient Session Expiration
25- HTTP Response Splitting	43- XML External Entities	48- Insecure Indexing
26- HTTP Request Smuggling	44- XML Entity Expansion	49- Insufficient Password Recovery
27- HTTP Response Smuggling	45- Fingerprinting	
28- Null Byte Injection	46- XQuery Injection	

3 Research Methodology

To accomplish the objective of this work, four specific actions were carried out. They were:

(1) Bibliographic research and selection of the main international organizations specialized in conceptualization, treatment and exploitation of information security, followed by the selection of works focused on the treatment of vulnerabilities in Web information systems.

(2) From the organizations selected (OWASP, CWE and WASC), a comparative analysis, mapping and union of their work results were made. It was created a category list of main risks, programming errors and security vulnerabilities in Web applications.
(3) From the result of item 2, a proposal of proactive treatment for vulnerabilities and risks by the developers of information systems in Web environments was created to support software development, based in proactive actions of previous works of the organizations selected.
(4) Finally, a use experience was carried out using a real information system of a Brazilian Public Institution, selecting a part of the system (functional modules) to test the efficiency of the application of the results of this work. The test was carried out by a company specialized in information security that provides services to public institutions and other companies.

4 Selection of Vulnerabilities and Risks

The purpose of this study is to provide a more comprehensive selection (categories) of the most critical factors to treat vulnerabilities in Web applications, to support software developers of different companies and institutions. They will not be limited to the implementation of controls based on vulnerabilities and risks identified by the rank of only one of these institutions, such as OWASP Top 10.

This work takes into account notes made by a larger group of institution that carry out these studies and selections, including also publications provided in time prior to the most recent ranks of some of these institutions.

It is worth mentioning that, although some vulnerabilities come out of the Top 10 and others have come in their place, does not mean that they are no longer important.

The result of OWASP work in previous publications caused developers to start to address a particular vulnerability, causing it to fall in the Top 10 rank after the calculations made by the methodology OWASP, which also considers the prevalence of occurrences.

So, some vulnerabilities, from previous periods, have also been included in the listing, because, although they are no longer within the most recently published listings, they are important.

Table 4 presents the proposed list of the main categories of selected vulnerabilities/risks, with the respective source in which they are cited, and respective actions of selected proactive controls will be mentioned in the subsequent sections.

It is important to highlight that the proposed table has focus on vulnerabilities whose main actions must be carried out by the information systems development team.

That is attacks, vulnerabilities or risks whose proactive action for treatment are not treated through safe coding practices, for example, training of employees to avoid social engineering, among other actions, are not object of study of the present work.

Finally, in Table 4, the field "CATEGORIES" comprises the risks, vulnerabilities, weaknesses and attacks.

Table 4. Proposed categories (Mapping)

CATEGORIES	OWASC TOP 10 (Year: Rank)	CWE TOP 25 (Rank)	WASC T. C (Rank)
Injection	2017: A1 | 2013: A1 2010: A1 | 2007: A2 2004: A6 | 2003: A1, A6	CWE: 01, 02	WASC: 12, 19, 23, 28, 29, 30, 31, 36, 37, 46
Broken Authentication (and Session Management)	2017: A2 | 2013 A2 2010: A3 | 2007: A7 2004: A3 | 2003: A3	CWE: 05, 07, 21	WASC: 01, 18, 37, 47, 49
Sensitive Data Exposure	2017: A3 | 2013: A6 2010: A7 | 2010: A9 2007: A8, A9 | 2004: A8 2003: A8	CWE: 08, 19, 25	WASC: 04
Cross Site Scriptiong (XSS)	2017: A7 | 2013: A3 2010: A2 | 2007: A1 2004: A4 | 2003: A4	CWE: 04	WASC: 08
Unvalidated Input/Output	2004: A1	CWE: 10	WASC: 20, 22
Security Misconfiguration	2017: A6 | 2013: A5 2010: A6 | 2004: A10 2003: A10	CWE: 11, 17	WASC: 14, 15, 16, 17, 48
Broken Access Control	2017: A5, A4 | 2013: A4, A7 2010: A4, A8 | 2007: A10 2004: A2 | 2003: A2	CWE: 6, 13, 15, 16	WASC: 02, 11, 21, 33, 34
XML External Entities	2017: A4	*	WASC: 23, 32, 35, 41, 43, 44
Cross Site Request Forgery (CSRF)	2013: A8 | 2010: A5 2007: A5	CWE: 12	WASC: 09

(*continued*)

Table 4. (*continued*)

CATEGORIES	OWASC TOP 10 (Year: Rank)	CWE TOP 25 (Rank)	WASC T. C (Rank)
Unvalidated Redirects and Forwards	2013: A10 | 2010: A10	CWE: 22	WASC: 38
Information Leakage and Improper Error Handling	2007: A6 | 2004: A7 2003: A7	*	WASC: 13
Malicious File Execution	2007: A3	CWE: 09	WASC: 05
Insufficient Logging and Monitoring	2017: A10	*	*
Denial of Service	2004: A9 | 2003: A9	*	WASC: 10
Buffer Overflows	2004: A5 | 2003: A5	CWE: 03, 23, 25	WASC: 03, 06, 07
Insecure Deserialization	2017: A8	*	*
Using Components with known Vulnerabilities	2017: A9 | 2013: A9	CWE: 14, 18	*
HTTP Request/Response Splitting	*	*	WASC: 24, 25

In order to help the selection of the categories and their importance for the composition of the proposal, a study and mapping between the results of the three institutions work products were carried out.

In the bibliographic researches [18, 19], it was verified that each institution carried out a mapping of its vulnerabilities/risks with ranks of other institutions. So, before the identification and selection of which categories would be included in the proposal, it was necessary to complement the mapping with information from the most recent ranks.

It should be highlighted that the indication of the sources is limited to vulnerabilities within the ranks or lists from each institution. The vulnerability may not appear in all ranks, because one institution may have considered the vulnerability at greater risk than the other.

Finally, two important categories may also be take into account in the software development process, which are inherent vulnerabilities in the programming language used for the development of a particular information system, and vulnerabilities inherent in the business logic in which due to insufficient validation in the process (WASC-40) can also generate openings for malicious agents.

These vulnerability categories were not included in Table 4 because they have proactive treatment controls depending on the choice of software developers or system business rules.

5 Selection of Proactive Controls

After selecting the main categories for vulnerability and risk management in Web information systems, it was also necessary to select which proactive controls that have to be implemented by the development team to minimize or avoid the action of malicious agents.

The treatment of a given vulnerability and/or risk is not necessarily unique. So they can have different solutions, and some vulnerabilities may be interrelated. For example, lack of protection or poor handling of a particular vulnerability may provide an opening for another vulnerability, even if the developer has implemented a right protection action.

After the study of proactive controls has been made on the body of knowledge of the institutions [20, 21], possible controls for use and implementation by developers of Web information systems were selected. They can treat the categories selected in the proposal (Table 5).

Table 5. Selection of proactive controls

Category	Actions/Proactive Controls
Injection	Primary Options: - Use of Prepared Statements (Parameterized Queries) - Use of Stored Procedures - Escaping all User Supplied Input Additional Defenses: - Enforce: Least Privilege - Input/Output Validation, Encoding
Broken Authentication (and Session Management)	- Multi-Factor authentication - Secure Password Storage (Use of Criptografy and Salt) - Secure Password Recovery Mechanism - Session Management, Generation and Expiration - Transport Layer Protection, Security Headers
Sensitive Data Exposure	- Transport Layer Protection, Security Headers - Secure Password and Criptografy Storage (Use of Criptografy and Salt)
Cross Site Scriptiong (XSS)	- Input/Output Validation, Encoding, Escaping, Sanitizing
Unvalidated Input/Output	- Input/Output Validation
Security Misconfiguration	- Secure configuration, Security Headers - Remove unnecessary features, components, frameworks, documentation and samples - Disable default accounts and passwords
Broken Access Control	- Disable web server directory listing and Disable Caching - Anti-automation measures - Authentication Validation (User validation in every page)
XML External Entities	- Input/Output Validation - XML validation against local Static DTD/ XML Schema/ XSD - Disable XML external entity and DTD processing

(*continued*)

Table 5. (*continued*)

Category	Actions/Proactive Controls
Cross Site Request Forgery (CSRF)	- Use of Synchronizer Token
Unvalidated Redirects and Forwards	- Input/Output Validation and Sanitizing
Information Leakage and Improper Error Handling	- Input/Output Validation; Secure configuration - Create a default error handler with sanitized error message
Malicious File Execution	- Input/Output Validation, Encoding, Escaping - Secure configuration
Insufficient Logging and Monitoring	- Input/Output Validation, Encoding, Escaping and Sanitizing
Denial of Service	- Input/Output Validation, Encoding, Escaping and Sanitizing - Anti-automation measures
Buffer Overflows	- Use components from official sources - Input/Output Validation
Insecure Deserialization	- Input/Output Validation
Using Components with known Vulnerabilities	- Remove unnecessary features, components, frameworks, documentation and files - Use components from official sources
HTTP Request/Response Splitting	- Renew the Session ID After Any Privilege Level Change - Input/Output Validation (Headers), Escaping and Sanitizing

As can be seen in Table 5, several actions deal with not only a particular category, but also impact on the treatment of other vulnerabilities, such as Input/Output Validation.

However, for example, although the action of validation serves to treat several categories (as well as other cases), the way the treatment will be implemented will depend on which category is being treated. For instance, the validation of information may have different sources such as HTTP headers, URLs, application forms, and so on, which have led to different types of implementations.

6 Experience of Use and Results

In order to analyze the efficiency of the proposal to the treatment or reduction of vulnerabilities in Web information systems, an experience of use was performed, testing modules of a real system of a Brazilian Public Institution.

6.1 Experience Configurations

For the accomplishment of the experiment, four software modules (functionalities) of a real system already existent in a certain Public Institution were selected, being: (1) Functionality of authentication and access control (Login); (2) Credentials recovery functionality (I forgot my password); (3) Enrollment functionality to enable access by completing forms with sensitive data (Request access); (4) Information query functionality in the local database (Individual query by name or personal ID).

After the selection and replication of the modules and databases for a test environment, they were subjected to vulnerability analysis and testing, and invasion testing by a company specialized in information security that provides services to the Public Institution in question.

To categorize the number of vulnerabilities found in criticality and risk levels, the security company adopted, according to Table 6, a terminology of levels for the definition of risk factors.

Table 6. Definition of risk factors

Risk	Definition of Factors
Critical Risk	Very easy-to-exploit security threats that aim to take complete control of the host, remote code execution, and denial of service
High Risk	Security threats that may compromise data integrity, exploit confidential information, or be used to render the systems unavailable
Middle Risk	Security threats that can breach the system for unauthorized persons, expose data, files and information, or cause some kind of downtime, usually in some specific application
Low Risk	Threats that do not normally result in unauthorized access or compromise of a system, but when combined with another type of vulnerability it contributes as a security breach to exploit a system

6.2 Results and Discussions

After completing the analysis and tests, the report available by the company pointed out 14 vulnerabilities in the selected modules, 3 of medium risk and 11 of low risk.

Then, the same modules were sent to a security company. However, all modules were corrected based on the categories and respective proactive controls presented in this work. It was also requested to the security company that all analyzes and tests must be performed in the same way as in the first test, so that a real comparative analysis of results could be possible.

The new result obtained was analyzed. So it could be seen that no vulnerability was found within any risk factor after the completion of the analyzes and tests by the security company.

We could verify the treatment of 15 out of 18 of the vulnerability categories discriminated in the proposal of the present work, remaining only the verification of the categories XML External Entities, Malicious File Execution and Insecure Deserialization.

This experience made it possible to have some evidence of the effectiveness of the proposal in the treatment or reduction of vulnerabilities for a real information system.

However, although the results obtained were promising, the experience did not have all the categories as object of analysis, because the actual system used as experience does not have functionalities implemented for testing the 3 remaining categories.

Another point to consider is that there may be so specific functionalities in information systems that are not implemented so frequently. And so, their vulnerabilities may not have been mentioned in the ranks of the institutions used by this work.

7 Conclusion

According to [22] it is important to emphasize that it is not possible to obtain absolute security, and it is only possible to seek and adopt security policies and practices with the aim of minimizing risks, since, at all times, new vulnerabilities and threats arise. Besides, it is essential, in the case of software, that security actions be applied since the beginning of the development.

This work presented a proposal to contribute to information security and software engineering area. It is related to the treatment of vulnerabilities in Web information systems, in order to minimize the performance of malicious agents. We could observe that it is fundamental to contract companies specialized in security services, since they provide more credibility to the tests. Tests carried out only by the development team of the institutions can introduce some kind of bias.

It is important to highlight that the work products and results of this work can generate impacts not only in Brazilian Public Institutions, but in companies and organizations of diverse sizes throughout the world.

For future work it is suggested to carry out new experience of use, including the implementation of the controls related to the remaining categories. Besides, tests in other companies and Public Institutions.

References

1. Rexha, B., Halili, A., Rrmoku, K., Imeraj, D.: Impact of secure programming on web application vulnerabilities. In: 2015 IEEE International Conference on Computer Graphics, Vision and Information Security (CGVIS), Bhubaneswar 2015, pp. 61–66. https://doi.org/10.1109/cgvis.2015.7449894
2. Symantec Corporation: Internet security threat report, vol. 21, April 2016. http://www.ijssst.info/info/IEEE-Citation-StyleGuide.pdf. Accessed 20 Nov 2017
3. OWASP: Segurança na web: Uma janela de oportunidades. Uma mensagem do OWASP Brasil ao Governo Brasileiro, March 2011. https://www.owasp.org/images/1/16/Seguranca_na_web_-_uma_janela_de_oportunidades.pdf. Accessed 5 Nov 2017
4. Holik, F., Neradova, S.: Vulnerabilities of modern web applications. In: 2017 40th International Convention on Information and Communication Technology, Electronics and Microelectronics (MIPRO), Opatija, pp. 1256–1261 (2017). https://doi.org/10.23919/mipro.2017.7973616
5. Espirito Santo, A.F.S.do.: Segurança da informação. Departamento de Ciência da Computação - Instituto Cuiabano de Educação (ICE), Cuiabá, MT, Brasil. http://www.ice.edu.br/TNX/encontrocomputacao/artigos-internos/aluno_adrielle_fernanda_seguranca_da_informacao.pdf. Accessed 20 Nov 2017, unpublished

6. Ferreira, F.N.F., Araújo, M.T.d.: Política de segurança da informação – Guia prático para elaboração e implementação (2nd. Ed Revisada). Editora Ciência Moderna Ltda, Rio de Janeiro, Brasil (2008)
7. Prazeres, A.P.: Princípios para o desenvolvimento de software seguro. Monografia (Especialização em Engenharia de Projetos de Software). Universidade do Sul de Santa Catarina, Florianópolis, SC, Brasil (2015)
8. Kumar, R.: Mitigating the authentication vulnerabilities in web applications through security requirements. Inf. Commun. Technol. (WICT) **60**(2), 651–663 (2016). https://doi.org/10.1109/WICT.2011.6141435
9. Holanda, M.T., Fernandes, J.H.C.: Segurança no desenvolvimento de aplicações: GSIC701. Curso de Especialização em Gestão da Segurança da Informação e Comunicações: 2009/2011. Departamento de Ciências da Computação da Universidade de Brasília, Brasília, DF, Brasil, November 2017
10. OWASP: About the Open Web Application Security Project, 25 November 2017. https://www.owasp.org/index.php/About_The_Open_Web_Application_Security_Project. Accessed 10 Nov 2017
11. Mon'teverde, W.A.: Estudo e análise de vulnerabilidades web. Trabalho de conclusão de curso (Curso superior de Tecnologia em Sistemas para a Internet). Universidade Tecnológica Federal do Paraná, Campo Mourão, PR, Brasil (2014)
12. Atashzar, H., Torkaman, A., Bahrololum, M., Tadayon, M.H.: A survey on web application vulnerabilities and countermeasures. In: 6th International Conference on Computer Sciences and Convergence Information Technology (ICCIT), Seogwipo, pp. 647–652 (2011)
13. OWASP: The Ten Most Critical Web Application Security Risks. OWASP Foundation, Final Release November 2017. https://www.owasp.org/index.php/Top_10-2017_Top_10. Accessed 25 Nov 2017
14. Mitre Corporation: CWE (Common Weakness Enumeration), 03 February 2018. https://cwe.mitre.org/. Accessed 06 Nov 2017
15. Mitre Corporation: CWE/SANS Top 25 Most Dangerous Software Errors, 06 June 2011. http://cwe.mitre.org/top25/. Accessed 25 Nov 2017
16. Web Application Security Consortion: WASC – The web application security consortion. http://www.webappsec.org/aboutus.shtml. Accessed 10 Nov 2017
17. Web Application Security Consortion: The WASC Threat Classification (2nd edn.), 01 January 2010. http://projects.webappsec.org/f/WASC-TC-v2_0.pdf. Accessed 10 Nov 2017
18. OWASP: Vulnerability Classification Mappings, 21 January 2010. https://www.owasp.org/index.php/Vulnerability_Classification_Mappings. Accessed 10 Nov 2017
19. Web Application Security Consortion: Threat Classification Taxonomy Cross Reference View (2013). http://projects.webappsec.org/w/page/13246975/Threat%20Classification%20Taxonomy%20Cross%20Reference%20View. Accessed 10 Nov 2017
20. OWASP: OWASP Top 10 Proactive Controls 2016. 10 Critical Security Areas That Web Developers Must Be Aware of. https://www.owasp.org/images/5/57/OWASP_Proactive_Controls_2.pdf. Accessed 10 Nov 2017
21. Mitre Corporation: 2011 CWE/SANS Top 25: Monster Mitigations, 23 June 2011. http://cwe.mitre.org/top25/mitigations.html. Accessed 10 Nov 2017
22. Santiago, H.L.P., Lisboa, G.d.S.: Segurança de sistemas da informação – O contexto da segurança dos sistemas de informação. Faculdade Atenas, Paracatu, MG, Brasil (2011). http://www.atenas.edu.br/Faculdade/arquivos/NucleoIniciacaoCiencia/REVISTAS/REVIST2011/6.pdf. Accessed 28 Nov 2017

An Approach of Filtering to Select IMFs of EEMD in Signal Processing for Acoustic Emission [AE] Sensors

Nur Syakirah Mohd Jaafar(✉), Izzatdin Abdul Aziz, Jafreezal Jaafar, and Ahmad Kamil Mahmood

Centre for Research in Data Science, Universiti Teknologi PETRONAS, Seri Iskandar, Malaysia
{nur_16001470, izzatdin, jafreez, kamilmh}@utp.edu.my

Abstract. The pipeline system is the important part in the media transportation for oil and gas transmission but due to weak maintenance, it leads to the corrosion, leakage stresses and mechanical damage of oil and gas pipelines. The signal processing is used to decompose the raw signal and analysis will be in time-frequency domain. Number of existing signal processing methods can be used for extracting useful information. However, the problem of signal processing method, essential to highlight the wanted information and attenuate the undesired signal is trivial. Several signal processing methods have been implemented to solve this issue. Research using Empirical Mode Decomposition (EMD) algorithm shows promising results in comparison to other signal processing methods, especially in the accuracy showing the relationship between signal energy and time – frequency distribution by represents series of the stationary signals with different amplitudes and frequency bands. However, this EMD algorithm will still have noise contamination that may compromise the accuracy of the signal processing to highlight the wanted information. It is because the mode mixing phenomenon in the Intrinsic Mode Function's (IMF) due to the undesirable signal with the mix of additional noise. There is still room for the improvement in the selective accuracy of the sensitive IMF after decomposition that can influence the correctness of feature extraction of the oxidized carbon steel. Using two data sets from the Acoustic Emission Sensors [AE], signal processing flows have been presented in this paper. Wave propagation in the pipeline is a key parameter in acoustic method when the leak occurs. More experiments and simulation need to be carried out to get more result for leakage signature and localisation of defect.

Keywords: Signal processing method · Empirical Mode Decomposition
Ensemble Empirical Mode Decomposition · Intrinsic Mode Function's
Noise contamination · Mode mixing · Selective accuracy of IMF

R. Silhavy et al. (Eds.): CoMeSySo 2018, AISC 860, pp. 100–111, 2019.
https://doi.org/10.1007/978-3-030-00184-1_10

1 Introduction

The pipeline operates for several years, and tend to leakage, corrode and mechanical damage. This damage can result in catastrophic failures. With growth of length and operation time of pipelines the number of their stops increases. Leakages, corrosion of pipelines will lead to damage of environments, financial losses in term of fluid, maintenance wastage and environmental impacts. In real environment, dealing with the dynamic signal is very challenging because not-frequent type of signal normally occurs for the domain. To partially solve the not-frequent type of signal, the urge to have pattern recognition in time domain is needed.

Improved signal denoising method is defined by reducing the unwanted signals from the raw signal which are collected from the sensors that installed outside of the pipeline network [1]. The ultrasonic inspection and magnetic flux leakage inspection technique are the common methods for evaluating and monitoring the state of pipelines through the collected signals from the interior of the offshore pipeline [2]. The signal characteristics will have derived the anomalies, signal amplitudes, signal phases and signal frequency to ascertain actions to be taken ahead. Recently, research works on the acoustics method are still ongoing by the researchers for multi-radius pipelines and variational work condition.

Lacking techniques for evaluating and monitoring the state of pipelines through the collected signals from the interior of the pipeline leads to inaccurate corrosion detection. This happens due to high background noise and noise condition which will affect the actual signal. For example, the signals which can reveal the anomalies can be overlapped. Thus, the recovery of a signal from observed noisy data is important. The analysis of techniques of signal processing is essential to highlight the wanted information and attenuate the undesired one [3]. Butterworth low pass filter, wavelet-based thresholding filter (Wavelet Transform) and Hilbert Huang Transform (HHT) are few examples of complex decomposition signal breaking into finite components [4]. These existed denoising methods in signal processing still have limitations in the removal of noise reduction to achieve the accurate reading of signal signatures. In general, most often signal processing methods are applied to solve the hidden results from the raw signal including removing noisy signals from the data collected through sensors which are installed outside of the pipeline network.

Empirical Mode Decomposition (EMD) algorithm in Hilbert Huang Transform (HHT) contains undesirable signal signatures. Undesirable signal signature contains anomalies such as noises interference, which leads to signal misinterpretation. Signal is decomposed into many IMFs by EMD in HHT. Based on the analysis on EMD algorithm, the mode mixing problems are encountered when signal contains intermittency. The mode mixing will affect the decomposition of signal into different IMFs on the time series as similar scale is residing in different IMF. Thus, this drawback creates a necessity for a proposal and development of an improved decomposition of signal using Ensemble Empirical Mode Decomposition (EEMD).

The problem found creates a necessity for a proposal and development of an improved method to decompose signals using Ensemble Empirical Mode Decomposition (EEMD). This is to improve the reconstruction of noise removed signal when the selection of the Intrinsic Mode Functions (IMFs) is relevant to the corresponding of the most important structures of the signal. The following research questions are extracted from the problem statement.

[RQ1]: What are the correct values for ensemble numbers and amplitude of the white noise that are relevant with signal processing method using Ensemble Empirical Mode Decomposition (EEMD)?

[RQ2]: To what extent does Ensemble Empirical Mode Decomposition (EEMD) that takes advantage of Intrinsic Mode Functions (IMF), able to analyses filtering in frequency-time domain?

[RQ3]: To what extent does the selection of the Intrinsic Mode Functions (IMFs), able to reconstruct the noise removed signal after the signal – filtering method been proposed?

The research aims are to improve the existing Empirical Mode Decomposition (EMD) algorithm using Hilbert Huang Transform (HHT). Through the preliminary research conducted, several research questions had been brought forward. To address the research questions arose; three research objectives have been devised:

[RO1]: To evaluate the correct values for each parameter; ensemble numbers and amplitude of the white noise that is relevant to Hilbert Huang Transform (HHT) using Ensemble Empirical Mode Decomposition (EEMD).

[RO2]: To compare and analyze filtering used in the frequency-time domain analysis.

[RO3]: To construct signal signature based on the selected principal Intrinsic Mode Functions (IMF) components to indicate the noise removed signal using data from oxidized carbon steel.

The study focuses on the existing EMD algorithm using HHT to recover a signal from the raw signals. The improved signal processing methods, which the extent of the EMD algorithm; are Ensemble Empirical Mode Decomposition (EEMD) algorithm are tested against the dataset obtained through field testing. The parameters tested for this research work are limited to the analysis of the EEMD algorithm which is many ensemble members, fixed sifting numbers and amplitude of the added white noise.

The algorithm used in this research is EEMD algorithm and for the representations in the frequency-time domain, HHT is used. The existing implementation of the EEMD algorithm is utilized to perform signal decomposition to enhance the thoroughness of denoising by fixing the parameters analysis of the EEMD. The algorithm is tested against two data sets obtained from the field testing. The study is limited to the signals that were corrupted by additive white Gaussian noise and is conducted based on extended numerical experiments. The algorithm is tested again two data sets obtained from the field testing by using the acoustic emission sensors R 15 D high frequency AE sensor. The two data sets are data sets for test cases of leak and no-leak test cases.

2 Sound Wave Propagation and Acoustic Method

Leak in the pipeline happens when the wall is having hole, crack, or ruptures. The pressure difference in pipeline always flows from higher pressure to lower pressure. The imbalance pressure detected and analyzed to detect the leak when the trials set are conducted, and the expected result is not consistent. When sound waves meet a surface of the wall of pipeline, some parts of the energy are reflected, transmitted through it and some is absorbed. The wave propagation in the pipeline is a key parameter in acoustic method when the leak occurs, and the pressure balance of the pipe will have generated by the friction between walls in the pipe.

Acoustic sensor normally installed outside the pipeline network. The leak generates noise and pick up by these acoustic sensors and the most important thing is to minimize the background noise. Thus, sensor is important to suit with the suitable of purpose of the research. The parameters that influence the selection of acoustic of sensors depends on such as sensitivity and frequency response over the operating range of the environment. The sensor itself should have design that isolates the sensing terminals electrically from the cavity. This electrical isolation makes the sensor particularly useful for applications where high background electrical noise is a major concern.

3 Signal Processing

The signal processing is used to decompose the raw signal and analysis will be in time-frequency domain. Number of existing signal processing methods can be used for extracting useful information. However, the problem of signal processing method, essential to highlight the wanted information and attenuate the undesired signal is trivial. Several signal processing methods have been implemented to solve this issue. The signal processing methods were used such as Fast Fourier transforms (FFT), Wavelet transforms (WT), and Hilbert Huang transforms (HHT) used for the leak detection issues, corrosion and mechanical changes issues. The raw signal from the dataset of field testing; signal processing is needed to reveal the hidden result and information. By having the data sets of test cases of leak and no-leak test cases, the leakage signatures can be detected by the acoustic signal extraction.

4 Experiment Setup

The objective of the experiment setup for this research is to compare leak and no-leak pipe operation using two sensors at two positions. Thus, to be more specific, to compare AE signatures between leak and no-leak condition at different distances. The placement of the sensor; sensor_1[s1] is placed on the leak position; while the sensor_2 [s2] were placed right side of the leak (downstream, pump side) at different distances.

Figure 1 shows the experiment setup for the lab scales experiment for the test-rig and sensor positioning layout. The data acquisition info consists of the analog filter; 20 kHz–3 MHz, digital filter; 100 kHz–400 kHz, the sampling frequency; 5 MHz and pre-amplifier gain: 40 dB. **d** shows distance from leak location.

<table>
<tr><td>Data acquisition info</td><td>Connection layout: Sensor (R 15 D) > Pre-amp (20 kHz – 1.2 MHz)> PCI-2 AE system
No. of sensors and position: 2, Sensor 1 – at leak location (leak at pipe bottom, sensor at top); Sensor 2 – right side of leak location (pump side) at different distances set by test no.
Analog filter: 20 kHz – 3 MHz
Digital filter: 100 kHz – 400 kHz
Sampling frequency: 5 MHz
Pre-amplifier gain: 40 dB
Data length:</td></tr>
<tr><td>Test-rig and sensor positioning layout</td><td>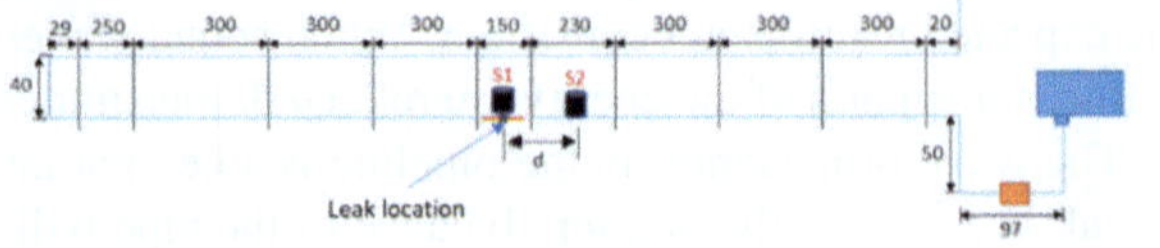

d= distance from leak location</td></tr>
<tr><td>Test conditions</td><td>Test 1: S1- 0m; S2- d=1m;
Test 2: S1- 0m; S2- d=2m;
Test 3: S1- 0m; S2- d=4m;
Test 4: S1- 0m; S2- d=6m;
Liquid pressure: No leak- 1 Bar (223 lpm); 2 Bar (302 lpm)
leak- 1 Bar (221 lpm); 2 Bar (302 lpm)</td></tr>
</table>

Fig. 1. Experiment setup

Test conditions consist from Test 1 to Test 4 as shown in the Fig. 1. The liquid pressure for no-leak for 1 bar pressure is (223 | pm); and for 2 bar pressure is (302 | pm). While for leak; for 1 bar pressure is (221 | pm); and 2 bar pressures (302 | pm).For this research, test condition from test 1, sensor_1[s1] is placed on the leak position; while the sensor_2[s2] were placed right side of the leak (downstream, pump side) at 1 m distance is analysed.

5 Literature Review

5.1 Selection of Signal Processing Method

Table 1 describes the existing method in signal processing methods. For this research, EEMD by HHT is selected. The main reason of EEMD by HHT is selected due to the feature that it is data driven basis and suitable for analysis on the signal interpretation algorithm. It is important that the signal processing method can decomposed the signal into mono component functions. EEMD by HHT works through performing a time-adaptive decomposition operation.

By having the decomposition signal without requires no prior knowledge of the target signal, this will be more efficient in filtering accuracy. Even though EEMD can filter between noise and denoised signal, it still lacks in accuracy to filter between selections of the IMFs. As EEMD, the method still encounters the problem of mode mixing when the signal contains intermittency when the parameter of the EEMD is not set up. The parameter of the EEMD is based on the knowledge of the researchers [11–13].

Wavelet Transform method are more effective in performance compared with the EMD. Still this method involves major drawbacks when it comes to the selection of the wavelet based [12, 13]. The next subsection discusses in brief on EEMD algorithm.

Table 1. Comparison between signal processing methods

Name	Advantages	Disadvantages
Fourier transform by (Fast Fourier Transform) [5]	Fast Fourier Transform has accuracy in analytical form of smoothness signal of the underlying function [5]	Limitations for Fast Fourier Transform for pipeline having no localization of signal discontinuity [6]
Wavelet based thresholding filter by (Wavelet Transform) [7]	Wavelets have allowed filters to be constructed for signal localization for both time and frequency domain [7]	Limitations for the Wavelet Transform in signal generated for length of the basic wavelet function, wavelet base, scale, threshold function and optimal threshold values are fixed [8, 9]
Empirical Mode Decomposition by Hilbert Huang Transform (HHT) [10]	No pre-determine filter or wavelet function for denoising the signal [11]	Mode mixing occur in the Empirical Mode Decomposition (EMD) [11]
Ensemble Empirical Mode Decomposition by Hilbert Huang Transform (HHT) [10]	No pre-determine filter or wavelet function for denoising the signal [12]	Parameters analysis of the Ensemble Empirical Mode Decomposition (EEMD) is not fixed for the decomposition of the signal leading to the mode mixing [11]

5.2 Ensembles Empirical Mode Decomposition (EEMD)

EEMD, means from the numbers of IMF's been extracted from the adapted of EMD to the original signal. The ensemble mean is calculated with a different addition of white noise.

Figure 2 shows the process of EEMD algorithm. Given the signal, EEMD can be summarized as follows. By having many amplitudes of added white noise to the original signal resulted as noise-added signal. Next, the EMD operation will be implemented with the noise-added signal. Lastly, the ensemble means of each IMF component will be calculated.

Figure 3 shows an example of an IMF (green curve) with corresponding envelopes. A signal consists of higher-frequencies waves riding upon lower-frequencies carrying waves. The characteristic of time scales is used to find the intrinsic modes.

The main idea of having the sifting process which is the IMF is to separate the data into a slowly varying local mean part and a fast-varying symmetric oscillatory part. The latter part becoming the IMF represented by the green envelope and the local mean represented by the black line defining a residue [14]. This residue serves as input for further decomposition, with the process being repeated until no more oscillations can be obtained [14].

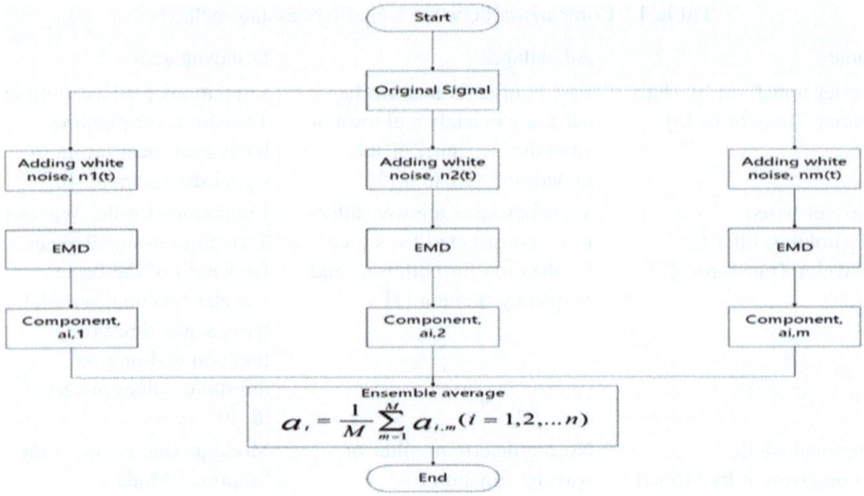

Fig. 2. Process of EEMD algorithm

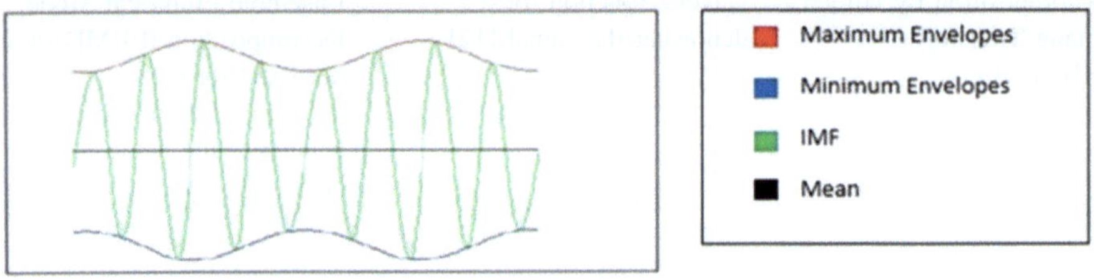

Fig. 3. Example of an Intrinsic Mode Function (IMF)

To overcome partially on the mode mixing phenomenon within the IMF, EEMD is proposed [15, 16]. This EEMD defines the true IMF components as the mean of an ensemble of trials, whereby each of the signals consist of the added white noise of finite amplitude. With this EEMD approach, the mode mixing scales in the IMF could be partially resolves based on the recent studies of the statistical properties of white noise [17]. The next section discusses existing works that utilize EEMD.

5.3 Existing Work for Ensembles Empirical Mode Decomposition (EEMD)

The content in the Table 2 explains the comparison between works implementing EEMD in decomposing the signal. Based on the existing works, all of them proved that EEMD could perform decomposition of signal. Either it is decomposition of signal in acoustic, ultrasonic guided waves, or any other types of signals, EEMD can decompose the signal into mono components to be more meaningful to be interpreted.

Table 2. Existing works of EEMD

Name	Type of signal	Findings	Way forward
Adnan et al. [8]	Acoustic signal	The signal processing is used to decompose the raw signal and show in time-frequency. Acoustic signal and HHT is the best method to detect leak in gas pipelines	This proposed approach could be applied for analysis of acoustic signal for the acoustic emission [AE] sensor in pipeline
Camarena-Martinez et al. [18]	PQ disturbances	An approach based on EMD for analysis of PQ signals has been presented. The method consists of iterative down sampling stage that helps to extract the fundamental component as the first IMF	This proposed approach could be applied for analysis of acoustic signal for the acoustic emission [AE] sensor in pipeline
Xu et al. [19]	Acoustic signal	An adaptive EMD threshold denoising method for acoustic signal has been presented. The method is introduced into the threshold determination process of the denoising approach	This proposed approach could be applied for analysis of acoustic signal for the acoustic emission [AE] sensor in pipeline
Siracusano et al. [20]	Acoustic emission signal	An approach on a framework based on the Hilbert Huang Transform for the analysis of the acoustic emission data. By having the EMD features, the acoustic signal could isolate and extract the signal to be interpreted such as the amplitude to be more meaningful	This proposed approach could be applied for analysis of acoustic signal for the acoustic emission [AE] sensor in pipeline

However, most of the existing works explained that there is minimal work in analyzing signal energy of each IMF's in the EEMD using HHT in identifying the accurate signal reading in frequency-time domain. For this research, the question is whether the improvement EEMD algorithm indeed helps in reaching the goal of the data analysis in extracting the signals. Thereby, leading to the identification of t signal signature when the carbon steel is oxidized. Thus, for this research, the number of ensemble numbers, and the amplitude of the added noise, will be the parameters to be studied in reaching the goal of the research.

6 Methodology

There are several activities which are carried out for this research. For example, Fig. 4 depicts the activities conducted for this research.

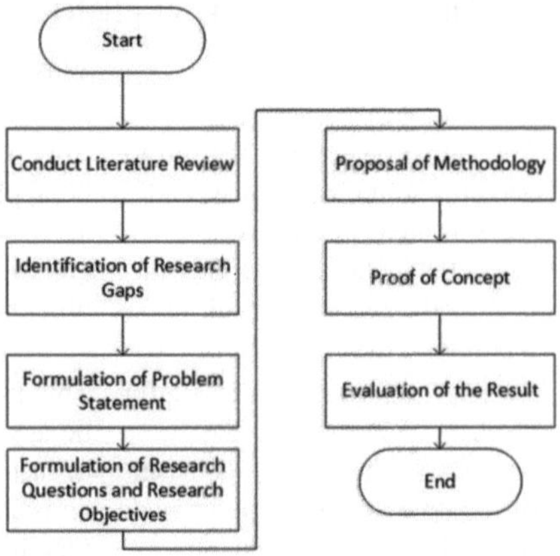

Fig. 4. Research activities

6.1 Solution Requirement

In this research, the parameter that is utilized is the accuracy. The parameter is selected based on the research gaps found from the existing implementations of the Ensemble Empirical Mode Decomposition (EEMD) algorithm. Accuracy is defined as context is defined as the closeness of the retrieve result to the intended result. Hence, accuracy of the algorithm is measured by the power spectral density [PSD] analysis comparison for the cases in 1 bar pressure for both cases which is leak and no leak cases for signal – reading in frequency – time domain.

Thus, from the PSD analysis comparison, further analysis of the dominant band in leak case that does not appear in the no-leak case will be taken ahead. The selected of the IMFs, the signal will be reconstructed indicated the noise removed signal of oxidized carbon steel. Based on the critical analysis performed, EEMD is selected as the decomposition of signal for this research.

Figure 5 shows the signal processing flow. The procedure of the signal processing flow is as follows:

Steps taken of the program will be to decompose the original signal by using the EEMD with the parameter that has been analyzed. The parameters that have been analyzed were the number of the ensemble numbers and the amplitude of the added noise. The decomposition of the signals by sifting process, IMF represents the series of stationary signals with different amplitude and frequency bands.

Frequency component, with different amplitude and frequency bands in each of the IMFs contained sampling of frequency. It resulted simple oscillatory modes as different frequency bands ranging from high to low.

First step of the signal processing flow is by slicing the data into shorter time windows once the parameters for the EEMD have been analyzed. The time windows that will be analyses are 20 ms, 50 ms, 100 ms. The slicing of the signal using a time-domain window to produce multiple samples. And, this will be calculated depending on the effective wave velocity in the guided media. Second steps will be band pass filter for the sliced of the signal in the time window based on the Table 3.

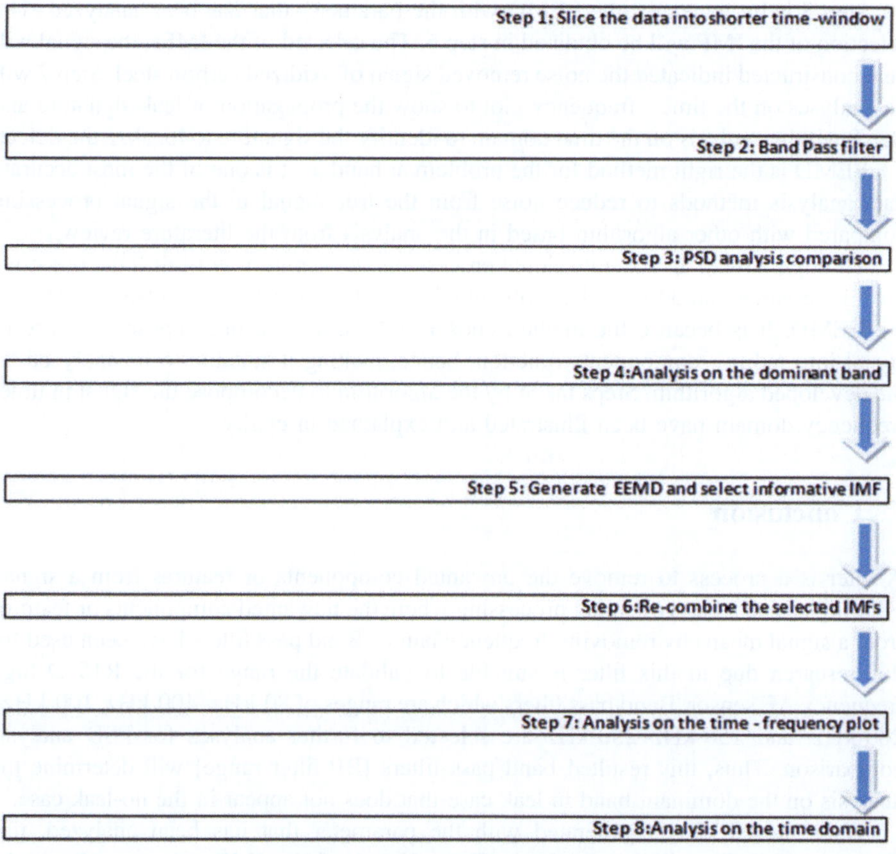

Fig. 5. Signal processing flow

Table 3. Range of Band Pass Filters

Band Pass Filters	Ranges
BP filter range	80 kHz–400 kHz
BP filter range	100 kHz–200 kHz
BP filter range	150 kHz–250 kHz

A filter is a process to remove the unwanted components or features from a signal. Filtering is also a class of signal processing where the unwanted components or features from a signal means by removing frequency bands. The range of the band pass filters are selected for the PSD analysis comparison as in Table 3. Step Three will be comparing for the PSD analysis for the cases in the 1 bar pressure for both cases which is leak and no leak cases. The three sets for both cases are three trials and expected to produce the same results. Next, step four will be analyses on the dominant band in leak case that does not appear in the no-leak case. This analysis will be in a type of signature in frequency domain.

Step 5 is by generates the EEMD with the parameter that has been analyzed. The selection of the IMF will be obtained in step 6. The selected of the IMFs, the signal will be reconstructed indicated the noise removed signal of oxidized carbon steel. Step 7 will be analyses on the time - frequency plot to show the propagation of leak signature and step 8 will be analysis on the time domain to identify the signature to localize the defect.

EEMD is the right method for the problem at hand as it is one of the most accurate data analysis methods to reduce noise from the true signal in the signal processing compared with other algorithm based in the analysis from the literature review.

Through having an accurate signal processing algorithm, it will fulfill the objective for this research and address the problem of accurately selecting the adaptive IMFs of the EEMD. It is because the method chosen will be able to decompose any type of signal into mono - components function, hence, making it suitable to be analyzed by the developed algorithm. Steps taken by the algorithm to decompose the signal in time-frequency domain have been illustrated and explained in clarity.

7 Conclusion

A filter is a process to remove the unwanted components or features from a signal. Filtering is also a class of signal processing where the unwanted components or features from a signal means by removing frequency bands. Band pass filters have been used for this research due to this filter is suitable to validate the range for the R15 D high frequency AE sensor. Band pass filters which are ranges of 80 kHz–400 kHz, 100 kHz–200 kHz, and 150 kHz–250 kHz are selected to further analyses for PSD analysis comparison. Thus, this resulted band pass filters [BP filter range] will determine the analysis on the dominant band in leak case that does not appear in the no-leak case.

Once the EEMD is generated with the parameter that has been analyzed, this selection of IMFs will be obtained. The selected of the IMFs, the signal will be reconstructed indicated the noise removed signal of oxidized carbon steel. By having the analyses on the time-frequency plot to show the propagation of leak signature. Thus, analysis on the time domain leads to identify the signature to localize the defect. Thus, further analysis needed to be conducted for more analysis on leak and no leak events to identify the signal signature. Different test cases supposed to transfer signal energy to a different band for a narrow frequency band (R15 D operating range). Once the signature of each case been identified, the IMF's for the different test cases could be analyzed and reconstruction.

Acknowledgments. This work was supported by Development of Intelligent Pipeline Integrity Management System (I-PIMS) Grant Scheme from Universiti Teknologi PETRONAS.

References

1. Underground pipeline corrosion
2. Shi, Y., Zhang, C., Li, R., Cai, M., Jia, G.: Theory and application of magnetic flux leakage pipeline detection. Sensors **15**(12), 31036–31055 (2015)

3. Gaci, S.: A new Ensemble Empirical Mode Decomposition (EEMD) denoising method for seismic signals. Energy Procedia **97**, 84–91 (2016)
4. Agarwal, M., Jain, R.: Ensemble empirical mode decomposition: an adaptive method for noise reduction. IOSR J. Electron. Commun. Eng. **5**, 60–65 (2013)
5. Karkulali, P., Mishra, H., Ukil, A., Dauwels, J.: Leak detection in gas distribution pipelines using acoustic impact monitoring. In: IECON 2016-42nd Annual Conference of the IEEE Industrial Electronics Society. IEEE (2016)
6. Datta, S., Sarkar, S.: A review on different pipeline fault detection methods. J. Loss Prev. Process Ind. **41**, 97–106 (2016)
7. Jiao, Y.-L., Shi, H., Wang, X.-H.: Lifting wavelet denoising algorithm for acoustic emission signal. In: 2016 International Conference on Robots and Intelligent System (ICRIS). IEEE (2016)
8. Adnan, N.F., Ghazali, M.F., Amin, M.M., Hamat, A.M.A.: Leak detection in gas pipeline by acoustic and signal processing-a review. In: IOP Conference Series: Materials Science and Engineering. IOP Publishing (2015)
9. Fang, Y.-M., Feng, H.-L., Li, J., Li, G.-H.: Stress wave signal denoising using ensemble empirical mode decomposition and instantaneous half period model. Sensors **11**(8), 7554–7567 (2011)
10. Yang, J., Wang, X., Feng, Z., Huang, G.: Research on pattern recognition method of blockage signal in pipeline based on LMD information entropy and ELM. Math. Probl. Eng. **2017** (2017)
11. Kevric, J., Subasi, A.: Comparison of signal decomposition methods in classification of EEG signals For motor-imagery BCI system. Biomed. Sig. Process. Control **31**, 398–406 (2017)
12. Rostami, J., Chen, J., Tse, P.W.: A signal processing approach with a smooth empirical mode decomposition to reveal hidden trace of corrosion in highly contaminated guided wave signals for concrete-covered pipes. Sensors **17**(2), 302 (2017)
13. Samadi, S., Shamsollahi, M.B.: ECG noise reduction using empirical mode decomposition based combination of instantaneous half period and soft-thresholding. In: 2014 Middle East Conference on Biomedical Engineering (MECBME). IEEE (2014)
14. Saeed, B.S.: De-noising seismic data by Empirical Mode Decomposition (2011)
15. Huang, Y., Wang, K., Zhou, Z., Zhou, X., Fang, J.: Stability evaluation of short-circuiting gas metal arc welding based on ensemble empirical mode decomposition. Meas. Sci. Technol. **28**(3), 035006 (2017)
16. Potty, G.R., Miller, J.H.: Acoustic and seismic time series analysis using ensemble empirical mode decomposition. J. Acoust. Soc. Am. **140**(4), 3423–3424 (2016)
17. Honório, B.C.Z., de Matos, M.C., Vidal, A.C.: Progress on empirical mode decomposition-based techniques and its impacts on seismic attribute analysis. Interpretation **5**(1), SC17–SC28 (2017)
18. Camarena-Martinez, D., et al.: Novel down sampling empirical mode decomposition approach for power quality analysis. IEEE Trans. Ind. Electron. **63**(4), 2369–2378 (2016)
19. Xu, J., Wang, Z., Tan, C., Si, L., Liu, X.: A novel denoising method for an acoustic-based system through empirical mode decomposition and an improved fruit fly optimization algorithm. Appl. Sci. **7**(3), 215 (2017)
20. Siracusano, G., Lamonaca, F., Tomasello, R., Garescì, F., La Corte, A., Carnì, D.L., Carpentieri, M., Grimaldi, D., Finocchio, G.: A framework for the damage evaluation of acoustic emission signals through Hilbert-Huang transform. Mech. Syst. Sig. Process. **75**, 109–122 (2016)

Influence Analysis of Selected Factors in the Function Point Work Effort Estimation

Zdenka Prokopova(✉), Petr Silhavy, and Radek Silhavy

Faculty of Applied Informatics, Tomas Bata University in Zlin, nam T. G. Masaryka 5555, Zlin, Czech Republic
{prokopova, psilhavy, rsilhavy}@utb.cz

Abstract. The paper presents an influence analysis of selected factors (independent variables – function point count approach, business area, industry sector, relative size) on the estimation of the work effort for which the function point method is primarily used. Productivity influencing factors and the productivity estimation capability in function point method is studied too.

Keywords: Function point method · Estimation · Work effort Productivity

1 Introduction

Software projects development is understood as a manufacturing process that requires human work. Projects work effort estimation often refer to the preliminary cost of the projects. Estimates are related to the planning, but they are not identical. According to McConnell [1], estimation should be an unbiased, analytical process. On the other hand, the planning should be a process of finding a goal. It is important to plan first and then make estimates.

It is necessary to understand the metrics that are used in the estimation of the projects. Metrics are ways to measure project properties that are used, for example, to estimate the cost and labor (work effort) of the project. There are many metrics, but we need to know when and for what metrics are appropriate.

As part of software measurement, two types of measurement are used [2]:

- Direct measurements - lines of code (LOC), calculation speed, memory size, number of errors for a time, etc.
- Indirect measurements - functionality, quality, complexity, effort, reliability, maintenance capability, etc.

To express the overall estimate, the terms person-hour or work effort are used. Already in the Albrecht publication [3], the number of person-hours is the time that is required to meet one goal by one employee. Person-hour is not a standardized unit, and it is appropriate to use historical data.

One of the methods that are used for work effort estimation is Function Point (FP) method. FP is a metric that is functionally oriented. This method was created as an alternative to the LOC method. The FP method was developed by Allan Albrecht

R. Silhavy et al. (Eds.): CoMeSySo 2018, AISC 860, pp. 112–124, 2019.
https://doi.org/10.1007/978-3-030-00184-1_11

(an IBM worker) in 1979. Its counting rules are established by the International Function Point Users Group (IFPUG). FP is used for a preliminary estimate using limited information. It is not necessary to have the source code to calculate it, which makes it possible to use this method in an earlier phase. Despite the potential for early use of the project, FP provides meaningful and fairly accurate results, what can be documented by some publications. The result, however, depends on the experience of the person using the method, because the functional points require a certain degree of subjectivity, which is one of the greatest drawbacks of this method. Comparison of the results of analyzes and other research intentions on a special dataset is addressed in a systematic mapping study of Gonzalez-Ladron-de-Guevara [4].

2 Problem Statement

The primary aim of our investigation is an influence analysis of selected factors on the determination of productivity, which could be used for work effort estimation. Our research is based on Functional Sized Measurement Method [5]. It was implemented on a special dataset - International Software Benchmarking Standards Group (ISBSG) that maintains a repository of data about completed software projects [6].

2.1 Research Questions and Goals

The research questions answered by this study are as follows:

RQ1: Is productivity dependent on selected factors (independent variables)?
RQ2: Which statistical value is better to use when estimating the work effort - the productivity mean or the median?
RQ3: Is any of the selected factors more appropriate for determining productivity regarding estimation accuracy?

2.2 Evaluation Criteria

There are several commonly used criteria, which are accepted as a standard evaluation for work effort estimation. Therefore, all performed simulations were evaluated according to four of these metrics. Their formulas are given as follows:

Mean squared error (MSE)

$$MSE = \frac{1}{n}\sum\nolimits_{i=1}^{n} (RealWorkEffort - EstimatedWorkEffort)_i^2 \quad (1)$$

The magnitude of relative error (MRE)

$$MRE = \frac{|RealWorkEffort - EstimatedWorkEffort|}{RealWorkEffort} \quad (2)$$

The magnitude of error relative (MER)

$$MER = \frac{|RealWorkEffort - EstimatedWorkEffort|}{EstimatedWorkEffort} \tag{3}$$

Balanced relative error (BRE)

$$BRE = \frac{|RealWorkEffort - EstimatedWorkEffort|}{min(RealWorkEffort, EstimatedWorkEffort)} \tag{4}$$

where n is number of observations.

3 Methods

The principle of FP is based on the division of a project into several smaller units, which may seem simple, but the problem is how to define these smaller units.

3.1 Methods Standards

The International Organization for Standardization and Electrotechnical Commission (ISO/IEC) has standardized five worldwide Functional Size Measurement (FSM) method standards [2]:

- IFPUG - International Function Point Users Group
- COSMIC - Common Software Measurement International Consortium
- FiSMA - Finnish Software Measurement Association
- Mark II - Method was primarily used in the UK
- NESMA - Netherlands Software Metrieken Gebruikers Associate

The most widely used method is the IFPUG method.

IFPUG

The IFPUG was founded in Toronto in 1986. It has developed the standardized IFPUG Function Point Method to count the functional size of a piece of software. The IFPUG has established that the method will focus on identifying five types of components (measurement parameters) [7–10]:

Function points related to transaction functions:

- External inputs (EI) - these components are responsible for changes in internal system data. These include, for example, screens, forms, dialogs, or control signals that allow the user or another program to perform data operations on the system. EI contains all inputs that have a unique format or processing logic.
- External Outputs (EO) - represents how the system presents its data. However, it is necessary to be able to distinguish between components belong to external outputs and which do external queries. EOs include, for example, screens, report charts, or control signals generated by the system for the end user.

- External queries (EQ) - methods used to read data without editing them. This concept comes from the area of database systems and can be imagined as a search for data. Nowadays, thanks to modern applications, it is difficult to distinguish the difference between EO and EQ. In simple terms, queries get data directly, while data outputs are still processing and formatting.

Function points related to data functions:

- External Interface Files (EIF) - files that are responsible for exchanging data with other systems. Often, these systems are controlled. These are each large set of data or informations that come in or out of the system.
- Internal Logical Files (ILF) - files that are used by the system itself. A logical file can consist of one file or one table in a relational database.

Once we have a system described using the above component groups, we can switch to the calculation of Unadjusted Function Points (UFP). Different component groups are additionally divided according to their complexity into three groups (low, medium and large). The following Table 1 shows weights for classification the complexity of individual component groups.

Table 1. Classification of the complexity of individual component groups [7]

Component	Complexity weights		
	Low	Medium	Large
EI	3	4	6
EO	4	5	7
EQ	3	4	6
EIF	5	7	10
ILF	7	10	15

For the calculation of UFP simply set the number of component in groups (multiplier), multiply them by complexity weights and make the sum of all fields in the table.

$$UFP = \sum_{i=1}^{m} \sum_{j=1}^{k} (multiplier_{i,j} \times weight_{ij}) \quad (5)$$

where m is a number of components and k is a number of complexity group.

In the next phase, it is necessary to calculate Value Adjustment Factor (VAF). This value is calculated based on 14 factors which affecting the system (see Table 2).

Table 2. Factors affecting the system

Factor	Description
F_1	Does the system require backup and recovery?
F_2	Are data required for communication?
F_3	Does the system include a distributed processing function?
F_4	Is critical performance required?
F_5	Will the system work during heavy loads?
F_6	Does the system require direct data input?
F_7	Do data inputs require multiple screens or operations?
F_8	Are the main files up-to-date?
F_9	Are inputs, outputs, files and queries intricate?
F_{10}	Is internal processing complicated and complex?
F_{11}	Is the code designed for reuse?
F_{12}	Are conversions and installation included in design?
F_{13}	Is the system designed for multiple installations in different locations?
F_{14}	Is the application designed to make it easy for users to make changes?

Each factor is rated by degree of influence on the application on the 0-5 scale. The following Table 3 represents the significance of the influence factors rating.

Table 3. Factors weights [7]

Influence	Rating
None	0
Not relevant	1
Mild	2
Average	3
Significant	4
Very strong	5

Then we can express:

$$VAF = 0.65 + 0.01x \sum_{i=1}^{14} (F_i \times rating) \tag{6}$$

The number of adjusted function points (AFP) is calculated as:

$$AFP = UFP \times VAF \tag{7}$$

FP calculation is time-consuming, as it requires detailed crawling of the project requirements specification and division into individual components. Several simplified methods have been proposed to minimize the effort required to calculate FP [11]. Of course, also new advanced methods have been developed to estimate the size of software projects [12].

3.2 Productivity Definition and Calculation

Several definitions of productivity exist in the literature. In our case, we understand productivity as the ratio of the software development process output (FP size) divided by the input resources (work effort in person-hours). ISBSG has obtained a productivity measure of the projects, which is called Product Delivery Rate (PDR), which is equal to the number of effort hours needed to release a Function Point [13].

Productivity should not be confused with PDR, which is a metric that reflects the number of input resources (work effort in hours) that it takes to produce one unit of output (FP). As such, PDR is measured in units of work effort in hours per FP. On the other hand, units for productivity (Prod) are FP per work effort in hours:

$$Prod = FP/Work\ \ Effort \tag{8}$$

The successful calculation of productivity depends on accurate measurement of the output software size, as well as the actual work effort expended.

According to several research studies, productivity rates vary widely across different sectors, business areas, as well as other factors. Research shows that some factors can influence the productivity rate and that, e.g. every organization has an optimum productivity range (illustration we can see in the Fig. 1) [14].

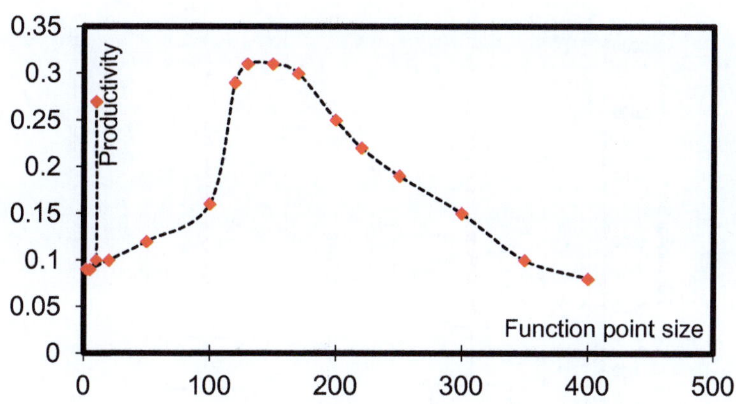

Fig. 1. Illustration of productivity characteristics [14]

4 Experiment Study

4.1 Dataset Description

As a dataset, we had used an ISBSG project repository Release 13. This dataset includes 6760 projects overall. The data in the dataset have come from over thirty countries, and the majority of the projects was completed in the last decade. Detailed information according to full dataset is in [6]. Dataset used in our experiments was obtained by selection of projects with four chosen factors; different count approaches, known business area, known industrial sector and not null functional size and normalized work effort. Based on these requirements we have got a working dataset DS of 1333 projects.

Next experiments will be evaluated by using two sub-datasets (training set DS1 and testing set DS2), which were created using the hold-out approach in ratio 2:1.

Table 4 shows characteristics of used sub-datasets. All values in Table 4 are based on Functional Size (FS) and Normalized Work Effort (NWE), where *n* indicates number of projects. As can be seen, both datasets have similar standard deviations (SD), but DS2 contains several projects with very high NWE (see Max NWE).

Table 4. Dataset's characteristics

	n	Mean FS	Median FS	Min FS	Max FS	SD FS	Mean NWE	Median NWE	Min NWE	Max NWE	SD NWE
DS1	934	345.62	191.50	5.0	2983	431.79	4679.3	2318.5	20.0	1.093e +5	8282.9
DS2	399	346.40	180.00	10.0	2922	475.65	4454.9	2010.0	97.0	78472.0	7435.9

In Fig. 2 a boxplot of the training (DS1) and testing (DS2) datasets is presented.

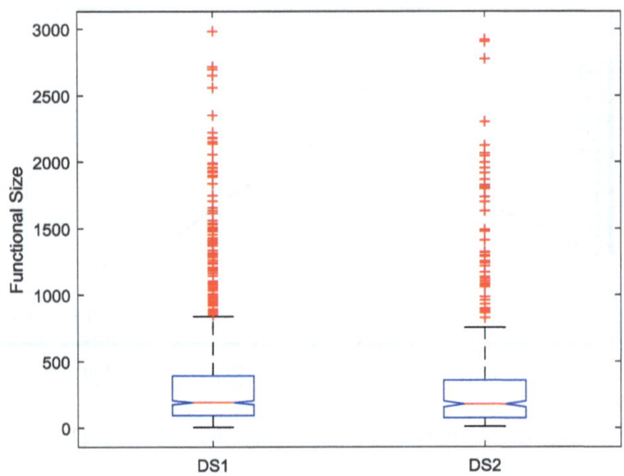

Fig. 2. Boxplot of the training and testing datasets

As we can see, there are many possible outliers in the datasets DS1 and DS2. However, in this study, the goal is not to deal with the cleaning or filtering of datasets. So all analyses were carried out on complete defined datasets DS1 and DS2, including possible outliners.

4.2 Analysis of Dependent and Independent Variables

According to Gonzales [4] "the four most used independent variables are FS, Development Type (DT), Language type (LT), and Development Platform (DP), which is consistent with the fact that these are the variables whose usage is recommended by ISBSG for data partitioning."

For our study, we have chosen other four independent variables, as are Count Approach (CA), Business Area Type (BAT), Industry Sector (IS) and Relative Size (RS). This selection was chosen to determine the influence of these variables on productivity and NWE. Productivity and NWE are understood as dependent variables.

Correlations between variables can be seen in Table 5. There is not a high correlation among examined variables.

Table 5. Correlation coefficients between pairs of variables

	CA	BA	IS	RS	Prod	NWE
CA	1	−0.065179	−0.027839	0.035837	0.16605	0.13415
BA	−0.065179	1	−0.2623	0.045001	0.0074739	0.047487
IS	−0.027839	−0.2623	1	0.0084151	0.072658	0.02247
RS	0.035837	0.045001	0.0084151	1	−0.36975	−0.54426
Prod	0.16605	0.0074739	0.072658	−0.36975	1	−0.54682
NWE	0.13415	0.047487	0.02247	−0.54426	−0.54682	1

4.3 Productivity Calculation Depending on FP Count Approach

The Count Approach variable can be used to select projects that were calculated using a consistent approach. Projects in the ISBSG used to calculate functional size by several FSM methods (COSMIC, IFPUG 4+, IFPUG old, FiSMA, NESMA, Mark II, Dreger, Automated, Feature Points, etc.). Table 6 presents summary statistics of training dataset DS1 organized by count approach.

IFPUG 4+ indicate version 4 or greater, IFPUG old indicate version before 4.

Table 6. Summary statistics organized by count approach

Count approach	Group count	Mean productivity	Median productivity
IFPUG 4+	751	0.10183	0.0808
IFPUG old	113	0.14865	0.1400
NESMA	40	0.12822	0.1041
COSMIC	17	0.14370	0.1376
Mark II	13	0.11729	0.1247

4.4 Productivity Calculation Depending on Business Area Type

Business area type identifies the subset within the organization being addressed by the project [6]. There are 80 business area types in the training dataset. Table 7 presents summary statistics of training dataset DS1 organized by business area type (17 most frequently occurring).

Table 7. Summary statistics organized by business area type

Business area type	Group count	Mean productivity	Median productivity
Telecommunications	185	0.10567	0.08060
Banking	91	0.10196	0.07600
Insurance	84	0.10419	0.08260
Manufacturing	74	0.10885	0.07100
Financial (excl. Banking)	59	0.12518	0.10620
Outbound Logistics	46	0.096274	0.08115
Sales	41	0.089015	0.06900
Inbound Logistics	39	0.12409	0.10680
Product. Process & Design	31	0.1205	0.10950
Accounting	26	0.10806	0.08500
Engineering	26	0.16815	0.20630
After Sales & Services	18	0.088894	0.05970
Quality	17	0.073735	0.05690
General	15	0.10745	0.11610
Inventory	14	0.13359	0.14635
Dealer Operations	13	0.11702	0.10000
Legal	11	0.14987	0.09530
Purchasing & Supplier...	11	0.099482	0.08400

4.5 Productivity Depending on Industry Sector

Industry sector attempts to summarize the type of organization that submitted the project into a single value of a defined set [6]. There are 15 industry sectors in the training dataset. Table 8 presents summary statistics of training dataset DS1 organized by industry sector.

4.6 Productivity Calculation Depending on Project Relative Size

Projects can be categorized into several numbers of size categories. The ISBSG dataset uses a categorization presented by Rule [15] and described in [2].

Table 8. Summary statistics organized by industry sector

Industry sector	Group count	Mean productivity	Median productivity
Manufacturing	351	0.10468	0.0836
Communication	210	0.10641	0.08075
Government	87	0.10587	0.0838
Banking	67	0.091978	0.064
Insurance	65	0.12008	0.0969
Financial	53	0.12727	0.0945
Utilities	36	0.1397	0.1041
Electronics & Computers	20	0.10333	0.08915
Service Industry	13	0.12338	0.1177
Professional Services	9	0.13882	0.1082
Construction	6	0.21312	0.2246
Mining	6	0.12265	0.108
Defence	4	0.078875	0.03415
Wholesale & Retail	4	0.17162	0.19575
Medical & Health Care	3	0.12957	0.1213

Table 9 presents summary statistics of training dataset DS1 organized by relative size.

Table 9. Summary statistics organized by relative size

Relative size	Group count	Mean productivity	Median productivity
M1	380	0.1058	0.0845
M2	241	0.125	0.1066
S	224	0.094816	0.07095
L	69	0.13683	0.0934
XS	16	0.064525	0.0584
XXS	4	0.081575	0.0269

5 Comparison and Discussion of Results

According to the results of the correlation analysis shown in Table 5, it can be seen that among the analyzed variables is not a high correlation. Only RS vs. NWE (−0.54426), which shows that the relative size of the project is an important factor for normalized work effort and Prod vs. NWE (−0.54682), which shows that the productivity is an important factor for normalized work effort. The other parameters are weakly (0.13415–0.36975) or not correlated.

The answer to RQ1: Productivity is weekly dependent on the examined factors. The highest dependency is on CA (0.11843) and RS (−0.11051).

In the following Table 10, we can see the resulting errors (MSE, MRE, MER, and BRE) that we obtained by comparing the estimate and the actual normalized work effort on the testing dataset. For estimates, productivity was calculated as mean or median, either from all data in the testing dataset (no category), or from data organized according to the examined factors.

The answer to RQ2: Based on these results presented in Table 10 we can see, that in all cases errors are smaller in the case of median than mean.

Table 10. Comparison of analysis results

Factor	MSE		MRE		MER		BRE	
	Mean	Median	Mean	Median	Mean	Median	Mean	Median
No category	2.90E+7	2.64E+7	0.52227	0.40323	0.89733	0.42085	1.02800	0.53180
Count approach	2.85E+7	2.61E+7	0.51736	0.39952	0.87371	0.40067	1.00790	0.51435
Business area	2.87E+7	2.58E+7	0.53822	0.43164	0.83765	0.41213	0.98649	0.55476
Industry sector	2.90E+7	2.60E+7	0.51520	0.39473	0.84633	**0.39580**	0.97766	**0.51189**
Relative size	3.35E+7	2.73E+7	0.51515	0.40273	0.85021	0.41785	0.98018	0.56710
Combination	2.96E+7	**2.54E+7**	0.50892	**0.38434**	0.84371	0.40227	0.96985	0.51319

The answer to RQ3: When we compare errors according to different criteria, and for different factors, it is not possible to specify the most appropriate factor that can be used to minimize the accuracy of the estimate.

The Fig. 3 shows the characteristics of NWE in the testing dataset and estimates work efforts through productivity. All estimates follow NWE, but none of them is significantly better than others are. In our opinion, biggest problems are caused by projects that are very different - outliners.

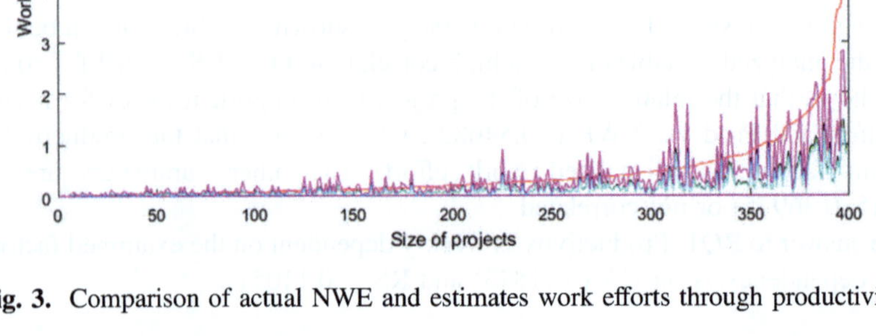

Fig. 3. Comparison of actual NWE and estimates work efforts through productivity

6 Conclusion

This paper presents the analysis of a selected factors influence on the work effort estimation using function points methods. For experimental purposes, an ISBSG dataset was chosen from which were filtered data corresponding to our requirements. We intended to find out whether it is possible to use the productivity gained from the training (historical) dataset to estimate the effort and whether the accuracy of the estimation is influenced by the selected factors. All simulations were calculated in the Matlab environment.

There are many possible outliers in the datasets, as can be seen in Fig. 2. In this study, all analyses were carried out on complete defined datasets DS1 and DS2 including possible outliners. The values of the resulting comparisons are affected by this fact. This fact also reflects the results of the errors presented in Table 10 and the estimates characteristics of the work efforts shown in the Fig. 3. However, it is very positive to find that it is preferable to use a median than the mean for these purposes.

For the future work, we plan to deal with outliners, data cleaning and estimating productivity by looking for its optimum value according to the curve of its course. We will also try to analyze some other factors in presented understanding.

References

1. McConnell, S.: Software Estimation: Demystifying the Black Art. Microsoft Press, Redmond (2006)
2. Bundschuh, M., Dekkers, C.: The IT Measurement Compendium: Estimating and Benchmarking Success with Functional Size Measurement. Springer, Heidelberg (2008)
3. Albrecht, A.J.: Measuring application development productivity. In: Proceedings of IBM Application Development Joint SHARE/GUIDE Symposium, Monterey, CA (1979)
4. Gonzalez-Ladron-de-Guevara, F., Fernandez-Diego, M., Lokan, C.: The usage of ISBSG data fields in software effort estimation: a systematic mapping study. J. Syst. Softw. **113**, 188–215 (2016)
5. ISO 2003 ISO/IEC 14143-3: 2003: Functional Sized Measurement Methods - Verification of Functional Sized Measurement Methods. International Organization for Standardization, Geneva 2003 (2003)
6. ISBSG: ISBSG Development & Enhancement Repository – Release 13. International Software Benchmarking Standards Group (ISBSG) (2015)
7. Abran, A., Robillard, P.N.: Function points analysis: an empirical study of its measurements process. IEEE Trans. Softw. Eng. **22**(12), 895–910 (1996)
8. Function Points: Functional Size Measurement and project estimation (2010). http://www.functionpoints.org. Accessed 22 Apr 2018
9. IFPUG: International Function Point Users Group. http://www.ifpug.org/. Accessed 22 Feb 2013
10. IFPUG: Function Point Counting Practices Manual, Release 4.1, International Function Point Users Group (IFPUG), Mequon, Wisconsin Release 4.1, January 1999
11. Bock, D.B., Klepper, R.: FP-S: a simplified function point counting method. J. Syst. Softw. **18**, 245–254 (1992)
12. Tichenor, C.: A new software metrics to complement function points: the software non-functional assessment process (SNAP). In: 25th Anniversary Issue, CrossTalk (2013)

13. Jones, C.: Applied Software Measurement, Assuring Productivity and Quality, 2nd edn. McGraw Hill, New York (1997)
14. Thomson, M., French, D., Netherland, B.: The use of function points in earned value management for software development. In: The 2013 ICEAA Professional Development and Training Workshop (2013). www.iceaaonline.com
15. Rule, G.: 'Small project', 'medium-size project' and 'large project': what do these terms mean? PowerPoint presentation. Software Measurement Services Ltd. (SMS), Copyright 2004–2005, 124 High Street, Edenbridge, Kent, UK (2005)

The Performance Evaluation for the Efficiency of Coastal Regional Innovation Network Based on DEA

Song Cheng[1(✉)], Hu Longying[1], and Yuan Haiyan[2]

[1] School of Management, Harbin Institute of Technology, Harbin 150001, China
472458635@qq.com
[2] Department of Mathematics, Heilongjiang Institute of Technology, Harbin 150050, China

Abstract. In order to evaluate the innovation efficiency of China's coastal regional innovation network accurately, this paper selects 11 coastal provinces, municipalities and autonomous regions, and simplifies the published financial data by the factor analysis method and the principal component analysis method. We build the innovation efficiency of China's coastal regional innovation network evaluation index system from the interaction between scientific research institutions and symbiotic environment, the interaction between enterprises and symbiotic environment and the interaction between scientific research institutions and enterprises. This paper analyzes the innovation efficiency of China's coastal regional innovation network evaluation and gives the input-output DEA model. It gives the coastal regional innovation network evaluation results also.

Keywords: DEA · Innovation efficiency · Coastal regional innovation network

1 Introduction

At present, as the environmental technology innovation becomes more and more complex, the cooperation and the collaborative innovation between enterprises becomes more and more so as to deal with the complex problems of innovation and the innovation risk, enterprises have begun to realize the complementary advantages, maximize the efficiency of innovation by through acquiring innovation network resources, the regional innovation network is bring out in this background. The regional innovation network refers to a region of many innovation subjects (enterprises, research institutions, universities, government, intermediary institutions and individuals in particular) which is in the certain innovation environment, and forms a relatively perfect and stable network system impacted by the formal or informal interactions. The regional innovation network plays an important role in promoting the regional economic growth, adjusting the regional industrial structure, stimulating the endogenous power of regional industry and promoting the overall competitiveness of the region. Therefore, how to evaluate coastal regional innovation network performance objectively, fairly and accurately is a multi-objective complex system evaluation problem which should consider all the impact factors. At present, there are many enterprises

R. Silhavy et al. (Eds.): CoMeSySo 2018, AISC 860, pp. 125–134, 2019.
https://doi.org/10.1007/978-3-030-00184-1_12

performance evaluation methods. All of them have their own characteristics. So it is necessary to choose a scientific and effective method with a complete theory and a relatively simple model for evaluating coastal regional innovation network performance, analyze the main factors affect the efficiency of innovation, find out the problems of regional innovation network in the process of innovation, put forward measures to improve the innovation ability. In the practical application, Berger [1], Humphrey [2] and Ruggiero [3] have put forward the DEA method for evaluating the non-parametric cross data and the multivariate input and output data effectively. It has been applied in the management science, systems engineering, decision analysis, and evaluation techniques and other areas. Therefore, we will use the DEA model to analyze the performance of the coastal regional innovation network performance in China.

2 The DEA Evaluation Model

DEA is short for Data Envelopment Analysis. It is a system analysis method based on the idea of relative efficiency evaluation improved by Chames et al. in 1978 (see [4]), it extends the idea of engineering productivity in single-input single-output system to a relatively efficient evaluation method in multiple input multiple output system.

Firstly, we simplify the data no losing or little losing the original information, compress and cut the dimension of the original indicators, change the original related variables into comprehensive and independent variables, eliminate the superiorities and inferiorities impacts of objective basic conditions, and reflect the original information as much as possible.

Suppose there are n samples, each has m data, define

$$X = \begin{bmatrix} x_{11} & \cdots & x_{1m} \\ \vdots & \ddots & \vdots \\ x_{n1} & \cdots & x_{nm} \end{bmatrix} = (x_1, x_2, \cdots, x_m),$$

After standardize the columns of X, we get a standardized matrix X^* and the correlative coefficient matrix R of index variables.

Where

$$X^* = \begin{bmatrix} x_{11} & \cdots & x_{1m} \\ \vdots & \ddots & \vdots \\ x_{n1} & \cdots & x_{nm} \end{bmatrix} = (x_1, x_2, \cdots, x_m), R = \begin{bmatrix} r_{11} & \cdots & r_{1m} \\ \vdots & \ddots & \vdots \\ r_{n1} & \cdots & r_{nm} \end{bmatrix} = \frac{1}{n} X^{*'} X^*.$$

Using the correlative coefficient matrix for principal component analysis, we can determine the principal component number.

We call $\lambda_k / [\sum_{i=1}^{p} \lambda_i]$ the kth principal component information contribution rate, denote it as β_k, and call $\sum_{i=1}^{k} \lambda_k / [\sum_{i=1}^{p} \lambda_i]$ the first k principal components cumulative information contribution rate. Choose all the principal components with all characteristic values greater than 1, if the first k principal components cumulative contribution rate is more than 80%, take the first k principal components instead of the original m indexes. Extract the common factor, build the factor analysis model and make the factor rotation for the loading matrix.

Then, define the rotated and simplified data as the decision making units. Suppose x_j, y_j the jth DMU input and output parameters, $X_j = (x_{1j}, x_{2j}, \cdots, x_{mj})$, m as the input index number, $Y_j = (y_{1j}, y_{2j}, \cdots, y_{sj})$ and s as the output index number. Define the impact factors indicators before input as input indicators, the funds technology index after input as output index, then, determine the performance of the specific decision unit by comparing the input, output vectors and the experience production possibility set PE.

$$\text{where } PE = \left\{ (Y^t, X^t) = \sum_{i=1}^{n} \mu_i (Y_i^T, X_i^t); \quad \sum_{i=1}^{n} \mu_i = 1, \mu_i \geq 0 \right\}.$$

Now we construct a hypothetical decision making unit composed by the linear combination of n decision making units. The i-th component is $\sum_{j=1}^{n} \lambda_j x_{ij}$, the r-th component gives $\sum_{j=1}^{n} \lambda_j y_{ij}$ with $\sum_{j=1}^{n} \lambda_j = 1$. If all the outputs of this hypothetical unit are no less than that of j_0, and all the input are less than that of the decision making unit, these mean that j_0-th decision unit is not on the production frontier.

So we have the following model

$$\min E$$
$$\text{st.} \begin{cases} \sum_{j=1}^{m} \lambda_j y_{ij} & r = 1, \cdots, s \\ \sum_{j=1}^{m} \lambda_j x_{ij} & i = 1, \cdots, m \\ \sum_{j=1}^{m} \lambda_j = 1 & j = 1, \cdots, n \end{cases}$$

It measures the past performance of evaluation object by the factor analysis method, obtains the reference index which can reflect the objective foundation conditions. Using the same method for the currently operating performance of the evaluation object, we can get the current index.

Suppose there are n enterprises, x_j, y_j are the j-th decision making unit (DMU) reference index and the current index, $(x_j, y_j) \in E$, (x_j, y_j) is the j-th index for the DMU performance management. E is a comprehensive index set for all the DMUs performance management given by the factor analysis method,

Define

$$T = \{(x,y) \left| \sum_{j=0}^{n} \lambda_j x_j \le x, \sum_{j=0}^{n} \lambda_j y_j \ge y, \lambda_j \ge 0, i = 0, 1, \cdots, n \right.\} \tag{1}$$

It consists of all (x_j, y_j), where $(x_0, y_0) = (0, 0)$. If we take the reference index as a input, the current index as an output, consider a convex exponent index set T [5] induced by A. Chames' DEA method, we build the BCC model [6]:

$$\begin{array}{ll} \max & Z \\ & \sum_{j=0}^{n} \lambda_j x_j \le x_{j0} \\ s.t & \sum_{j=0}^{n} \lambda_j y_j \le Z y_{j0} \\ & \sum_{j=0}^{n} \lambda_j = 1, \\ & \lambda_j \ge 0,\ j = 0, 1, 2 \cdots, n \end{array} \tag{2}$$

If the optimal value of linear programming (2) is $Z = 1$, then the innovation performance of the DMU is called on the frontier of the index set T; If $Z^* > 1$, then the innovation performance of the DMU is called not on the frontier of the index set T.

Let $\bar{x}_{j0} = x_{j0}$ and $\bar{y}_{j0} = Z^* y_{j0}$, if $(\bar{x}_j, \bar{y}_j)$ is not on the frontier of the index set T, then $(\bar{x}_j, \bar{y}_j)$ is called the projection of innovation performance of the DMU j_0 index (x_{j0}, y_{j0}) onto the frontier of the index set T. As shown in Fig. 1.

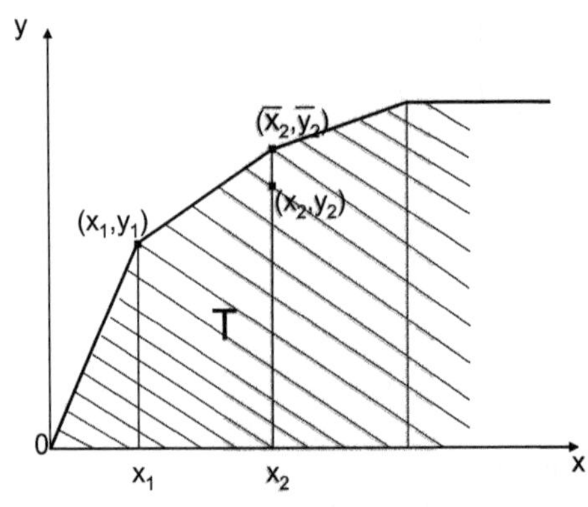

Fig. 1. The index set

The shadow of Fig. 1 is the index set T, (x_1, y_1) and (x_2, y_2) are the index of innovation performance of the DMU (1) and (2), where the index (x_1, y_1) of innovation performance of the DMU (1) is on the frontier of $T(Z^* = 1)$; the index (x_2, y_2) of innovation performance of the DMU (2) is not on the frontier of $T(Z^* > 1)$, $(\bar{x}_2, \bar{y}_2)$ is called the projection of index (x_2, y_2) onto the frontier of the index set T. If Z^* is the optimum value of (2) in the evaluation, then $\eta = 1/Z^* \times 100\%$ is the j_0-th innovation performance value of the DMU.

By $1/Z^* = y/\bar{y}$, we have that the evaluation value of the DMU's innovation performance is the maximum percentage of the current index for each participating DMU's innovation performance on the same conditions [7].

3 Build the Index System

Building a scientific evaluation index system is the premise and foundation of the DEA evaluation method. The coastal regional innovation network performance evaluation system is constructed on the basis of the DEA modeling [8–14] following the relation between nodes and nodes which form the innovation network. The model index system consists of the primary indicators and the secondary indicators, 3 first class indexes and 11 secondary indicators, see the following Table 1.

Table 1. Coastal regional innovation network performance evaluation index system

The primary indicators	The secondary indicators
Y_1 interaction between scientific research institutions and symbiotic environment	X_1 the non corporate parts of R & D institutions and university technology innovation funding X2 R & D institutions and university technical innovation personnel X3 R & D institutions and university international scientific papers X4 R & D institutions and university invention patent applications
Y_2 interaction between enterprises and symbiotic environment	X5 employment of industrial enterprises above designated size X6 total assets of industrial enterprises above designated size X7 new product sales revenue of industrial enterprises above designated size X8 total profits of industrial enterprises above designated size
Y_3 interaction between scientific research institutions and enterprises	X9 the enterprise part R & D institutes and university technology innovation funds X10 technology market contract value X11 new research and development personnel above designated size

4 Empirical Analysis

This paper considers Liaoning, Hebei, Tianjin, Shandong, Jiangsu, Shanghai, Zhejiang, Fujian, Guangdong, Guangxi, Hainan, the coastal 11 provinces, autonomous regions and municipalities directly under the central government as the research object, use the "China Statistical Yearbook" and "Statistical Yearbook" and "Chinese technology university technology statistical data collection" data to evaluate the coastal the innovation performance of regional innovation network. The research data comes from the scientific research institutions and universities' data which is added up using the principal component analysis method, and the correlation index of enterprise is measured by the data industrial enterprises above the scale published. Since the innovation activities have a certain delay from input to output [15], this paper chooses 5 years as an evaluation stage, takes 2006–2010 years' data as the reference index and takes 2011–2015 years' data as the reference index as current index. It simplifies the data using SPSS, and decides whether the data can be used as a factor analysis. See the following Table 2.

Table 2. KMO test and Bartlett sphericity test results

Kaiser-Meyer-Olkin measure of sampling adequacy		.736
Bartlett's test of sphericity	Approx. Chi-Square	243.908
	df	55
	Sig.	.000

In Table 2, we can see that KMO is 0.736, by the rule given by Kaiser, 0.736 is greater than 0.5, it means "ok" and suits for factor analysis. The Bartlett sphericity inspection accompanied probability is 0.000, less than 0.05, so it suits for the factor analysis by the Bartlett sphericity zero hypothesis.

From Table 3 we can see there are three factors with eigenvalue λ greater than 1, the cumulative variance explained by the first 3 common factors is 86.854%, considering it is more than 85%, so most of the information can be summarized and the relative data can be explained by the first three factors.

Table 3. Total Variance Explained

	Total	% of variance	Cumulative %
1	5.164	37.929	38.929
2	3.071	27.505	57.434
3	1.952	13.420	86.854
4	0.836	7.410	88.264
5	0.691	3.816	89.080
6	0.616	2.709	89.789

SPSS software is used to do the factor analysis and principal component analysis of the listed data, we get three new variables which concentrate most of the original variables' information. As the data analyzed by data envelopment analysis satisfy the production possibility set invalid axiom, it is possible to increase input or decrease output on the basis of the original production activities and it is rational to determine the reference performance index and the current performance index.

Then, according to the transformation invariance of DEA model, the effectiveness of the decision-making unit has not changed when every decision-making unit plus or minus the same positive number such that $(x, y) > 0$. That is the classification of the decision-making unit and the efficient frontier surface would not be affected by the inputs and outputs data and the DEA efficient decision-making unit would not be invalid when the data changed. So, we verify and correct the reference index and the current performance index for each sample using Lindo software to make sure the demands of DEA model for the input and output data. Define the data $Y_i (i = 1, 2, 3)$ adjusted in 2011–2015 as the performance index, the data $Y'_i (i = 1, 2, 3)$ adjusted in 2006–2010 as a reference performance index. The results should are shown in Table 4.

Table 4. The current and reference performance index after revise

Area	Data in 2011–2015 (The current performance index)			Data in 2006–2010 (The reference performance index)		
	Y_1	Y_2	Y_3	Y'_1	Y'_2	Y'_3
Liaoning	0.29086	0.14287	0.04119	0.11959	0.05001	0.51455
Hebei	0.34309	0.10622	0.70133	0.26329	0.02558	0.63954
Tianjin	0.11929	0.48323	0.40458	0.40842	0.41132	0.71969
Shandong	0.30676	0.45526	0.44486	0.25294	0.24067	0.36225
Jiangsu	0.12029	0.24306	0.67703	0.24776	0.27071	0.60612
Shanghai	0.17426	0.27823	0.20372	0.13621	0.26901	0.31424
Zhejiang	0.47763	0.07397	0.28194	0.41445	0.23721	0.45871
Fujian	0.41125	0.09258	0.87282	0.67774	0.64093	0.43969
Guangdong	0.58275	0.22859	0.86652	0.68513	0.26973	0.38008
Guangxi	0.41516	0.38202	0.45877	0.56554	0.42186	0.46766
Hainan	0.30365	0.64774	0.59732	0.12101	0.45724	0.22342

Input the relevant data to the data envelopment analysis output-oriented BCC model, cross calculated each variable value by 05SR2.0 evaluation system software. At the same time, we obtain the composite score of every decision-making unit by the distance from every unit to the frontier surface, and comparative analyze every two of the decision-making unit, form different visual effects, determine the optimal performance unit and the performance evaluation value, the results are shown in Table 5.

In Table 5, we can see that, the evaluation value in Zhejiang, Jiangsu, Shanghai etc. is greater than 0.85, they are classed as first-class, the evaluation value in Liaoning,

Table 5. The relative evaluation results of innovation performance of coastal regional innovation networks

Area	Reference index	Current index	Frontier index	Two times relative advantage	Ranking
Zhejiang	0.2013	0.2037	0.2037	1	1
Jiangsu	0.2103	0.1445	0.1552	0.9313	2
Shanghai	0.4406	0.3568	0.4179	0.8539	3
Guangdong	0.102	0.2225	0.2699	0.8245	4
Tianjin	0.3826	0.2063	0.2699	0.7645	5
Shandong	0.2198	0.3094	0.4174	0.7413	6
Hainan	0.3821	0.1957	0.2874	0.6808	7
Fujian	0.2391	0.2249	0.3412	0.6594	8
Liaoning	0.2983	0.2216	0.3543	0.6235	9
Hebei	0.3423	0.2291	0.3813	0.6009	10
Guangxi	0.3905	0.2547	0.425	0.5992	11

Hebei, Guangxi is less than 0.65, they are classed as low-grade, the evaluation value in Guangdong, Tianjin, Shandong, Hainan and Fujian is among 0.65–0.85, they are classed as medium.

5 Conclusions

On the basis of studying the results in this paper we know that: Firstly, the provinces which have scientific research institution and higher enterprise innovation efficiency always have higher innovation performance evaluation value, both their input innovation network resources into output technology results are efficient, and converting technical achievements into economic benefits are also efficient. Such areas have more developed economy, innovative ideas, more perfect system to protect intellectual property rights and encourage technological innovation, and have higher technology market operation capability. We suggest they maintain such advantages, and increase investment in innovation resources and produce more economic results. Secondly, the provinces which have higher enterprise innovation efficiency but the commercialization of research findings is inefficiency always have medium innovation performance evaluation value, their Innovation capability are at the bottleneck, they do not strengthen the transformation of scientific and technological achievements, so the development of science and technology and the economic development has the gap, Shandong province is an example. For these areas, we suggest they establish the mechanism combining production and research, and take the enterprise as the core, so as to speed up the transformation of technologies into economic efficiency. The other situation is that the results are higher, but the efficiency of technology research and development is low. The transformation mechanism of scientific and technological achievements in such provinces is in good operation and the transformation is efficient, but they do not pay attention to R & D and rely on the introduction of external scientific and technological achievements, Guangdong province is an example. Therefore, we suggest they consider the learning and re innovation as an important way to enhance self innovation on the basis of introduction of external scientific and technological achievements, improve the ability of integrated innovation and short the cycle of innovation.

The use of DEA to evaluate the innovation performance of the coastal regional innovation network takes the regional innovation system as a "black box", without considering the influence of the actual operation of internal system and the overall efficiency of the system. This method is more inclusive of input and output indicators, it can accept indicators that are difficult to quantify in general sense. In light of the operational results, we can see the innovation performance of China's coastal innovation networks are consistent with the conclusions shown in this paper, it is verified that the DEA is useful for the performance evaluation of the coastal regional innovation network.

Acknowledgements. This work was supported by the Creative Talent Project Foundation of Heilongjiang Province Education Department (UNPYSCT-2015102).

Declare. The authors declare that there is no conflict of interests regarding the publication of this article.

References

1. Berger, A.N., Humphrey, D.B.: Efficiency of financial institutions: international survey and directions for future research. Eur. J. Oper. Res. **98**, 175–212 (1997)
2. Berger, A.N., Mester, L.J.: Inside the black box: what explains differences in the efficiencies of financial institutions. J. Bank. Financ. **21**, 895–947 (1997)
3. Ondrich, J., Ruggiero, J.: Efficiency measurement in the stochastic frontier mode. Eur. J. Oper. Res. **129**(2), 434–442 (2001)
4. Charnes, A., Cooper, W.W., Rhodes, E.: Measuring the efficiency of decision making units. Eur. J. Oper. Res. **2**, 429–444 (1978)
5. Gao, W.Z., Feng, Y.Z.: Research of management efficiency characters and its measurements. J. HIT (Soc. Sci. Ed.) **9**(5), 103–106 (2007)
6. Wei, Q.L.: Data Envelopment Analysis, pp. 146–169. Science Press, Beijing (2004)
7. Zhuang, S.Y., Feng, Y.J.: An evaluation of management performance about company core competence based on DEA. J. Syst. Sci. Inf. **3**(2), 363–370 (2005)
8. Braglia, M., Zanoni, S., Zavanella, L.: Measuring and benchmarking productive systems performance using DEA: an industrial case. Prod. Plann. Control **14**(6), 542–554 (2003)
9. Glaister, K.W., Buckley, P.J.: Strategic motives for international alliance formation. J. Manag. Stud. **33**, 301–332 (1996)
10. Hui, D.: Research on the impact of regional innovation network on enterprise innovation ability. Jiang Nan University, pp. 7–8 (2014)
11. Toshiyuki, S.: Production analysis in different time periods: an application of data envelopment analysis. Eur. J. Oper. Res. **86**, 216–230 (2010)
12. Banker, R.D., Potter, G., Srinivasan, D.: An empirical investigation of an inventive plan that includes nonfinancial performance measures. Acc. Rev. **75**(1), 65–92 (2000)

13. Suwignjo, P., Bititci, U.S., Carrie, A.S.: Quantitative models of performance measurement system. Int. J. Prod. Econ. **64**(1–3), 231–241 (2000)
14. Mike, B., Bourne, M., John, W.M.: Designing implementing and updating performance measurement systems. Oper. Prod. Manag. **20**(7), 754–771 (2000)
15. Shi, X., Zhao, S.: Research on regional innovation efficiency and its spatial difference in China. Quant. Tech. Econ. Res. **3**, 44–55 (2009)
16. Sheng, Z.H., Zhu, Q., Wu, G.M.: DEA Theorems, Methods and Application, pp. 353–360. Science Press, Beijing (1996)
17. Zhu, J.: Multi-factor performance measure model with an application to Fortune 500 companies. Eur. J. Oper. Res. **123**, 105–124 (2000)

Models and Algorithms of Vector Optimization in Selecting Security Measures for Higher Education Institution's Information Learning Environment

Berik Akhmetov[1], Valeriy Lakhno[2], Bakhytzhan Akhmetov[3], Yuri Myakuhin[2], Asselkhan Adranova[3](✉), and Lazat Kydyralina[3]

[1] Caspian State University of Technologies and Engineering Named After Sh.Yesenov, Aktau, Kazakhstan
Berik.Akhmetov@yu.edu.kz
[2] European University, Kiev, Ukraine
[3] Kazakh National Pedagogical University Named After Abay, Almaty, Kazakhstan
aselhan.adranova@mail.ru

Abstract. The paper looks at the features of the application of Edgeworth-Pareto method in an effort to address the multicriteria discrete optimization problem, associated with the assessment of the security in a higher education institution's (HEI) information learning environment (ILE) and its technical means of protection (TMP). Modification to the method, based on a combination of discrete optimization of Edgeworth-Pareto and Podinovski's lexicographic method, is proposed. The algorithm of choosing the rational decision from the suggested TMP and ILE of HEI is proposed. In order to highlight a variety of Pareto-optimal solutions the vector criterion for evaluating output, which considers two conditions for the optimality as its elements: cost assessment of the analyzed TMP and its technical efficiency evaluation.

Keywords: Higher education · Information security · Discrete optimization

1 Introduction

With the growing quantity and sophistication of cyber attacks on state and private entities, providing proper cybersecurity (CS) of the information learning environment (ILE) of the HEI has become one of the key objectives of the education informatization in all over the world [1–3]. Technical capacity of today's HEI's ILE includes various area networks (in particular, corporate ones (CAN), information learning platforms, portals, resources for educational activities, as well as technical means of network communications. However, when designing the ILE of a HEI, the existing methodological approaches of comprehensive information security (IS) and cybersecurity (CS) are not taken into account [2–4].

R. Silhavy et al. (Eds.): CoMeSySo 2018, AISC 860, pp. 135–142, 2019.
https://doi.org/10.1007/978-3-030-00184-1_13

The existing international experience [5, 6] confirms that the extensive approach to solving the problems of IS and CS by expanding means and measures for the protection of information (PI), is often little efficient. Such approach does not solve the problems associated with the PI in the HEI's ILE and its cybersecurity. Despite the global trend of development of HEIs' network ILE as well as the widespread implementation of informatization means in educational activities, this field still lacks innovative approaches in solving problems of IS and CS management.

Therefore, the subjects in this paper devoted to the development of the method and algorithm of the multi criteria discrete optimization of the TMP selection for HEI's ILE are relevant.

2 Literature Review and Formulation of the Problem

The growth in the number of threats to ILEs information security resulted in extensive research in the field of developing effective TMP [7, 8] and information security systems (ISS) [9, 10]. A separate direction of research concerns the development of decision-making support systems (DSS) [11] and expert systems (ES) [12] in order to solve the problems when choosing compatible hardware and software for TMP and ISS. The research is still in progress.As the authors note [13, 14] many models of selecting TMP and ISS, particularly based on the method of F. Edgeworth-Pareto when addressing multiobjective discrete optimization problem associated with the assessment of the security level, do not take into account ILE's specific features, in particular, the specifics of the HEI. In addition, as shown in [15], the sophisticated algorithms of selecting the TMP and ISS are difficult to implement in software, which in turn undermines their practical use in HEI's ILE. Thus, considering debate in the existing literature [12, 13, 15], an objective to improve the method and algorithm of the multi criteria discrete optimization problem associated with the assessment of the security of HEI's ILE seems urgent. This, in particular, may be achieved through improvement of models and algorithms of selecting by using the combination of F. Edgeworth-Pareto and Podinovski's lexicographic method.

3 The Purpose of Research

To develop an algorithm of discrete optimization in decision-making about organizational and technical aspects when selecting the means of information security for HEI's ILE.

To fulfil this purpose, the specific objectives of the study are:

- to improve the method of multi criteria discrete optimization to meet the challenges of the HEI's ILE security assessment using the combination of F. Edgeworth-Pareto and Podinovski's lexicographic methods;
- to develop an algorithm of optimal choice when designing an ISS for the HEI's ILE, taking into account all possible options of implementing information security policy (ISP).

4 Materials and Methods

A modified mathematical method for optimal choice of technical means of protection for HEI's ILE using the multi criteria vector optimization method is proposed.

It is therefore proposed to use the combination of mathematical paradigms based on the F. Edgeworth Pareto [13, 14] and Podinovski methods (lexicographic multi criteria optimization method) [15]. The proposed approach has two criteria: cost criterion and the criterion of effectiveness of the application. It is assumed that the criterion of 'effectiveness' is more preferable than the criterion of 'cost'. It is possible that the criteria have numerical and weight ratios, the sum of which always equals 1.A graphic representation of the optimality areas and computing process of finding solutions in all areas, that have their own priorities shown in Fig. 1. The priority number corresponds with the area. Finding the optimal solution for the area I is a global optimum, and for all other areas – local optimum. Area II is not optimal from the value optimality point of view. Area III is not optimal by "application effectiveness" criterion. The area IV is not optimal by both criteria.

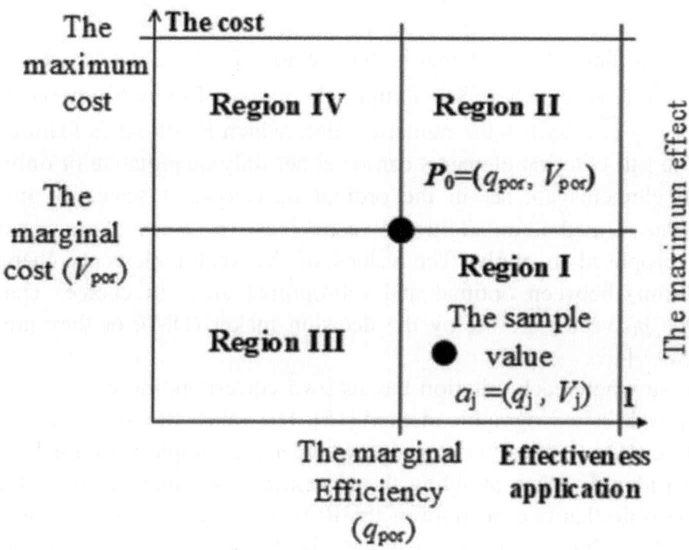

Fig. 1. Schematic representation of the regions of optimality

Finding maximum multiple solutions for the 'application effectiveness' criterion for the objects of study in the given areas is proposed. Multiples solutions refers to the objects that are different in structure, but have same quantitative values. These values characterize the same technical specifications (TS) or have same numerical values on the same criteria. If at least one research object (RO) has numerical multiple values, comparison of the two numeric values for the respective samples is undertaken, but only by the second criterion. It is important to note that the RO with the lowest value by the ‘cost’ criterion should be chosen. This RO is included in the list of the optimal options, and the other is excluded from the list of optimal solutions. Moreover, it is possible that for the two RO values for the taken criteria are equal. Then, both of these RO are included in the list of optimal solutions.

During the computing process each area will have one of three possible outcomes: (1) the considered area doesn’t have optimal RO; (2) the area has one the optimal RO; (3) the area has at least two optimal RO.

Each RO is represented as j-th vector-column of an A-type matrix (1). The values of these vectors’ elements characterize specific properties of criteria. Each vector-line of the A-matrix is represented as a vector, the elements of which are presented as numeric values. The number of vectors-lines (N) is equal to the number of criteria or TS.

$$A(\overline{a}_{\mathrm{ij}}) = \begin{pmatrix} \overline{a}_{11} & \overline{a}_{12} & \cdots & \overline{a}_{1\mathrm{j}} & \cdots & \overline{a}_{1\mathrm{S}} \\ \overline{a}_{21} & \overline{a}_{22} & \cdots & \overline{a}_{2\mathrm{j}} & \cdots & \overline{a}_{2\mathrm{S}} \\ & \cdots & \cdots & \cdots & \cdots & \\ \overline{a}_{\mathrm{i}1} & \overline{a}_{\mathrm{i}2} & \cdots & \overline{a}_{\mathrm{ij}} & \cdots & \overline{a}_{\mathrm{iS}} \\ & \cdots & \cdots & \cdots & \cdots & \\ \overline{a}_{\mathrm{N}1} & \overline{a}_{\mathrm{N}1} & \cdots & \overline{a}_{\mathrm{Nj}} & \cdots & \overline{a}_{\mathrm{NS}} \end{pmatrix} \tag{1}$$

For the i-th line of the A-matrix the vector $\bar{Q}_i = \left(\bar{a}_{i1}, \bar{a}_{i2}, \ldots, \bar{a}_{\mathrm{ij}}, \ldots, \bar{a}_{is}\right), S = \dim(\bar{Q}_i)|\forall i \in \overline{1, N}$ is set, where S is a multitude of given RO by N criteria. The element $\bar{a}_{ij}|i = \overline{1, N};\ j = \overline{1, S}$ defines the numeric value, which is j-th RO’s feature by the i-th criterion. The i-th criterion elements can be either only quantitative or only qualitative values. The elements are set in the ordinal measurement scale. To implement F. Edgeworth Pareto method an additional vector $\bar{P}_0 = (p_1, p_2, \ldots, p_i, \ldots, p_N)$ with $N = \dim(\bar{p}_0)$ is proposed to apply. The values of the vector elements characterize the threshold (limit) between optimal and sub-optimal areas of choice. The numerical values of the $\bar{p}_0$ vector are set by the decision maker (DM), or they are calculated mathematically [13].

In the case where each criterion has its own corresponding weight, the method is called Podinovski’s lexicographic method [15]. The values of the vector-column for the corresponding RO allow us to determine whether the sample is optimal or not by the criterion considered. When applying the aforementioned method of vector optimization, it is possible that one or more of the RO on the criterion under consideration is sub-optimal. In such case, the DM may include this RO in the list of the optimal ones. Then, the list of the optimal objects will end up having several RO with identical values in all criteria. In this case, the DM suggests experts to evaluate Ros and choose the best.

The developed computational algorithm for optimal selection of samples of the access control and management to the HEI’s ILE system using “effectiveness-cost” criteria implements this method and takes the specifications mentioned above into account. A block diagram of this algorithm is shown in Figs. 2a and b.

Execution of the algorithm begins with entering the baseline data on S researched objects. Dataset is represented as two vectors $\{\bar{Q}, \bar{V}\}$. Next, another two values {qopt, Vopt} – threshold of optimal areas (index-opt) are entered. Thus, $\bar{P}_0$ is set. In order to highlight the importance of the criteria weighting factors of these criteria should be set.

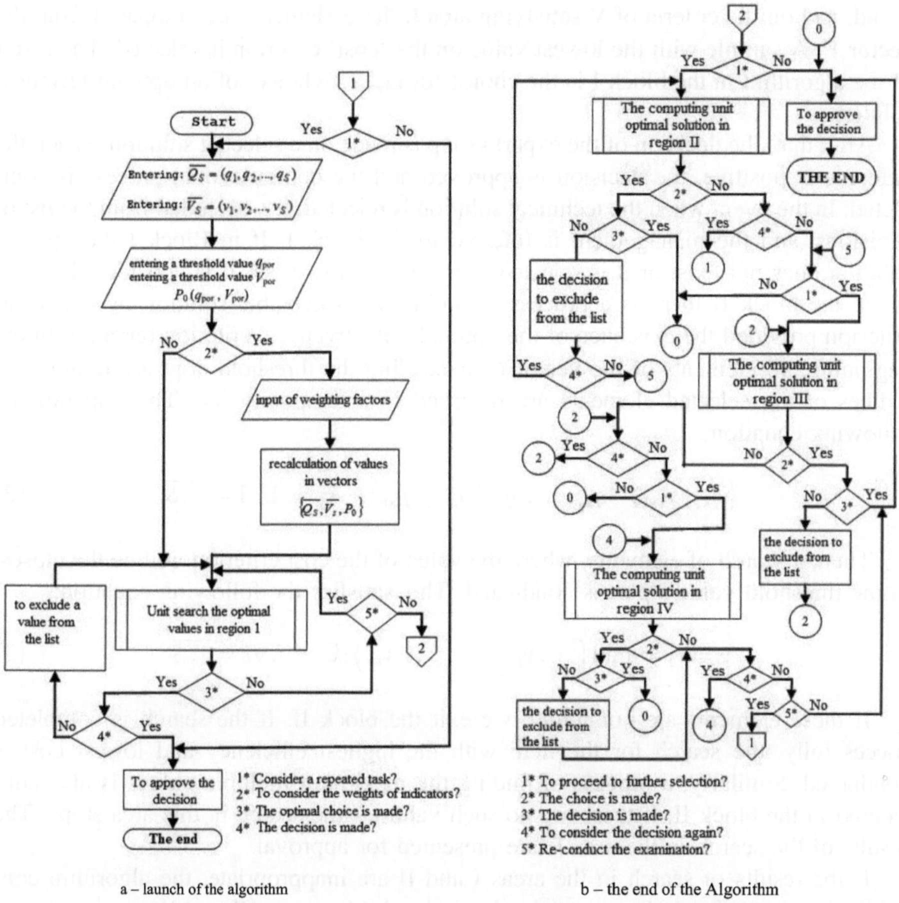

Fig. 2. The algorithm of the optimal choice of access control and management system on the "efficiency–cost" criterion

This, in turn, will recalculate the vector elements $\{\bar{Q}, \bar{V}, \bar{P_0}\}$:$\{\alpha_1 \cdot \bar{Q}, \alpha_2 \cdot \bar{V}, \alpha_1 \cdot q_{\text{opt}}, \alpha_2 \cdot V_{\text{opt}}\}$, with $\{\alpha_1, \alpha_2\}$ as values of the weighting factors.

Then, the management is transferred to the block I–to search solutions in the area I. In block I the search is conducted by the element $\bar{Q}$ vector, the value qj of which is maximum and exceeds the limit qopt. If such element is found the optimality on the second criterion is checked.

The value of the element vj from the vector $\bar{V}$, standing in the same position as the qi, with the value of Vopt are compared. If the sought value does not exceed Vopt, the specified element is put in the optimal samples' list. If the latter condition is not met, the element is excluded from the list and the program returns to the beginning of the block I. The search for the next element from the remaining list is conducted. The algorithm can check and find the presence of the maximum multiple values on the "effectiveness" criterion. If the elements with multiple values from the vector $\bar{Q}$ are found, and on the criteria of V satisfying area I, these elements are compared from the vector $\bar{V}$. A sample with the lowest value on the 'cost' criterion is selected. The result of the algorithm in the block I is the choice (or lack of choice) of an optimal technical solution.

After that, the decision of the expert group is made on a selected solution. When the outcome is positive, the decision is approved and the computational process is completed. In the event when the technical solution is rejected, it's excluded from the list of decisions, and the management is referred to the block I. If in Block I the optimal solution does not exist or it is rejected, transition to the block II is considered.

In the block II there is a process of finding an acceptable solution by the 'cost' criterion provided the existence of the optimal 'effectiveness' criterion (area II). In the beginning, all elements of the vector $\bar{Q}$, exceeding the threshold qopt are found. The indices of the selected elements are recorded in a temporary list. This satisfies the following equation:

$$\left\{q_{\mathrm{k}_1}, q_{\mathrm{k}_2}, \ldots q_{\mathrm{k}_g}, \ldots, q_{\mathrm{k}_f}\right\} | q_{\mathrm{i}} \geq q_{\mathrm{por}}; \quad \mathrm{k}_{\mathrm{g}} = \mathrm{i}, \forall \mathrm{i} = \overline{1, S} \tag{2}$$

Then, a search of elements, where the value of the cost criterion will be the closest to the threshold value Vpor, is conducted. This satisfies the following equation:

$$v = v_{\mathrm{i}} = \min\left\{v_{k_1}, v_{k_2}, \ldots v_{k_g} \ldots, v_{k_f}\right\} | k_g = i, \forall i = \overline{1, S} \tag{3}$$

If these elements are not found, we exit the block II. If the search is completed successfully, the search for the item with the highest efficiency and lowest cost is conducted. Similarly, a process of finding the maximum multiple values is also considered in the block II. If there are no such values – the search in this area stops. The results of the search in the area II are presented for approval.

If the results of search in the areas I and II are inappropriate, the algorithm considers the search for a close-to-optimal solution in the areas III and IV.

In the event of the rejection of proposed options the algorithm provides the transition to find optimal solutions (local optimum) by one criterion (the effectiveness or cost), in respective areas II or III and IV, in which decisions are globally sub-optimal. If the re-consideration is rejected, or further selection of options is halted, the algorithm forms a report on the conducted process of finding optimal solutions from the proposed options of technical systems.

5 The Experiment

The computational experiments (Fig. 3) are performed.

The results obtained on the graph are an expanded area for finding optimal solutions for the proposed method and the "efficiency-cost" criterion. During the testing of the program and the algorithm, the correctness of the results obtained was established.

The choice of variants of the TMS and IS, performed according to the proposed algorithm, made it possible to shorten the selection time by 15–20% in comparison with the methods described in [4, 10, 12].

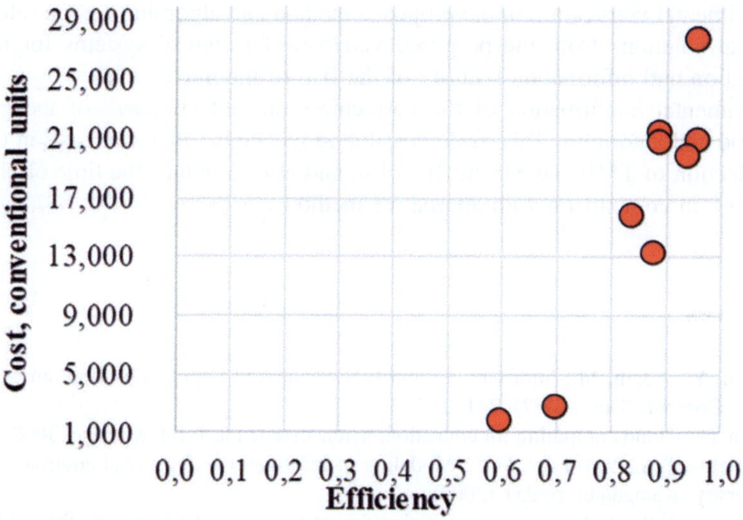

Fig. 3. Extended search area for finding optimal technical samples

6 Discussion of Results and Prospects for Further Research

Improved method for multi criteria discrete optimization to meet the challenges of the assessing HEI's ICS level of security, has allowed to reduce the costs of the TMP and ISS by 16–19% compared with alternative methods [10, 12, 13].

A certain shortcoming of the proposed model is the need to involve several independent experts to select the compatible versions of TMP and ISS for HEI's ILE in the initial phase of the study.

The further development of this work can be improvement of interaction of selection algorithm with modules of the 'DMSSCIS' [11, 12].

Overall, based on this research, the proposed method and algorithm are deemed to be effective.

7 Conclusions

The main results of the study were:

1. An improved method of multi-criteria discrete optimization for solving the problems of assessing the security of the institution's ILE. The solution is based on a combination of the Edgeworth-Pareto discrete optimization method and the Podinovsky lexicographic method. It has been developed vector criterion for estimating the solution, which includes two optimality conditions as components: the cost estimate of the considered version of the HEI and the evaluation of the technical efficiency of the modification for a specific ILE of the university.
2. A new algorithm for the optimal choice in the design of the TMP for the ILE of the university, taking into account all possible options for implementing the IS system.

3. Experimental verification of the proposed method and algorithm for the selection of optimal solutions from the proposed versions of technical systems for technical protection and information security of the ILE of the university.
4. Experimental confirmation of the correctness and effectiveness of the proposed method and algorithm. This made possible to reduce the financial costs at the stage of selection of TMP and SIS by 16–19%, and also to reduce the time of choice by 15–20% in comparison with alternative methods.

References

1. Rezgui, Y., Adam, M.: Information security awareness in higher education: an exploratory study. Comput. Secur. **27**(7), 241–253 (2010)
2. Sultan, N.: Cloud computing for education: a new dawn? Int. J. Inf. Manage. **30**(2), 109–116
3. Akhmetov, B.S., Yavorsky, V.V.: Modeling of information educational environment of the university, Karaganda, p. 251 (2006)
4. Schneider, F.B.: Cybersecurity education in universities. IEEE Secur. Priv. **11**(4), 3–4 (2013)
5. Conklin, A.: Cyber defense competitions and information security education: an active learning solution for a capstone course. In: Proceedings of the 39th Annual Hawaii International Conference on the System Sciences, HICSS 2006, vol. 9. IEEE (2006)
6. Schuett, M., Rahman, M.: Information Security Synthesis in Online Universities. The arXiv preprint arXiv:1111.1771 (2011)
7. Mariusz, N., Benton, M.: Cybersecurity cost of quality: managing the costs of cybersecurity risk management (2017). [El. resource]. https://arxiv.org/ftp/arxiv/papers/1707/1707.02653.pdf
8. Jalali, M., Siegel, M., Madnick, S.: Decision making and biases in cybersecurity capability development: evidence from a simulation game experiment (2017). [Electronic resource]. https://arxiv.org/ftp/arxiv/papers/1707/1707.01031.pdf
9. Gordon, L.A., Loeb, M.P., Zhou, L.: Investing in cybersecurity: insights from the Gordon-Loeb model. J. Inf. Secur. **7**(02), 49 (2016). https://doi.org/10.4236/jis.2016.72004
10. Goztepe, K.: Designing fuzzy rule based expert system for cyber security. Int. J. Inf. Secur. Sci. **1**(1), 13–19 (2012)
11. Akhmetov, B., Lakhno, V., Boiko, Y., Mishchenko, A.: Designing a decision support system for the weakly formalized problems in the provision of cybersecurity. East.-Eur. J. Enterp. Technol. **1**(2(85)), 4–15 (2017)
12. Lakhno, V., Boiko, Y., Mishchenko, A., Kozlovskii, V., Pupchenko, O.: Development of the intelligent decision-making support system to manage cyber protection at the object of informatization. East.-Eur. J. Enterp. Technol. **2/9**(86), 53–61 (2017)
13. Lotov, A.V., Bushenkov, V.A., Kamenev, G.K.: Interactive Decision Maps: Approximation and Visualization of the Pareto Frontier, vol. 89. Springer Science & Business Media. (2013)
14. Fischione, C., Jakobsson, M.: The optimality of radio power control via Fast-Lipschitz optimization. IEEE Trans. Commun. **64**(6), 2589–2601 (2016)
15. Podinovski, V.V., Kuosmanen, T.: Modeling weak disposability in data envelopment analysis under relaxed convexity assumptions. Eur. J. Oper. Res. **211**(3), 577–585 (2011)

Implementing DevOps in Legacy Systems

Adriano Bessa Albuquerque(✉) and Vinicius Lima Cruz

Universidade de Fortaleza, Fortaleza, Ceará, Brazil
adrianoba@unifor.br, vinnycruz@gmail.com

Abstract. Legacy systems are a challenge for the operations of modern organizations as they limit their changing ability and grow business. However, they are often systems suited to their purpose, they deliver the expected value and investment in them is justifiable. The concept of DevOps has come to reduce the separation between existing development and operations teams in software development companies, decreasing the product lifecycle. With this concept comes the practice of continuous delivery, allowing teams to deliver and deploy any version of their software in any computing environment at any time through a fully automated process. This practice improves the feedback of the development process, so that problems are identified and resolved as early as possible. In this paper, we present a case study of deploying DevOps, and propose a formal process for deploying DevOps and its modifications necessary to adapt to legacy systems so that the delivery process of the legacy systems has a short and high quality lifecycle, delivering frequent versions in an automated way.

Keywords: Legacy systems · DevOps · Continuous Integration
Continuous Deployment · Deployment Pipeline

1 Introduction

With increasing competition in the software market, organizations are allocating resources to develop and deliver high-quality software at a rapid pace [15]. But even with all the advancement in technology in recent years, most organizations still have to deal with legacy systems, which are often old systems but very important to the business, difficult to be evolved, and with a number of risks associated with any change in these systems [10, 11, 17].

There are several agile initiatives that are being described in the literature and widely adopted in the software industry, to deliver high value products to the market, quickly and in line with new trends [8, 12]. The concept of Continuous Integration (CI) was a first step, inspired by agile techniques, that significantly accelerated the product lifecycle, encouraging developers to integrate changes more frequently into shared storage [9]. This helped to detect problems soon after any change. Detecting problems early means reducing integration time, developing software faster, and increasing productivity.

Most projects fail due to people problems, not due technical problems. At no point in the development process is this as valid as in the part that involves deployment in testing and production environments [12]. Almost all medium and large companies

R. Silhavy et al. (Eds.): CoMeSySo 2018, AISC 860, pp. 143–161, 2019.
https://doi.org/10.1007/978-3-030-00184-1_14

separate development and infrastructure activities (or operations) into two different groups. Each team has to be specialized in its own tools and manage a different and isolated phase of the product life cycle. Moving from one phase to the next often requires complex processes, manual steps, and approvals, making deliveries too slow for business expectations.

The need to improve the lifecycle in all the different phases and teams and, especially, the great competitiveness of the market, having to launch products quickly aligned to the trends, was what originated the concept of Continuous Delivery that, as an extension of Continuous Integration, emphasizes the automation of tests, launching processes and deployment stages [12]. For that, it had to be reduced the separation of the roles between the different teams and led to the emergence of the DevOps movement. The DevOps movement is focused on one goal: to encourage collaboration among all those involved in the process of delivering higher value software with more confidence and speed.

In this work, we present a DevOps case study, and we propose a formal process for deploying DevOps and its necessary modifications to adapt to legacy systems, aiming to improve the system development process through the implementation of the continuous delivery practice. Thus, we intend to improve the maintenance and evolution of these systems, through the automation of the tests, launching processes and the deployment stages, detecting problems soon after any alteration, reducing the errors caused by manual steps, reducing the necessary approvals for deployment, reducing integration and delivery time, making maintenance faster and increasing team productivity.

This article consists of six sections: Sect. 2 presents the bibliographic review with the main concepts of legacy systems and DevOps. Section 3 presents the case study and lessons learned. Section 4 describes the proposed process. Section 5 presents the evaluation of the process, and finally Sect. 6 presents the final considerations.

2 Bibliographic Review

2.1 Legacy Systems

The authors present various definitions to the legacy system. A legacy system is a sociotechnical system that is useful or even essential to an organization, but which has been developed using obsolete technologies or methods. Because legacy systems often perform business-critical functions, they need to be maintained [23]. The term legacy system often refers to an old system that remains in operation within an organization. Legacy systems were typically developed using obsolete development and technology practices and have long life, needing extensive change [26]. Systems become legacy when they begin to resist to modifications and evolution [22].

Legacy systems are not a transient issue but a continuous issue: "Today's new system will become the legacy system in the next, and inevitable, stage of changes" [5]. Legacy systems are those that do not have automated testing (useful and simple definition, although controversial) [7].

Legacy systems are a known challenge to the operations of modern organizations as they limit their ability to change and increase their businesses [10, 11]. Government agencies continue to invest substantially in legacy systems, with an estimated $ 35.7 billion spent by the United States federal government in 2010 [25].

Despite the well-known disadvantages, such as being inflexible and difficult to maintain, legacy systems are still vitally important to business as they support complex business processes and store critical business logic of great value to the company, so they cannot simply be removed [14].

Knowledge embedded in the legacy system is a significant organizational asset, and if the system provides value to the business, then it must be modernized or replaced by a more modern system [22]. Many legacy systems support vital business functions and are indispensable. If legacy software meets the needs of its users and works reliably, it is not broken and does not need to be fixed [17].

Most of the work on legacy systems addresses the reengineering of legacy systems, either by transforming them into services or by migrating to more current platforms [1, 13, 24]. Most legacy systems modernization efforts fail because of complexity of system modernization, lack of understanding of the technology in which the software was built, lack of risk management, difficulties from the adoption of commercial components, and lack of alignment with the business objectives [22].

The specialists argue that often the migration or transformation of the legacy systems is not possible, but by improving the development, testing, and delivery process, it would facilitate the maintenance and evolution of these systems, reducing many of the existing problems.

2.2 DevOps

Companies typically separate development and operations activities into two different teams with conflicting objectives. While the development team is encouraged to deliver software faster, the operations team seeks stability in the production environment. But everyone involved has a common goal: to make the delivery of value-added software to users a low-risk activity. To help this challenge the DevOps movement was created. The DevOps movement is focused on one goal: to encourage greater collaboration among all those involved in the process of delivering higher value software with more confidence and speed. There is no precise definition of the term DevOps in the literature, nor a specific process to be followed, that is, there is no single solution that can be used by all companies [6].

The DevOps movement can be considered an extension of the agile movement, that arose in 2001 due to new values demanded by software development, that were compiled in the “Agile Manifesto” [27]. These values encourage rapid response to change and team collaboration. Among the agile methodologies we can highlight Scrum and Extreme Programming (XP), which encourage the need for small development cycles and multifunctional teams. XP was the first methodology that fully embraced the concept of Continuous Integration [2], defending the need to deliver and test small changes, reducing the risk of errors and facilitating debugging [9].

With the need for continuous improvement of the lifecycle in different phases of development and to reduce the delivery time of the products to the market, the concept of Continuous Integration was extended and the concept of Continuous Delivery was created. Continuous Delivery is the focal point of the DevOps movement, and advocates the automation of testing and deployment steps. Continuous Delivery is expected to offer several benefits, such as: (1) obtaining greater and faster feedback from customers and the software development process; (2) with frequent and reliable releases, lead to improve the customer satisfaction and product quality; (3) the link between development and operations teams is reinforced and manual tasks can be eliminated [4, 14, 20].

Many industrial cases reported indicate that Continuous Delivery is being increasingly used in software engineering practices in companies of various sizes and domains [15, 16, 18]. There is a study in a company that adopted Continuous Delivery and had the release lifecycle of a software evolution reduced from several months to 2 to 5 days, from the creation of user stories to the release of the production version [4]. But few works address the benefits brought in the application of Continuous Delivery to the processes of maintenance, evolution and implantation of legacy systems. Even if legacy systems do not need to have frequently released versions, they may also benefit from slowing down the evolution lifecycle.

2.3 Deployment Pipeline

The main idea of Continuous Delivery is to build a deployment pipeline that allows each change in the version control system to trigger an automatic standard process, to compile all parts of an application, deploying it in any environment, testing it, and generate a potentially candidate version to be deployed in production [Fowler2013].

The deployment pipeline is, in essence, an automatic process to delivery software. It does not imply that there is no human interaction with the system during the delivery process; it ensures that complex and error-prone steps are automated, repeatable, and reliable. It basically has three goals. Firstly, it makes every part of the build, deployment, test, and release delivery process visible to all involved by promoting collaboration. Second, it improves process feedback, so that problems are identified and resolved as soon as possible. Finally, it allows teams to deliver and deploy any version of the software at any time through a fully automated process, enabling deliveries to be made in any required environment (test, development, or production) at the push of a button [14].

Humble and Farley synthesize the steps of a typical deployment pipeline: The process begins when a developer makes a new check-in in the version control system. At this point, the continuous integration system responds to the check-in by creating a new instance of the pipeline. The first stage of the pipeline compiles code, runs unit tests, performs code analysis, and builds installers. If the unit tests are successful and the quality of the code is good, it is compiled into binaries that are stored in an artifact repository. The second stage consists of automated acceptance tests that take longer to execute. This stage will be called automatically if the first stage is successful. At this point, the pipeline can be split into automatic deployments of code compiled in various environments. The testers should be able to see which candidate versions are available and their situation. They should then be able to push a button and deploy a selected compiled code by running the deployment script in the appropriate environment.

Despite the typical pipeline presented, there is no standard or single pipeline for all companies [Phillips], the modeling and implementation of a deployment pipeline in a given enterprise can be influenced by a number of factors, such as skills, experience and structure of the team, structure and budget of the organization, client environments, and project domain [Bass].

2.4 Related Works

One of the works that addresses the challenges of DevOps in companies that work with legacy systems was developed by Schaller [21], which addresses the challenges in the areas of security, organizational culture, large-scale legacy systems architecture, large-scale automation. Schaller did an extensive research in the literature on the experiences of companies that have adopted DevOps and pointing out the specific challenges encountered in the literature when dealing with legacy systems. Another work was developed by Rao [19], where modifications are discussed in a typical DevOps script to adapt to legacy systems. According to the author, a typical DevOps script involves building a deployment pipeline, which is an automated implementation of the process of compiling all parts of an application, deploying it in any environment, testing it, and delivering it.

The deployment pipeline has three goals. Firstly, it makes every part of the build, deployment, test, and release delivery process visible to all involved by promoting collaboration. Second, it improves process feedback, so that problems are identified and resolved fast. Finally, it enables teams to deliver and deploy any version of their software at any time through a fully automated process, enabling deliveries to be performed in any required environment (test, development, or production) at the push of a button [12].

By adopting DevOps in a new project, teams will find it easier than applying to legacy systems [19], because in a new project, Continuous Delivery would be considered in the analysis, design, and architecture definition stage. In legacy systems, which have evolved over time without regard to automation, the adoption of DevOps can lead to the need to redesign the system and large-scale code refactoring. So, introducing Continuous Delivery and a deployment pipeline into legacy systems can be a major challenge.

This work differs from the work developed by Schaller [21] and Rao [19], as we present a DevOps process deployment use case and propose a formal process to deploy DevOps, with modifications needed to adapt to legacy systems, detailing activities and necessary workflows.

3 A Case Study Implementing DevOps in Legacy System

A case study investigates a contemporary phenomenon within its real-life context, especially when the boundaries between phenomenon and context are not clearly evident [28]. The case study was performed in a financial institution not identified in this work and had the objective of deploying DevOps and learning from the case study, since there was no formal process defined for deploying DevOps in legacy systems. It was carried out for 2 years, beginning in March 2016 and being implemented and refined until the production of this work. First of all, it was identified that the deployment process executed until the beginning of this work required a lot of effort and there was a lot of bureaucracy. A team was organized in the Software Architecture Sector of the institution to improve and automate the actual process. For this, the benefits of applying agile practices and Continuous Delivery began to be studied.

3.1 Continuous Integration and Deployment Automation

At first, tools to automate the process of Continuous Integration were searched, including the compilation and building of the system, and the automatic deployment of it. Initially, a tool, called Jenkins, was identified to be the continuous integration server. This tool was then installed by the team. Jenkins is an open source highly customizable and flexible platform, due to its plug-in architecture and its huge Open Source community.

A simple application in Java was chosen as a pilot project, and it was researched what it would take to automate the compilation and build process (which we will call the build process) of that application. For the automatic building of Java projects, it was seen that it would be necessary to use Maven. Maven is a build automation tool used in Java projects that uses an XML file (POM - Project Object Model) to describe the software project being built, its dependencies on external modules and components, the order of compilation, directories, and plug-ins required. The chosen Java application was then modified to use Maven with the help of documentation and the responsible development team. Then all necessary and registered dependencies were gathered in the Maven XML files and a script was created to perform the automatic construction. This script has been tested and approved. After the success of this execution, this automatic build script was already capable of being triggered and automatically run by Jenkins.

The next step was to verify how the automatic deployment of this Java application could be performed in the development environment. Java applications were deployed to an IBM Websphere Application Server (WAS) server following some manual steps that were performed by Java experts. Then it was found out, on the internet, a script that could use Maven to perform the automatic deployment of Java applications on WAS servers. This script was then extended and it was created a plugin for the Jenkins platform. Once tested and approved, this plugin has become the default script for automatic deployments of Java applications on WAS servers.

From the success of the construction and automatic deployment to a Java project, initiatives have emerged to use the Jenkins platform to automate the construction and automatic deployment of systems from other languages. The Jenkins platform was supported by managers as a corporate solution that would be used throughout the institution. Then the development team began an analysis of systems that had the most evolutionary demands and a system queue, by priority, was established for migration to the Jenkins platform. The plugins for other technologies were being created.

At the same time, the version control tool has also undergone changes. As the configuration management and version control tool IBM Rational Clearcase, had to be updated, it was decided to migrate to the IBM Rational Team Concert (RTC) tool. First, the new version control tool was installed and, in a second step, the source code and artifacts of all systems and projects were migrated to the new tool. The third step was training the entire development team in the new tool. And then, all systems were migrated to new platform, which began to be widely used. With the adoption of the RTC tool, a Jenkins plugin had to be developed to access it. So the system building scripts were modified to be integrated to the new version control tool (RTC).

After the installation and configuration of the version control tool, a regularization work was started on the branches of the systems. It was identified that some systems had many branches, which made it difficult to merge the various development branches of the system, when it was necessary to generate a version to be tested. Then, a standardization of the branches of all systems was defined in three types of them: integration, which contained the current system code used in production; project, which contained the candidate code for the new version; development, which contained the code of each evolution. Next, a baseline was defined for all systems, which would be the code of each system in production environment, and would be in the integration branch, being used as a milestone for the creation of any project from then on.

Finally, the managers of the IT Area of the institution decided to establish rules for the applications implementation in the environments. All deployments should be done automatically through Jenkins, and no more requests could be made by analysts, attaching the system build and documents needed for deployment. A one-month deadline was defined to adequate all applications to the new deployment rule.

This decision was important to ensure the integrity of the systems and all applications evolved from then on would execute the entire automated process. This has reduced errors caused by builds with wrong dependencies, incorrect deployment documents, and human errors in general, since only automated processes would deploy in the environments. As a result, any changes made to the systems and people who operated at each stage of the deployment pipeline could be tracked, verified, optimized, and audited.

3.2 Tests Automation

The verifications were performed manually, through requests from the development team to the quality team, who applied some source code verification scripts. For automation of the tests, the SonarQube tool was installed. SonarQube is an open source code control platform, which is basically a plugins-based web tool that manipulates the results of various code analysis tools, generating results in the form of a report. The quality team specialists responsible for verifications defined the quality metrics (code duplication, coding standards, code coverage and potential bugs) to be used in the SonarQube tool. A plugin was created for the Jenkins platform to integrate with SonarQube. Then it became possible to perform code quality analysis automatically even for cobol, powerbuilder, and other common languages in legacy systems.

Analyzing the defects pointed out in the automatic verification performed by SonarQube, development teams could check out the source code of their applications and adjust them to the defined source code standards. Each project manager defined a deadline to correct the defects pointed out by the SonarQube tool in their systems.

Then, the IBM Rational Quality Manager (RQM) tool was deployed. It is a tool for planning testing and managing test assets. An integration between Jenkins and RQM plugin was downloaded, and then it was possible to create functional tests to run automatically.

3.3 Monitoring Automation

The monitoring automation initiated independently of the automation initiatives carried out by systems development teams. First, 24 critical business systems were identified, so a team was organized in the Infrastructure Sector to deploy a more effective monitoring for these systems.

After identifying critical systems, a CA toolkit known as CA Service Operations Insight (SOI) was purchased and installed. In order to start the process, it was requested systems documentation, verified with the analysts of each system what functioning of each system (language, integrations, database, etc.) should be monitored and verified. Initially, each system was monitored for 15 days to verify the average performance. So, based on the measured performance and the expected performance by the analysts and users, the limits and alerts were created in the tool, and the rules to treat the occurrences were set up.

Most critical systems were legacy, so the adequacy was much more difficult as there were no scripts to monitor legacy systems. It was decided to monitor the infrastructure used by the applications, and specific scripts were created to simulate and monitor the user experience in legacy systems, such as: time to begin the system, response time, query time, and so on. The first scripts created could serve as a standard for that type of system, and were reused in other systems of the same type or language.

After the implementation of the monitoring for all 24 critical systems, a queue was created to organize the inclusion of all systems used in the institution in the monitoring automation, according to the amount of demand and how critical the systems was.

3.4 Infrastructure Automation

For infrastructure automation, the System Center Orchestrator platform was used, which provides tools to develop, test, debug, deploy and manage task automation using a graphical interface. First, the types of servers in the production infrastructure were identified and also which servers followed the same pattern of creation and provisioning. Server patterns were then defined (database server, Web server, component server) and templates were created in the VMware tool to provide these servers. Finally, the complete tasks in System Center Orchestrator were create to generate virtual machines. Then, the server creation process was automated.

3.5 Obtained Results

With the case study and the reading of articles, we were able to understand some relevant points, among others:

(a) DevOps is a culture and does not refer only to the use of tools. Everyone should feel responsible for changing the mentality, contributing to the improvement of the process as a whole and making software delivery faster and better;
(b) The gradual implementation of DevOps culture and tools is interesting for everyone be adapted to the impact of the change and do not generate resistance by some team;

(c) The training of the team in the use of the tools should be done just before they begin to use them. If it is much earlier, when they use it on a day-by-day basis they may have forgotten what they learned in training. If it is later, the teams will have the tools available but will not know how to use them;
(d) IT managers has to support and be committed to the implementation of the process, establishing controls, when the process is deployed and mature, to guarantee that applications can be deployed only if the whole process is executed;
(e) Many plugins, standard compilation, construction and tools integration scripts are available on the internet, facilitating the implementation of the process;
(f) As the introduction of the tools is repeatable, it is interesting that people who have already participated in the implementation of the integration process or Continuous Delivery participate in the process implementation team of the company that will implement DevOps;
(g) There will be people who are resistant to change to the new DevOps mentality, who will not be able to see the benefits of implementing the process. However, after the implementation of the process and daily use, probably they will notice the agility acquired when using the processes;
(h) Initially there will be the impression that some legacy systems and technologies will not be able to be automated or benefitted by the Continuous Delivery process, but in the case of the institution of the case study, even COBOL and Powerbuilder systems have benefited from process automation.

4 Process to Implement DevOps in Legacy Systems

As a result of the bibliographic review and the case study, a process to introduce legacy systems into DevOps was defined aiming to assist other institutions that work with software development and legacy systems.

The proposed workflow, represented graphically in Fig. 1, can be divided into five subprocesses, which can be executed in parallel, when carried out by different teams:

- Continuous Integration - a subprocess in which, when the code is delivered by the developers, the processes of integration of generated code, configuration management and build generation will be automated.
- Deployment Automation - a subprocess in which the implementation of any build generated in any of the environments (development, test and production) will be automated.
- Tests Automation - subprocess in which the applications of the unit tests, functional tests and any other type of verification and validation of the generated build will be automated.
- Monitoring automation - a subprocess in which the monitoring, alerts and treatment rules of the systems executed in the production environment will be automated.
- Infrastructure Automation - a subprocess in which the creation and provision of servers from all environments will be automated.

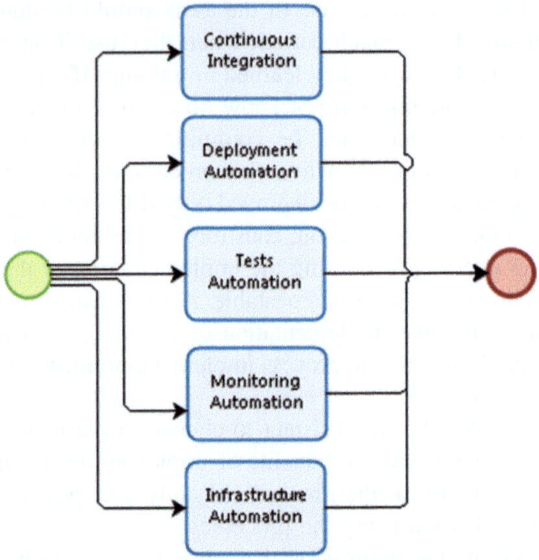

Fig. 1. Graphical representation of the proposed process

4.1 Continuous Integration

The continuous integration subprocess, shown in Fig. 2, begins with the configuration of the script automation tool. This is the main tool of the process because all other tools will integrate with it to automate all the steps of Continuous Delivery. The other scripting activities will be based on it. After configuring the scripting automation tool, the version control tool compatible with continuous integration will be configured and the branches of the systems will be defined. In addition an integration baseline will be created to be the starting point for the evolution of systems.

Next, the compilation and construction tools will be configured, the system building scripts and standards will be created, as well as other tools that will optimize the system build process will be set. Finally, all scripts created will be automated in the scripting automation tool.

Finally, the integrity constraints will be implemented in the process. In this activity is where IT managers have to determine that all systems that are deployed in production will need to follow the automated process that has been defined. Then, it is possible to track which changes have been introduced to a system version, who was the requestor, the responsible party, the modified artifacts, generated assets, deployed assets, and those responsible for the deployment. With this, we guarantee the integrity and tracking of the entire process by which the implemented systems have passed.

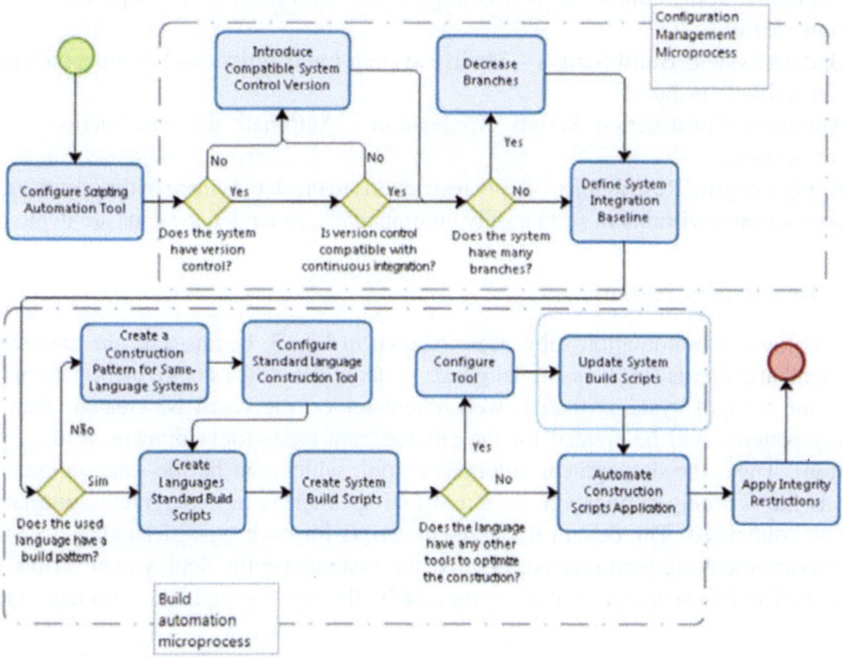

Fig. 2. Graphical representation of the Continuous Integration subprocess

Next, we will describe the activities of the subprocess "Continuous integration" and its main objectives:

- Configure Scripting Automation Tool - Introduce and configure a scripting automation tool to be used by all systems.
- Introduce Compatible System Control Version - Update the control version tool to a version compatible with the IC process.
- Decrease Branches - Check out which development branches are active and decrease the number of branches.
- Define System Integration Baseline - Define the last valid system version, to serve as the baseline in the main branch.
- Create a Construction Pattern for Same-Language Systems - Verify how systems of the same language build and create a standard way of building, to be used by everyone.
- Configure Standard Language Construction Tool - Introduce and configure a standard build tool to be used by all systems in the same language.
- Create Languages Standard Build Scripts - Encode the build pattern you created earlier and create patterns building scripts for languages.
- Create System Build Scripts - Modify the standard build scripts to languages and define the system build scripts.

- Configure Tool - Introduce and configure any additional tool to optimize system construction.
- Update System Build Scripts - Modify system build scripts and add dependencies and artifacts in build
- Automate Construction Scripts Application - Automate the construction scripts application.
- Apply Integrity Restrictions - Apply restrictions to the deployment of the system in the production environment so that only automatically generated systems are deployed.

4.2 Deployment Automation

The deployment automation subprocess, shown in Fig. 3, begins with the creation of deployment patterns for the same language systems. Next, the deployment patterns for the same type of systems (batch, web, client-server, etc.) will be created, then the deploy patterns will be created for the environment types (development, testing, production). Then, the deployment automation tool, which can be the same automation tool as the scripting application, or a tool that specializes in deployment automation, will be configured. The default deployment scripts for each type of language, system, and environment are then created. Finally, the system-specific deployment scripts will be created and these scripts will be automated by the script application automation tool.

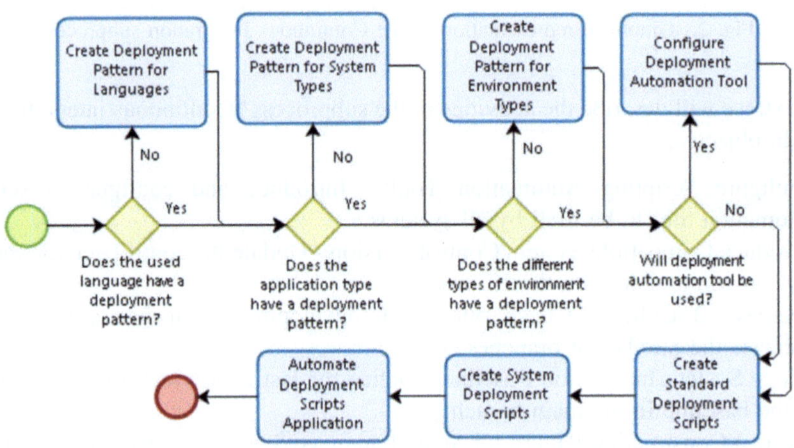

Fig. 3. Graphical representation of the Deployment Automation subprocess

Following, we describe the activities of the automation subprocess of the deployment and its main objectives:

- Create Deployment Pattern for Languages - Verify how to deploy the applications of a given language and define a pattern to be used.
- Create Deployment Pattern for System Types - Standardize how a system type is deployed (web application, batch, client server).

- Create Deployment Pattern for Environment Types - Standardize how deployments of each technology are made in development, testing and production environments.
- Configure Deployment Automation Tool - Configure a deployment automation tool to be used by all deployments.
- Create Standard Deployment Scripts - Based on the deployment patterns, create the default deployment scripts for the language, by system type, and by environment type.
- Create System Deployment Scripts - Modify the default deployment scripts and produce the system deployment scripts for each environment.
- Automate Deployment Scripts Application - Automate, with the help of the scripting automation tool, the application of the system deployment scripts for each environment.

4.3 Tests Automation

The automation subprocess of the tests, presented in Fig. 4, begins with the definition of the baseline of metrics to be analyzed for each language. Next, the automated code analysis tool will be configured. Then the standard static code analysis scripts will be created by language, then the static analysis scripts for each system. Unit testing tools and automated functional tests will then be configured. Finally, the test scripts created will be automated by the scripting automation tool.

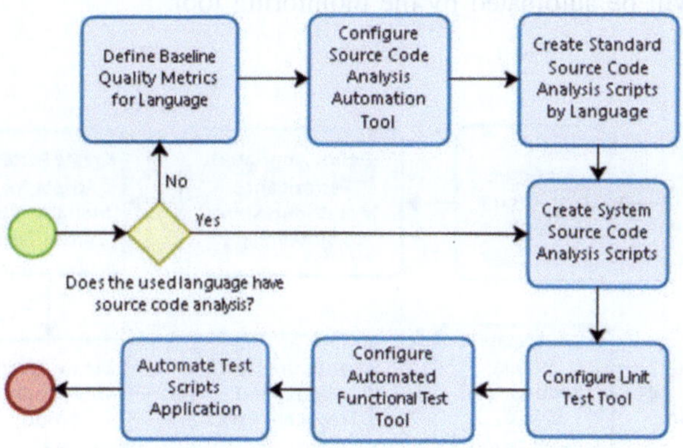

Fig. 4. Graphical representation of the Tests Automation subprocess

Following we will describe the activities of the automation subprocess of tests and its main objectives:

- Define Baseline Quality Metrics for Language - Define, according to the system language, what metrics will be applied to static source code quality analysis (Examples: Code Duplication, Coding Standards, Code Coverage and potential bugs).

- Configure Source Code Analysis Automation Tool - Configure a source code analysis tool to automate the quality verification.
- Create Standard Source Code Analysis Scripts by Language - Based on the baseline of project quality metrics for the language and in the source code analysis tool of the chosen code, create the standard source code analysis scripts of the language.
- Create System Source Code Analysis Scripts - Modify the static analysis scripts of the language source code and produce the static analysis scripts of the system source code.
- Configure Unit Test Tool - Configure the unit test tool that will be used by all languages compatible with this tool.
- Configure Automated Functional Tests Tool - Configure the automated functional test tool that will be used by all languages supported by this tool.
- Automate Test Scripts Application - Automate, with the help of the script automation tool, the application of automated system test scripts.

4.4 Monitoring Automation

The monitoring automation subprocess, presented in Fig. 5, begins with the configuration of the monitoring tool. Then, the metrics to monitor the performance of applications by system type will be defined aiming to know what should be monitored and what is the performance expected. Then the pattern monitoring scripts will be created by system type, followed by the monitoring scripts for each system. Then, the limits, alerts and treatment rules for each system will be configured. Finally, the monitoring scripts created will be automated by the monitoring tool.

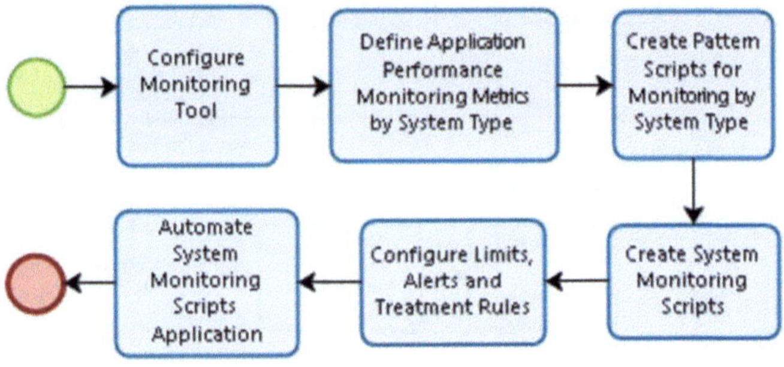

Fig. 5. Graphical representation of the Monitoring Automation subprocess

Below, we will describe the activities of the monitoring automation subprocess and its main objectives:

- Configure Monitoring Tool - Configure the monitoring tool that will be used by all systems in the production environment.

- Define Application Performance Monitoring Metrics by System Type - Define, according to the type of system, which application performance metrics will be monitored (Examples: page load times, error rates, slow transactions, etc.)
- Create Pattern Scripts for Monitoring by System Type - Create the pattern scripts for monitoring by system type, based on metrics to monitor the application performance by system-type.
- Create System Monitoring Scripts - Modify the default monitoring scripts for the system type and produce the system-specific monitoring scripts.
- Configure Limits, Alerts and Treatment Rules - Configure in the monitoring tool, for each system, the limits of each defined metric, the alerts to be generated and treatment rules for each alert.
- Automate System Monitoring Scripts Application - Automate, with the help of the monitoring tool, the application of system monitoring scripts.

4.5 Infrastructure Automation

The infrastructure automation subprocess, shown in Fig. 6, begins identifying the current configuration of the infrastructure of the environments. Next, infrastructure requirements will be standardized into categories, separating servers and components of the same type into standards. Then, the infrastructure automation tool will be configured. Patterns scripts to server creation will be created, and based on the pattern scripts, the system environment creation scripts will be created so that each environment in which the system will run is created automatically. Finally, the infrastructure creation scripts will be automated by the automation tool.

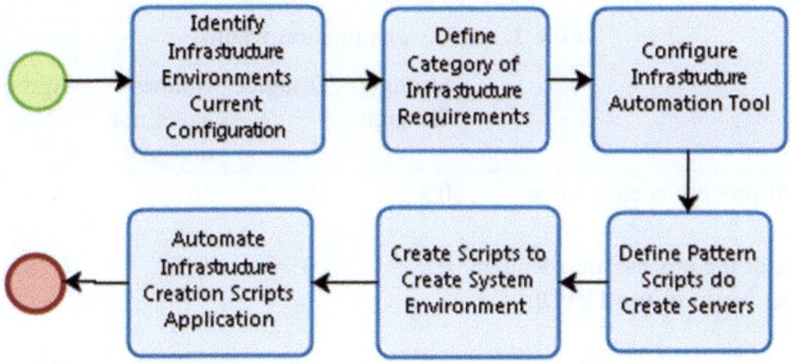

Fig. 6. Graphical representation of the Infrastructure Automation subprocess

Below, we will describe the activities of the subprocess of infrastructure automation and its main objectives:

- Identify Infrastructure Environments Current Configuration - Identify, with the help of the development team, the current configuration of the infrastructure of each environment, based on architecture documents, network topology, environment management tools, monitoring and version control.

- Define Category of Infrastructure Requirements - Define server categories and the instance sizes (for example: application server and database server with small, medium, and large instances) aiming that all applications have a pattern of servers to use.
- Configure Infrastructure Automation Tool - Configure an infrastructure automation tool that will be used to manage and create environments.
- Define Pattern Scripts do Create Servers - Define the pattern scripts to create servers that will be used for all environments based on the identified server pattern.
- Create Scripts to Create System Environment - Modify the pattern scripts to create servers and produce the define the scripts for creation of the system environment.
- Automate Infrastructure Creation Scripts Application - Automate, with the help of the infrastructure automation tool, the application of the scripts to create infrastructure.

5 Process Evaluation

The methodology used to validate the process was the peer review, which consists of the inspection of the work products by a pair, that is, a collaborator with skills similar to the author of the artifacts [3]. In this work, the peer review was carried out by 10 (ten) participants who have training in information technology. The peer review was divided into three stages: 1 - Presentation of the process to those involved; 2 - Filling in the questionnaire by the interviewees; and 3 - Analysis of the results by the researcher. The result is presented in Table 1.

Table 1. Process evaluation results.

Question	Strongly disagree	Disagree	Neither agree nor disagree	Agree	Strongly agree
The proposed process is easy to be understood	0%	10%	0%	80%	10%
The activities of the process are adequate for companies begin to use DevOps practices	0%	0%	10%	60%	30%
The amount of documents of the process are adequate, without excess	0%	0%	20%	80%	0%
The process is adaptable to software development companies	0%	0%	10%	70%	20%
The process can be executed in different scenarios of companies that have legacy systems	0%	0%	0%	70%	30%
The process is efficient and can be executed demanding just a little of effort	0%	40%	30%	30%	0%

Some suggestions were proposed by the evaluators. Among them, the most outstanding point was the ease of understanding the process organized in workflow and separated into activities. Another point commented upon by most was the possible need for effort to deploy the DevOps process because it is directly tied to the company culture, requiring management support, executor skills, and team training, especially when dealing with legacy applications.

6 Final Considerations

With the theoretical review and the case study of implementation of DevOps we have learned that it is challenging to improve the software development process, especially if we are dealing with legacy systems, which have many risks involved and any changes that lead to errors can cost companies a lot. However, applying DevOps concepts and practices such as Continuous Delivery means that evolutions will be delivered faster, with faster user feedback, faster identified errors, and higher quality deliveries because of integration processes and deployment automation.

We have also been able to create an incremental process for introducing DevOps for legacy systems, which can be used as a step-by-step in companies that decide to adhere to the practices of agile software development, benefiting from the improvements in the way in which the releases will be published and in the improvement of the relationship between development and operations teams.

As future work, we identify: (i) Apply the introduction of the DevOps process to legacy systems in other companies to identify which activities can be modified or extended; (ii) Write the detailed process for some legacy technology, indicating specific technology tools that can be used. (iii) Conduct studies to measure the product life cycle time, and the time from the beginning of the evolution to the product delivery, before and after the process introduction.

References

1. Aversano, L., Canfora, G., Cimitile, A., De Lucia, A.: Migrating legacy systems to the Web: an experience report. In: Proceedings Fifth European Conference on Software Maintenance and Reengineering, Lisbon, pp. 148–157 (2001). https://doi.org/10.1109/.2001.914979
2. Beck, K., Flower, M.: Planning Extreme Programming, 1st edn. Addison-Wesley Longman Publishing Co., Inc., Boston (2000)
3. Beizer, B.: Software System Testing and Quality Assurance. Van Nostrand Reinhold Co., New York (1984)
4. Chen, L.: Continuous delivery: huge benefits, but challenges too. IEEE Softw. **32**(2), 50–54 (2015). https://doi.org/10.1109/ms.2015.27
5. Coakes, E., Elliman, T.: Focus issue on legacy information systems and business process engineering: the role of stakeholders in managing change. Commun. AIS **2**(1), Article 4 (1999)

6. Erich, F., Amrit, C., Daneva, M.: Report: DevOps literature review. University of Twente, Technical report (2014)
7. Feathers, M.C.: Trabalho eficaz com código legado, 3rd edn. Porto Alegre, Bookman (2013). 600 p
8. Fitzgerald, B., Stol, K.-J.: Continuous software engineering: a roadmap and agenda. J. Syst. Softw. **123**, 176–189 (2017). https://doi.org/10.1016/j.jss.2015.06.063
9. Fowler, M.: Continuous Integration (2006). http://martinfowler.com/articles/continuousIntegration.html. Accessado emfev 2018
10. Gong, Y., Janssen, M.: From policy implementation to business process management: principles for creating flexibility and agility. Gov. Inf. Q. **29**(Supplement 1), S61–S71 (2012). https://doi.org/10.1016/j.giq.2011.08.004. ISSN 0740-624X
11. Halachmi, A.: Imagined promises versus real challenges to public performance management. Int. J. Prod. Performance Manag. **60**(1), 24–40 (2011). https://doi.org/10.1108/17410401111094295
12. Humble, J., Farley, D.: Entrega Contínua: Como Entregar Software de Forma Rápida e Confiável. Bookman Editora, Porto Alegre (2013). 464 p
13. Jeyaraman, G., Krishnamurthy, K., Raveendra, V.V.S.: Reengineering legacy application to E-business with modified rational unified process. In: Proceedings of the Seventh European Conference on Software Maintenance and Reengineering (CSMR 2003), p. 143. IEEE Computer Society, Washington, DC (2003). https://doi.org/10.1109/csmr.2003.1192421
14. Laguna, M.A., Crespo, Y.: A systematic mapping study on software product line evolution: from legacy system reengineering to product line refactoring. Sci. Comput. Program. **78**(8), 1010–1034 (2013). https://doi.org/10.1016/j.scico.2012.05.003
15. Leppänen, M., et al.: The highways and country roads to continuous deployment. IEEE Softw. **32**(2), 64–72 (2015). https://doi.org/10.1109/ms.2015.50
16. Olsson, H.H., Alahyari, H., Bosch, J.: Climbing the "Stairway to Heaven" – a multiple-case study exploring barriers in the transition from agile development towards continuous deployment of software. In: Proceedings of the 2012 38th Euromicro Conference on Software Engineering and Advanced Applications (SEAA 2012), pp. 392–399. IEEE Computer Society, Washington, DC (2012). https://doi.org/10.1109/SEAA.2012.54
17. Pressman, R.S.: Software Engineering: A Practitioner's Approach. Palgrave Macmillan, London (2005)
18. Rahman, A.A.U., Helms, E., Williams, L., Parnin, C.: Synthesizing continuous deployment practices used in software development. In: Proceedings of the 2015 Agile Conference (AGILE 2015), pp. 1–10. IEEE Computer Society, Washington, DC (2015). https://doi.org/10.1109/Agile.2015.12
19. Rao, G.: DevOps for legacy systems – The demand of the changing applications landscape. Infosys limited (2015). https://www.infosys.com/it-services/application-development-maintenance/white-papers/documents/DevOps-legacy-systems.pdf. Accessed Feb 2018
20. Roche, J.: Adopting DevOps practices in quality assurance. Commun. ACM **56**(11), 38–43 (2013). https://doi.org/10.1145/2524713.2524721
21. Schaller, A.E.: DevOps transformation challenges facing large scale legacy systems. Doctoral dissertation, Utica College (2016)
22. Seacord, R.C., Plakosh, D., Lewis, G.A.: Modernizing Legacy Systems: Software Technologies, Engineering Processes, and Business Practices. Addison-Wesley Professional, Boston (2003)
23. Sommerville, I.: Engenharia de Software. Pearson Prentice Hall, São Paulo (2011)

24. Sosa, E., Clemente, P.J.J., Conejero, M., Rodríguez-Echeverría, R.: A model-driven process to modernize legacy web applications based on service oriented architectures. In: Proceedings of the 2013 IEEE 15th International Symposium on Web Systems Evolution (WSE) (WSE 2013), pp. 61–70. IEEE Computer Society, Washington, DC (2013). https://doi.org/10.1109/WSE.2013.6642418
25. Tobin, M., Bass, B.: Federal Application Modernization Road Trip: Express Lane or Detour Ahead. UNISYS and MeriTalk, January 2011
26. Warren, I.: The Renaissance of Legacy Systems: Method Support for Software-System Evolution. Springer, London (2012). https://doi.org/10.1007/978-1-4471-0817-7
27. Waters, K.: What is Agile? (10 Principles of Agile), All About Agile (2007). http://www.allaboutagile.com/what-is-agile-10-key-principles. Accessed Feb 2018
28. Yin, R.K.: Case Study Research: Design and Methods, 2nd edn. Applied Social Research Methods Series, vol. 5 (1994)

Development of Sectoral Intellectualized Expert Systems and Decision Making Support Systems in Cybersecurity

Bakhytzhan Akhmetov[1], Valeriy Lakhno[2], Berik Akhmetov[3], and Zhuldyz Alimseitova[4](✉)

[1] Kazakh National Pedagogical University named after Abay, Almaty, Kazakhstan
[2] European University, Kyiv, Ukraine
[3] Caspian State University of Technologies and Engineering named after Sh. Yesenov, Aktau, Kazakhstan
[4] Satbavev University, Almaty, Kazakhstan
zhuldyz_al@mail.ru

Abstract. The paper considers the prerequisites for the integration of various expert and decision support systems for information security and cybersecurity. Analyzed the possibility of sectoral pooling and sharing of knowledge bases of such intellectualized systems. The model of the knowledge management subsystem in sectoral expert and decision making support systems for information security and cybersecurity of critical computer systems is described. Also article presents the results of studies of the knowledge management system of existing local expert systems and decision making support systems for information protection. Algorithms for the distribution of requests between similar systems are specified.

Keywords: Cybersecurity · Decision making support systems
Expert systems · Clusters

1 Introduction

At the initial stage of implementation of decision making support systems (DMSS) [1, 2] and expert systems (ES) [3], in particular adaptive [4], in information protection (IP) and cybersecurity (CS) problems, their architecture matched to the classical version [5]. However, as the tasks related to the provision of CS have become more complicated, problems have arisen that can no longer be solved within the framework of classical architectures of the DMMS and ES for IP. For example, when in the process of logical inference it becomes necessary to use the conclusions of several independently functioning DMSS or ES. If the DMSS or ES was initially developed only for the solution of narrow-profile tasks, such as supporting solutions for choosing hardware-compatible rational variants of complex information protection systems [3] or expert analysis of complex features detected during the targeted attack [4], in such case to extend the range of solutions for other than the initial options is not possible. The relevance of this study is determined, first of all, by the state of the problematic of

R. Silhavy et al. (Eds.): CoMeSySo 2018, AISC 860, pp. 162–171, 2019.
https://doi.org/10.1007/978-3-030-00184-1_15

the integrated implementation of the critical important computer systems (CICS) of DMSS and the ES, combined into sectors (clusters [6]) into the contours of IP and CS. This causes the further development of the methodological apparatus, and the principles for the creation of sectoral intellectualized DMSS and ES for IP and CS CICS tasks.

2 Analysis of the Literature Data and the Formulation of the Problem

The increasing complexity of cyberattacks, especially targeted ones on CICS, aroused interest in the development of intellectualized DMSSES in the field of CS [7, 8]. The need for prompt decision-making related to the provision of CS of CICS made promising studies on the development of sectoral DMSSES able to combine their knowledge in the framework of the solution of the tasks [7].

The studies [9, 10] analyze the experience of using ES and DMSS in the problems of risk assessment for CS CICS. The scope of knowledge base (KB) of these systems was limited only to risk assessment.

The studies [11, 12] describe the DMSS using for decision-making in insufficiently structured situations for estimating the CS CICS KVKS. However, the research [12] is not brought to hardware and software implementation yet. The experience of using intelligent DMSS and ES in the tasks of management of IP and CS at individual enterprises is presented in studies [13, 14]. The scope of these systems was limited only to the management of information security.

The studies [15, 16] analyze the experience of commercial DMSS and ES implementation on IP and CS. The authors note that commercial systems have a closed nature, and their acquisition by individual enterprises is associated with significant financial costs.

The study [17] has shown that the problems of introducing DMSS and ES in the context of their multitasking were not systematically considered.

The conclusions of studies [7, 8, 12, 15] allow us to speak about the problem of the systemic introduction of adaptive DMSS and ES into the contours of IP and CS CICS. At the same time, the problems of theoretical substantiation and development of the basic principles for the creation of sectoral intellectualized ES and DMSS in the tasks of CS CICS remain unresolved.

3 Purpose and Objectives of Research

The goal of the work is the development of the basic principles and methodology for the creation of sectoral intellectualized ES and DMSS for the tasks of IP and CS CICS.

In order to achieve the aims of the work there is a need to solve the following tasks:

- to develop a model for the functioning of the subsystem of coordination and control of the knowledge output in sectoral DSSES for IP tasks;
- to test the control system and integration of the DB of local DSS and ES by IP.

4 Models and Methods

In sectoral DSSES the sector (cluster) is considered as an information unit. The sector integrates the capabilities of several independent (i.e. local) ES or DSS, Fig. 1. We will assume that the sector nodes can exchange specific information related to different tasks of the IP and CS. But the end user, for example, the analyst of IS, the cluster is available as a single resource. Therefore, the task of the sectoral system is the distribution of user requests and their conversion for specific local DSS or ES on IP and CS.

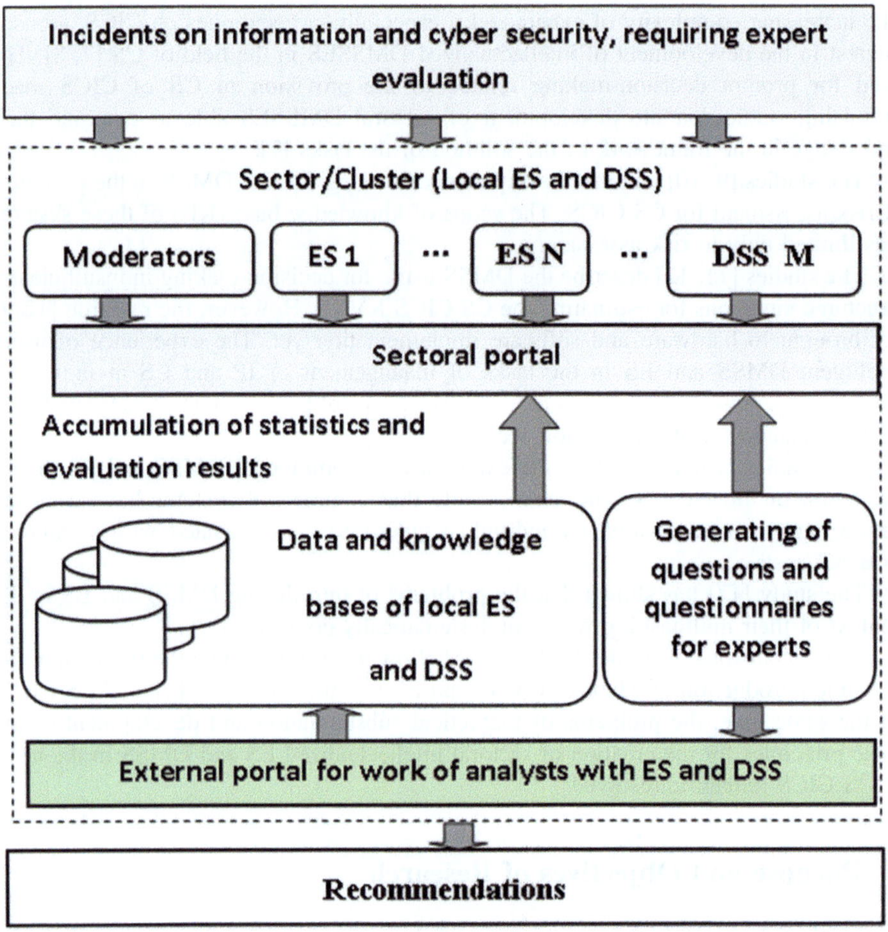

Fig. 1. Structural diagram of the sectoral (cluster) platform for combining local DSS and ES on cybersecurity

Let introduce the following notations: the initial (input) data – X; the resulting (output) data – Y; W – representation of the sets of input variables into the set of resulting data. Then $X = \langle x_1, \ldots, x_m \rangle$, $Y = \langle y_1, \ldots, y_n \rangle$, $W = \langle w_1, \ldots, w_n \rangle$. Where $X \cap Y = \varnothing$.

Sectoral DSSES will be considered as resulting if the following requirements are met:

1. $Y = w\langle X \rangle$.
2. $$(\forall x_i \in X)(\exists q_1 \in x_1, \ldots, q_i \in x_i, q_i \in x_i, \ldots, q_m \in x_m) \times [w(q_1, \ldots, q_i, \ldots, q_m) \neq w(q_1, \ldots, q_i, \ldots, q_m)].$$

At the first stage of the request processing there is specified a set of variables to the sectoral DSSES – $x_i = f_i(x_1, \ldots, x_n)$. Values x_i should to be obtained during the implementation of $Y^r = \langle y_1^r, \ldots, y_n^r \rangle$.

Then, $y_i^r = z_i(x_{i1}, \ldots, x_{im})$.

The set of functions $Z^1 = \langle z_1, \ldots, z_m \rangle$ forms a list of $X^1 = \bigcup_{i=1}^{m} x_i$ in a resulting list $Y^r = Z^1(X^1)$, where X1 – list of variables of the final rank of the functional structural module of the sectoral DSSES. Then $u = (X, Y, W)$: $X^1 = X^{u1} \bigcup Y^1$, where $X^{u1} \bigcap Y^1 = \varnothing$.

Let consider that X^1 contains variables that are the initial data for the module U^1. In list Y^1 there are included the parameters which should be calculated during the operation of the module U^1.

The list Y^1 is presented through a new list. For example: $Y^1 = Z^2(X^2)$, where $X^2 = X^{u2} \bigcup Y^2$, $X^{u2} \bigcap Y^2 = \varnothing$.

Operations continue as long as the list of variables not belonging to X^1 for the rank k is not exhausted.

The described model can be represented in the form of a graph of variables $x_i = f_i(x_1, \ldots, x_n)$, Fig. 2. The nodes of the graph, denoted by red dots, correspond to the resulting variables.

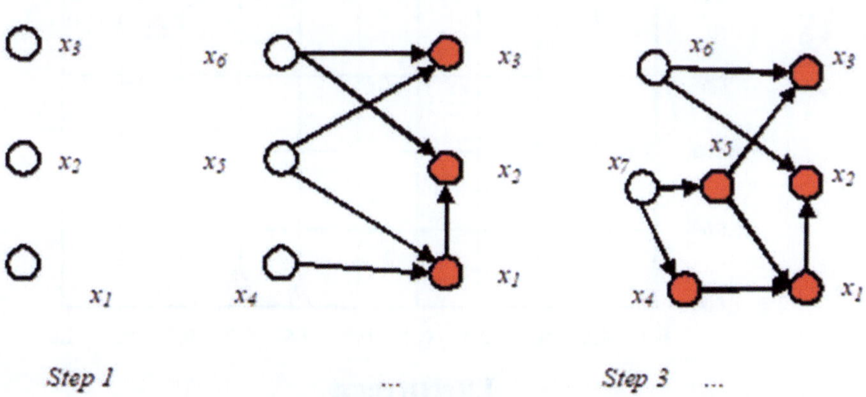

Fig. 2. Solution steps

At the decision stages, i.e. at each step of the graph construction, there are drawn arrows to non-initial nodes. We assume that the process is complete in the case when all non- initial nodes are connected by arrows. Correct is the result if there are no loops in the resulting graph. The proposed approach makes it possible to link knowledge from local DSS and ES for the related tasks and subject areas (for example, cybersecurity, the choice of technical means of protection, the cost of ISS, risk assessment for IS, etc.). In addition, it is possible to compile a tree of logical output for a sectoral DSS or ES on IP and CS of CICS.

5 Experiment

The proposed model, taking into account the results of the works [1, 4, 12], was implemented in the sectoral (cluster) system on cybersecurity – Cluster system of cybersecurity ("CLSC"). The knowledge base of "CLSC" integrated the DSS DB and ES of DMSSCIS and "DMSSCSE", respectively to the works [4, 18]. CLSC was tested in the computer centers of several companies [1, 4].

The effectiveness of the CLSC sectoral (cluster) platform, which integrated local DSS and ES on cybersecurity, was also tested during computational experiments (Figs. 2 and 3). As initial data there were taken the parameter values generated by the program according to the data from [1, 4, 12].

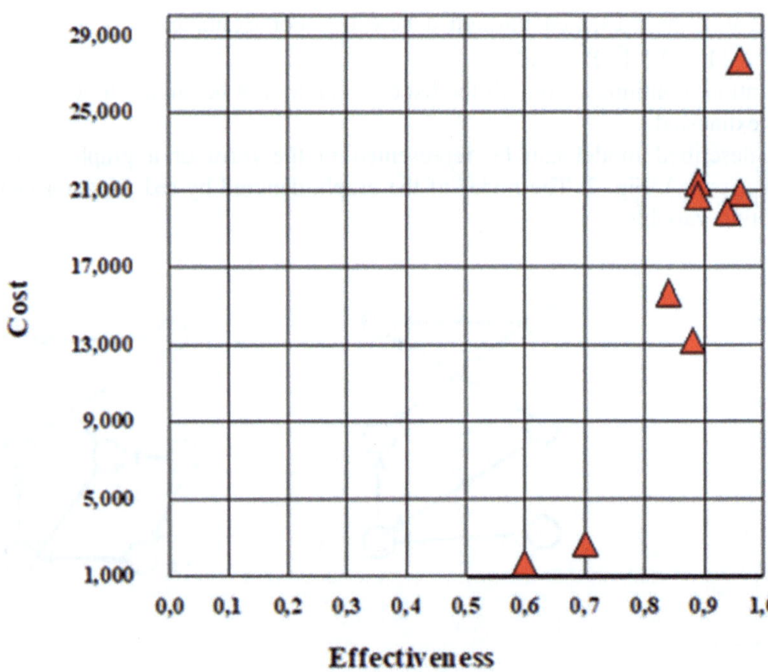

Fig. 3. Extended search area for the optimal samples of TMP for CICS

During the experiment there were compared the possibilities of sectoral DSS ("CLSC" and local DSS and ES (DMSSCIS and "DMSSCSE", [4, 18]) at evaluating the decisions regarding: the selection of compatible software and hardware protection tools, optimization of the cost of IISS for CICS, evaluation of risks for IS, evaluation and predicting the transformation of the situation for identified anomalies or threats in CICS, etc.

For example, during the computational experiments there was checked the correctness and adequacy of the algorithm (model) of the functioning of the subsystem for coordinating and control of knowledge output in the sectoral intellectualized DSS ("CLSC", for finding optimal solutions among the proposed variants of technical means of protection (TMP) of CICS, as well as of IISS for CICS.

During the computational experiment there were obtained two-dimensional arrays, which are presented in the form of a point diagram for the TMP, see Fig. 3, and for IISS for CICS, Fig. 4.

The triangles on Fig. 3 – coordinates of the effectiveness and cost values for the considered samples of TMP. Squares, Fig. 4 – coordinates of the effectiveness and cost values for the considered samples of complex means of information security and cybersecurity for CICS.

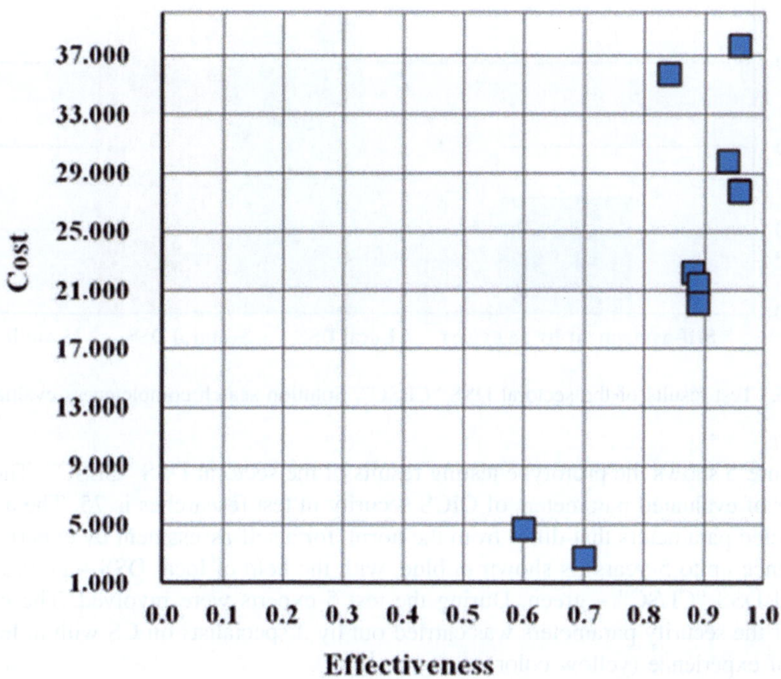

Fig. 4. Extended search area for the optimal samples of IISS for CICS computer network segment

For TMP and IISS as a protection object there was considered a segment of a computer network or a local CICS network [1–4]. The cost parameters were calculated in conventional monetary units.

The results obtained on the graph are an expanded search area for optimal solutions using the "effectiveness-cost" criterion [12].

The implementation of the research results on real CICS, Fig. 5, showed that the proposed solutions allow to increase the degree of protection of CICS, in particular due to the complex integration of local DSS and ES into cybersecurity tasks.

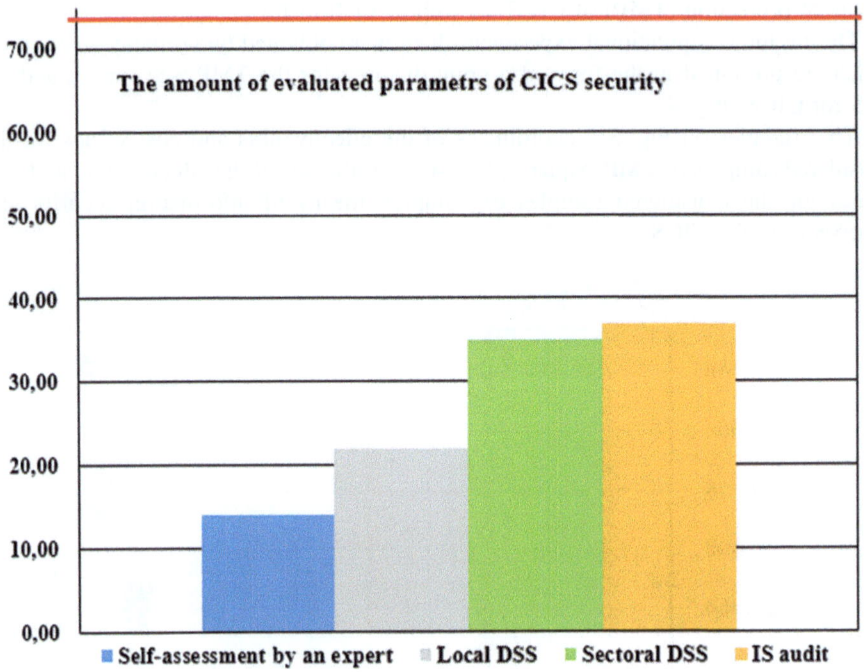

Fig. 5. Test results of the sectoral DSS "CLSC". Solution search completeness evaluation

Figure 5 shows the prototype testing results of the sectoral DSS "CLSC". The total number of evaluated parameters of CICS security in test researches is 75. The amount of detected parameters that differ from the norm, for a self-assessment by experts (with experience up to 5 years) is shown in blue, with the help of local DSS – gray, for the sectoral DSS "CLSC" – green. During the test 5 experts were involved. The control audit of the security parameters was carried out by 3 specialists on CS with at least 10 years of experience (yellow color).

Calculation of the amount of rules involved in the process of logical inference showed that in the sectoral DSS "CLSC" their amount is 5–7% less than in local DSS and ES [4, 18]. This is due to the fact that only the necessary rules for a specific task are sampled.

The sectoral DSS "CLSC" according to the main parameters (decision search time, prediction accuracy, search depth, etc.) exceeds solutions described in [4, 12, 18]. For example, the test showed that decisions support related to the choice of rational variants of TMP and ISS, the costs at the protection tools selection step are reduced by 16–19%.

6 Discussion of the Obtained Results and Prospects for Further Researches

The conducted researches are the development of work [19]. The proposed approach has advantages in comparison with traditional local DSS and ES in the tasks of IP and CS. Firstly, sectoral integration of systems makes it possible to combine the accumulated information of local DSS and ES in the DB. Secondly, sectoral (cluster) systems account better the specifics of the functioning of specific CICS, for example, in the tasks of selecting rational hardware compatible TMP or IISS [14] or at the decisions support on the operational control of cybersecurity, in conditions of uncertainty, inconsistency and incompleteness of knowledge about the state of the object [18, 19].

The potentially eliminated disadvantage of sectoral DSS or ES, and in particular CLSC, is the need to employ sufficiently qualified specialists, that is not always possible in real operating conditions. Also, to the disadvantages there can be attributed the additional financial costs for the development and implementation of coordination and control subsystems of the knowledge output.

The development of this research area can be the improvement of the interaction of the algorithm and program modules of the sectoral DSS "CLSC" with modules of the "DMSSCIS" system [17–19].

In general, on the basis of the conducted researches, it is possible to state the effectiveness of the proposed method and algorithms, as well as the software package for the sectoral DSS "CLSC".

7 Conclusions

1. There are considered prospects of integration of various expert and decision support systems for the tasks of information protection and cybersecurity ensuring on the basis of their sectoral aggregation and joint use of knowledge data bases. Proposed the model of the knowledge output control subsystem in sectoral ES and DSS for the IP and CS tasks of CICS.
2. Proposed a model and conducted computational research experiments of the knowledge control system of the existing local DSS and ES on IP, integrated into the cluster. Tested the prototype of the sectoral (cluster) decisions support system on CICS cybersecurity ensuring. Evaluated the effectiveness of the sectoral DSS "CLSC" use in comparison with local ES and DSS used in the tasks of decision-making support on IP and CS.

References

1. Akhmetov, B., Lakhno, V., Boiko, Y., Mishchenko, A.: Designing a decision support system for the weakly formalized problems in the provision of cybersecurity. Eastern-Eur. J. Enterp. Technol. **1**(2), 4–15 (2017). Article no. 85
2. Rees, L.P., Deane, J.K., Rakes, T.R., Baker, W.H.: Decision support for cybersecurity risk planning. Decis. Support Syst. **51**(3), 493–505 (2011)
3. Chang, L.Y., Lee, Z.J.: Applying fuzzy expert system to information security risk assessment-a case study on an attendance system. In: IEEE 2013 International Conference on Fuzzy Theory and Its Applications (iFUZZY), pp. 346–351 (2013)
4. Lakhno, V., Tkach, Y., Petrenko, T., Zaitsev, S., Bazylevych, V.: Development of adaptive expert system of information security using a procedure of clustering the attributes of anomalies and cyber attacks. Eastern-Eur. J. Enterp. Technol. **6**(9), 32–44 (2016)
5. Goztepe, K.: Designing fuzzy rule based expert system for cyber security. Int. J. Inf. Secur. Sci. **1**(1), 13–19 (2012)
6. Mahmood, T., Afzal, U.: Security analytics: big data analytics for cybersecurity: a review of trends, techniques and tools. In: 2013 2nd National Conference on Information Assurance (NCIA), pp. 129–134 (2013)
7. Kim, K., Kim, I., Lim, J.: National cyber security enhancement scheme for intelligent surveillance capacity with public IoT environment. J. Supercomput. **73**(3), 1140–1151 (2017)
8. Medhat, K., Ramadan, R.A., Talkhan, I.: Security in mission critical communication systems. In: Multimedia Services and Applications in Mission Critical Communication Systems, p. 270 (2017)
9. Radziwill, N., Benton, M.: Cybersecurity Cost of Quality: Managing the Costs of Cybersecurity Risk Management, [Electronic resource] (2017). https://arxiv.org/ftp/arxiv/papers/1707/1707.02653.pdf
10. Jalali, M., Siegel, M., Madnick, S.: Decision Making and Biases in Cybersecurity Capability Development: Evidence from a Simulation Game Experiment, [Electronic resource] (2017). https://arxiv.org/ftp/arxiv/papers/1707/1707.01031.pdf
11. Fielder, A., Panaousis, E., Malacaria, P., Hankin, C., Smeraldi, F.: Decision support approaches for cyber security investment. Decis. Support Syst. **86**, 13–23 (2016)
12. Lakhno, V., Petrov, A., Petrov, A.: Development of a support system for managing the cyber security of information and communication environment of transport. In: International Conference on Information Systems Architecture and Technology, pp. 113–127 (2017)
13. Benaroch, M.: Real options models for proactive uncertainty-reducing mitigations and applications in cybersecurity investment decision-making, p. 39 (2017)
14. Wagner, N., Şahin, C.Ş., Winterrose, M., Riordan, J., Pena, J., Hanson, D., Streilein, W.W.: Towards automated cyber decision support: a case study on network segmentation for security. In: IEEE Symposium Series on Computational Intelligence (SSCI), pp. 1–10 (2016)
15. Atymtayeva, L., Kozhakhmet, K., Bortsova, G.: Building a knowledge base for expert system in information security. In: Soft Computing in Artificial Intelligence, pp. 57–76 (2014)
16. Silva, M.M., de Gusmão, A.P.H., Poleto, T., e Silva, L.C., Costa, A.P.: A multidimensional approach to information security risk management using FMEA and fuzzy theory. Int. J. Inf. Manag. **34**(6), 733–740 (2014)
17. Tamjidyamcholo, A., Baba, M.S.B., Shuib, N.L.M., Rohani, V.A.: Evaluation model for knowledge sharing in information security professional virtual community. Comput. Secur. **43**, 19–34 (2014)

18. Lakhno, V., Zaitsev, S., Tkach, Y., Petrenko, T.: Adaptive Expert systems development for cyber attacks recognition in information educational systems on the basis of signs' clustering. In: Advances in Computer Science for Engineering and Education. ICCSEEA 2018. Advances in Intelligent Systems and Computing, vol. 754, pp. 673–682 (2018)
19. Akhmetov, B., Lakhno, V.: System of decision support in weakly formalized problems of transport cybersecurity ensuring. J. Theor. Appl. Inf. Technol. **96**(8), 2184–2196 (2018)

Optimal Multi-robot Path Finding Algorithm Based on A*

Alexander Erokhin[1], Vladimir Erokhin[2(✉)], Sergey Sotnikov[2], and Anatoly Gogolevsky[2]

[1] T-Systems CIS LLC, St. Petersburg, Russia
erohin_a_v@mail.ru
[2] Mozhaisky Military Space Academy, St. Petersburg, Russia
erohin_v_i@mail.ru, sergeyy_sotnikov@lenta.ru, gogolevski@bk.ru

Abstract. This paper presents an optimal path finding algorithm for a group of robots. Presented approach is based on A* graph traversal algorithm and contains modifications that make it possible to apply strengths of an original algorithm to solve multi-robot path planning problem. Proposed modification is to dynamically change costs of nodes used in path of one robot while planning route for another. In another words to give the cost meaning of a time which robots need to wait to be able to take the node. Although presented approach was used to solve multi-robot path finding problem on unknown map it can be successfully applied to solve navigating problems of group of mobile agents on known maps. Computer simulation showed reliable results for maps of different configuration.

Keywords: Multi-robot · Path finding · Graph traversal · A*

1 Introduction

Navigation and guidance of autonomous mobile vehicles is a fundamental problem in robotics, which has attracted intensive research in recent decades [1]. A lot of attention is focused on multi-robot path planning methods as a group of robots has the potential of solving problems, which single robot could only handle in a very inefficient way or could not handle at all. Usage of a group of robots has many advantages. The most important are shortening of the total path or mission time and increased robustness [2]. But there are also some complications that come along with advantages of using a group of robots.

For instance, we can no longer rely on a path finding algorithm for single vehicle when dealing with a group. What can be an optimal solution in path planning for a single mobile robot, can be totally unsuitable when solving problem for two or more robots moving as a group or having common goal. Furthermore, there are some new class problems we have to deal with when working with group,

R. Silhavy et al. (Eds.): CoMeSySo 2018, AISC 860, pp. 172–182, 2019.
https://doi.org/10.1007/978-3-030-00184-1_16

such as paths intersections, robot collisions, deadlock and lot more. Known path finding algorithms could no longer be applied to solve such tasks without major modifications.

Let's consider problem where group of autonomous mobile vehicles needs to reach certain area of a goal target. This problem is considered to be solved then and only then each vehicle in the group has reached the target area. Thus it is worthwhile to talk about optimal path planning for a group as a whole.

2 A* Graph Traversal Algorithm Reference

A* is one of the most commonly used path finding and graph traversal algorithms. It was originally proposed by P. Hart, N. Nilsson and B. Raphael in 1968 as an extension of Edsger Dijkstra's algorithm [3]. Nowadays due to numerous modifications A* is widely used to solve problems in different areas of applied science. It is successfully applied in such tasks as path planning in navigation services [4], path planning in game development [5], for affordable choice of wiring route in the electrical and telecommunications networks in the residential areas [6] and even classroom scheduling [7] to name a few.

A* is a best-first search and solves problems by searching among all possible paths to the solution for the one that has the smallest cost. In general it is guaranteed that A* will find a path to goal if there is one. It is formulated in terms of weighted graphs: starting from a specific node of a graph, it constructs a tree of paths starting from that node, expanding paths one step at a time, until one of its paths ends at the predetermined goal node.

At each iteration of its main loop, A* needs to determine which of its partial paths to expand into one or more longer paths. It does so based on an estimate of the cost (total weight) still to go to the goal node. Specifically, A* selects the path that minimizes:

$$f(n) = g(n) + h(n) \tag{1}$$

where n is the last node on the path, $g(n)$ is the cost of the path from the start node to n, and $h(n)$ is a heuristic that estimates the cost of the cheapest path from n to the goal. The heuristic is problem-specific. For the algorithm to find the actual shortest path, the heuristic function must be admissible, meaning that it never overestimates the actual cost to get to the nearest goal node. Moreover, if heuristic rarely overestimates actual cost, then A* will generate a path that is rarely overestimates the optimal solution. In fact, you can generally trust A* to generate the fewest nodes necessary to find a solution [8].

Considering all of the above lets decide what modifications should be made to the original A* in order it to become applicable in a multi-robot path finding problem.

3 Necessary Modifications to A* to Expand It on Multi-robot Path Finding Problem

First of all, it should be noted that original A* is a graph traversal algorithm and deals with vertexes and edges. Robot path finding problem is posed on terrain modeled with passability matrix. So first of all a way to map a weighted graph to a passability matrix model is needed. It could be easily done in a definite manner with a matrix cell containing information about cost of transition to it from another cell. Values of $g(n)$ and $f(n)$ will be determined based on this information. Second and most important task to be solved is a way to extend application area of A* algorithm to cover multi-robot path finding problem (Fig. 1).

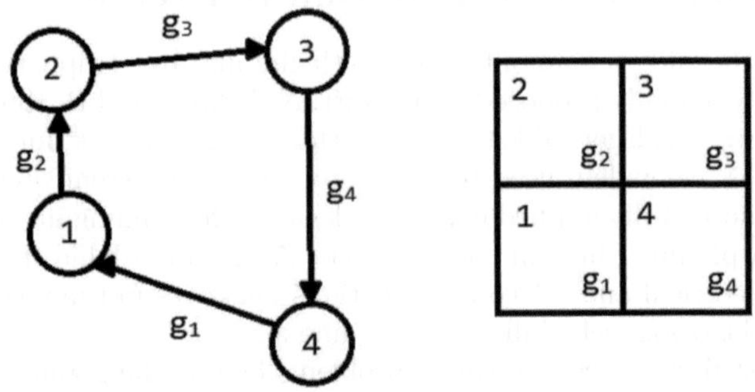

Fig. 1. Weighted graph to matrix mapping

Let's consider difficulties which arise when dealing with this task. Firstly, shared map structure means that data contained in it will be accessed and modified by all of the group members. So it is very important that map data integrity is properly maintained. Secondly, all path intersections should be controlled to exclude situations when robots are trying to use the same map cell at once, what will cause a collision. Best way to deal with situations like this would be intersection detection on a path planning phase before actual path following. This would prevent possibility of collision and save time needed for path adjustment otherwise. Thirdly, effectiveness of an algorithm should be estimated for a group as a whole and not for individual robots. Some robot individually can find most optimal path. But what would be optimal for a group: to wait in a queue of robots and to follow that same path when it is available or to plan an independent less optimal path but to move along without any delay.

Easiest way to maintain data integrity is to serialize access to it. Access order could be determined based on the priority of group members of on their serial number. For a better performance access to map cells can be parallelized. This will allow different robots to explore different map zones simultaneously. But each robot have to obtain an exclusive lock on a cell it is about to modify data in.

As for path intersection and robot collision, these problems could and should be solved on the path planning stage. This could be done by a proper A* algorithm modification. There are already some existing methods of collision free path planning for multiple objects [8], including such methods as complete and partial path blocking and on-the-fly adaptation.

Path blocking methods are based on approach in which each robot marks cells involved in its path as impassable for another robots either temporary or permanently. Advantages of such approach include good performance, ease of realization and the fact that collision situations are prevented on a path planning stage. The obvious drawback is that with the more robots participating in path planning map becomes more complex for each subsequent robot to plan a path, which becomes longer and more unusual.

On-the-fly adaptation methods are based on idea of dealing with collisions when there is one. All paths for all robots are planned independently and without considering other members of a group. Then on an action stage of path following, when collision is about to happen, methods of collision prevention are used. Most common are path corrections by repeating path planning procedure or passing by approach in which all robots wait for others to pass the disputed area based on their priority. Possibility of usage of any known path finding algorithm without any modification could be considered as advantage of this approach. But there are also many drawbacks including necessity of path replaning and the fact that in case of presence of one optimal path group members will be forced to wait in queue for available cells to make each step.

Our goal is to propose such modification that would allow avoiding collision possibility on path planning stage and would provide optimal paths for a group as a whole.

As were stated earlier A* selects the path that minimizes function (1) based on overall cost. The important thing is that the nature of that cost was not specified. Most natural way to estimate path cost is by time needed to travel that path. So time needed to travel vertically or horizontally from one map cell to another is equivalent as long as all cells have equal passability and velocity is constant. This means that time needed to travel diagonally equals

$$d = \sqrt{h^2 + v^2} \tag{2}$$

where h is time needed to travel one cell horizontally and v is time needed to travel one cell vertically.

When talking about a group, passability is no more the only factor affecting the cost of a cell. A cell already involved in a path of any group member means a new member planning a path will be forced to wait for a cell to become available to continue its travel. This would affect the cost of all path timewise. At some point it will become more effective to choose a path that seems longer in distance but is shorter in terms of time as expects less collision avoidance and wait time during the travel.

Let's consider a simple example. Suppose a group of three robots is deployed on a terrain described by a map showed in Fig. 2. Empty cells represent passable areas, filled cells mark obstacles, circle markers stand for a robot positions and goal cell contains a golden star. As could be easily seen the shortest path in terms of distance lies right between two obstacles. If robots would plan their paths independently all will choose that same shortest path considering their start positions. This will lead to situation in which all group members are forced

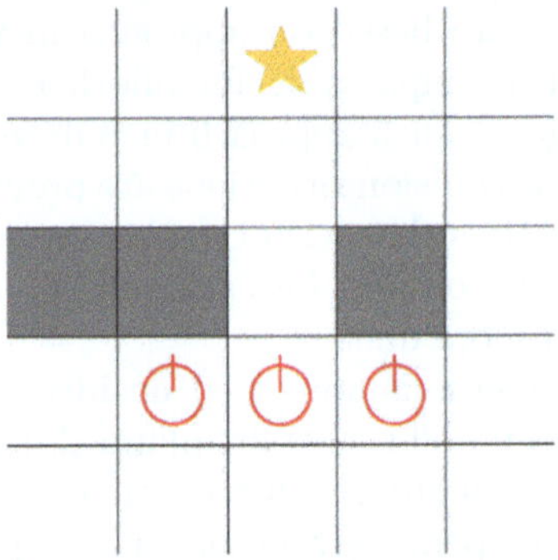

Fig. 2. Initial position of the robots

to move sequentially waiting for a member with a higher priority to release cells shared in all robots' path and would lead to overall increase of the time needed for all members of a group to reach the goal. It would have been so much better if on path planning stage robot knew that it might wait longer for a particular cell to become available. In other words it means that cells already involved in other robots paths put on cost by some constant value C_n. The more robots are planning to use a cell the more costly it becomes to every next robot planning its path. This way A* will automatically avoid using this cells to lower possible waiting time and will choose more optimal path.

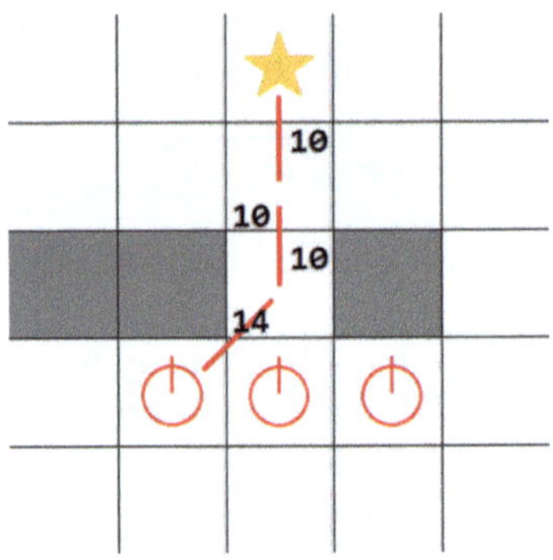

Fig. 3. First robot's path

Let's try to apply described approach in our example and assume for convenience that horizontal or vertical one cell step costs 10 units of time, diagonal step costs 14 units of time and constant addition cost value C_n equals 10. We also assume that robots are planning their paths sequentially ordered from left to right. First most left robot plans it path as shown in Fig. 3. Left bottom value in each cell involved in the path denotes initial cost of a cell. Right top value denotes additional cost of a cell after it has been used in path. Thus total cost of the first robot's path equals 24 units of time.

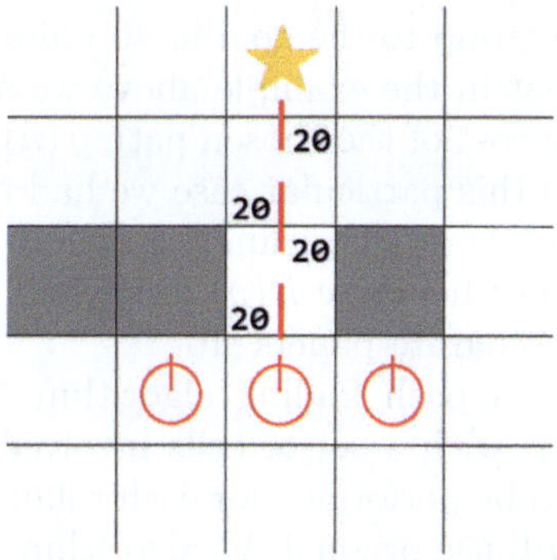

Fig. 4. Second robot's path

Next goes robot in the middle. Its path would lie between the obstacles as well because it is still the most optimal path, despite additional costs (Fig. 4). The total path cost for a second robot is 40 units of time: $10 + 10$ for two vertical moves and $10 + 10$ additional costs for every cell from the first robot. This means that two cells between obstacles are now used in two paths and thus have additional costs of 20 units of time each.

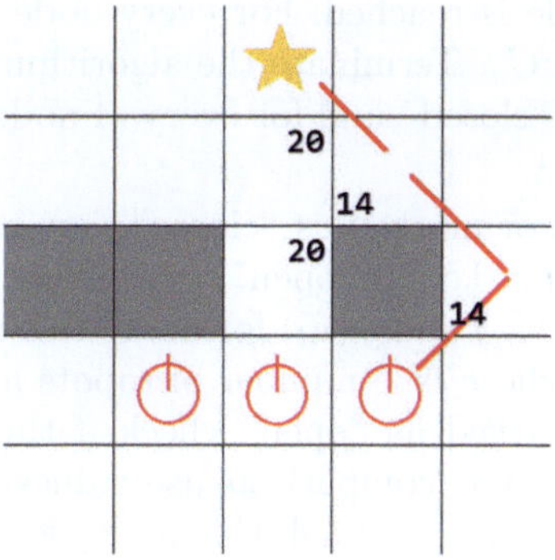

Fig. 5. Third robot's path

Now it is finally turn of the last most right robot. If it planed its path independently it would definitely lie between obstacles as the shortest possible one. But now if it chooses this path it would be forces to wait for first and second robots to release cells in order to start moving. This brings us total cost of the path of 64 units of time (diagonal move of 14, vertical move of 10 and 20 additional cost twice for each cell).

On the other hand if it chooses to move across an obstacle it wouldn't need to wait for cells to become available at all. With only two diagonal moves total cost of the path in this case would be 28 units of time (Fig. 5). That's 36 units of time cheaper.

This brings the whole group to the goal in 40 units of time instead of 64.

Important notice is that in the example above we only considered one component that makes up the cost of the chosen path $g(n)$, thus second component was taken as $h(n) = 0$. In this particular case we had no priority in considering cells when planning path and our algorithm degraded to Dijkstra's algorithm [9]. The task of choosing proper heuristic $h(n)$ is beyond the scope of the current paper and is a subject of separate papers [10].

To sum up lets define a path finding algorithm for a group of robots as the sequence of operations with a set of cells involved in path as output. The following procedure should be performed for each robot sequentially on the same map. As a template we will use original A* algorithm description [2] so instead of word "cell" we will be using word "node" to describe an invariable part of the map that can only be occupied by one robot at a time.

Let s be starting position of a robot, T – set of nodes denoting goal position area, $f(n)$ is a function (1). Then modified A* algorithm can be defined as follows.

Multi-robot path finding algorithm A*:

1. Mark s "open" and calculate $f(s)$.
2. Select the open node n whose value of f is smallest. Resolve ties arbitrarily, but always in favor of any node $n \in T$.
3. If $n \in T$, mark n "closed". Compile path P moving from node n to its parent and so on, until s node is reached. For every node $n \in P$ increase constant cost addition by value C_n. Terminate the algorithm for robot.
4. Otherwise, mark n as "closed" and for every of node ancestors n_i do subroutine.
   (a) If n_i is impassable or marked as "closed", ignore it.
   (b) If n_i has not been marked as "open", mark it as "open". Set n as a parent for n_i. Calculate $h(n_i)$ and $g(n_i)$ for node with respect to additional cost value of $N \cdot C_{n_i}$, where N – number of robots already using the node.
   (c) If n_i is already marked as "open" check if the path cheaper if it goes through this node. For comparison use values $g(n_i)$. If $g(n_i) < g(n)$, meaning path is cheaper through this node, set node's parent to n recalculate values $h(n_i)$ and $g(n_i)$ with respect to $N \cdot C_{n_i}$.
5. Go to step 2.

4 Computer Simulation and Results

Proposed approach was implemented within the computer model simulation. It is important to notice that this specific computer model was used in simulations of a robot group movement on an unknown map. This implies that robots were

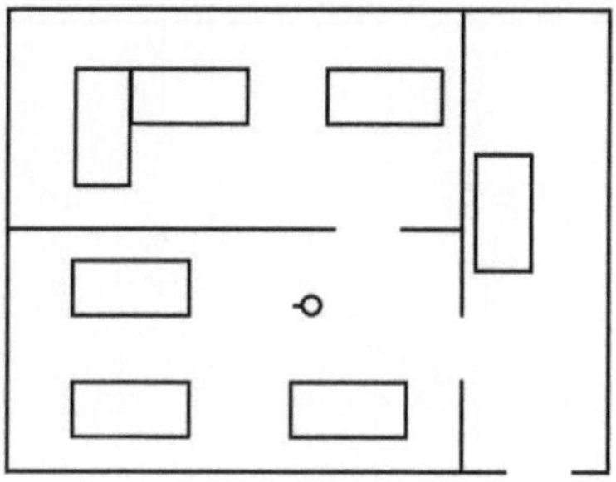

Fig. 6. Example of an "open map" configuration

forced to rebuild their planned roads in case new information about obstacles came along from scanning devices.

During simulation robots take turns to perform one of two types of activity: try to make a step or rebuild path and try to make a step. By making a step we mean moving to the next cell along the path. If cell is occupied by another robot, current robot remains in idle state and passes a turn. We call a full cycle when every robot in a group has taken its turn. As a measure of efficiency we count number of cycles needed for group to reach a goal.

Simulation was performed on maps of two different configuration types. Maps of first type we conventionally call "open maps" and consider a map as open when obstacles are compact and localized and spaces between them are much bigger than size of a robot (Fig. 6).

We reference to maps of second type as "closed maps". A map is considered closed when obstacles are elongated and interconnected. Spaces between obstacles are narrow and compared to robot size. Good example of "closed map" is a labyrinth (Fig. 7).

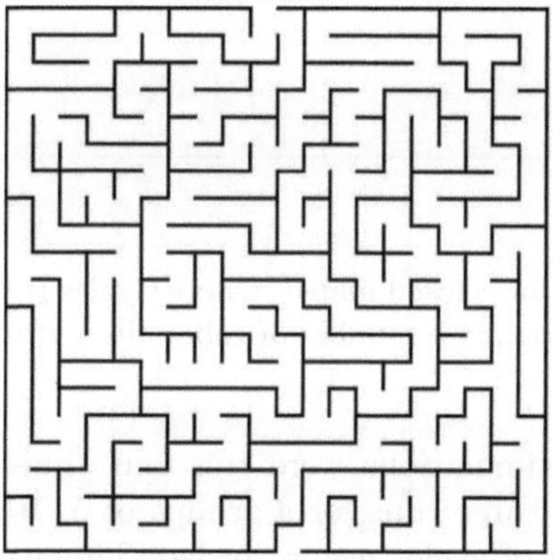

Fig. 7. Example of a "closed map" configuration

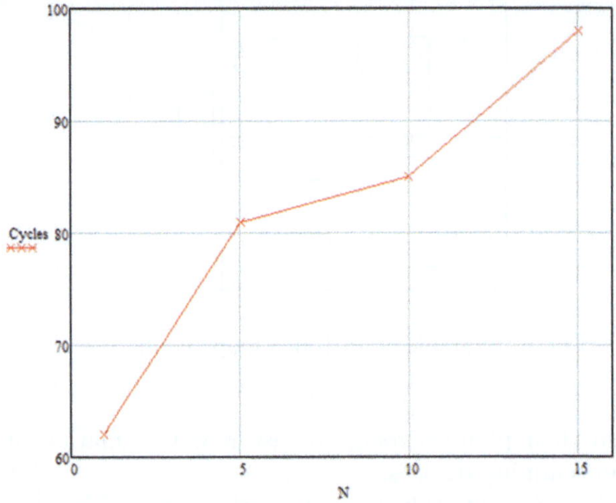

Fig. 8. Simulations results for "open maps", where Cycles is a number of cycles needed for a group to reach the goal, N – number of robots in a group

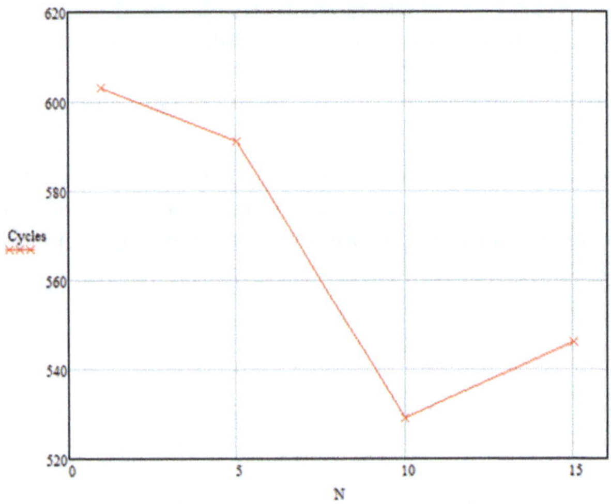

Fig. 9. Simulation results for "closed maps", where Cycles is a number of cycles needed for a group to reach the goal, N – number of robots in a group

On each map type simulations were run four times. First calibration run with a single robot and three runs for a groups of 5, 10 and 15 robots. During each simulation number of cycles was recorded. The average graphs are shown in Figs. 8 and 9.

As the graph at Fig. 8 demonstrates single robot shows good results in terms of cycles on open maps. As the group size grown number of cycles grows accordingly. It can be explained by existence of the single optimal path, which can be freely used by one robot but causes competition inside a group. As a group size continues to grow map becomes oversaturated with robots and they are forces to compete for every route possible. That explains sudden increase of number of cycles needed for group to reach the goal.

On the other hand when talking about "closed maps" we can observe an improvement in needed number of cycles as the group size grows. This happens because "closed maps" have a vast number of possible routs available for only one robot at a time. Modifications made to A* make robots choose their own path and thus explore map more effectively. As a group size keeps growing map becomes oversaturated, so we can talk about an optimal number of robots on a map, which equals to the number of possible routes to explore.

5 Conclusion

A* graph traversal algorithm had proven its effectiveness over years. It was applied to successfully solve lots of different tasks.

One of these tasks is robot navigation and path finding. With resent trends a problem of optimal multi-robot path finding is acute. In current paper we focused on a multi-object path planning problem and proposed modification to A* algorithm based on additional node costs. This modification makes possible finding optimal paths in terms of group movement. Moreover, proposed approach allows excluding potential collision situations on a path planning stage.

Proposed algorithm was successfully applied in a group navigation on unknown map problem. It showed reliable and interesting results on maps of different configurations.

References

1. Hoy, M.: Methods for collision-free navigation of multiple mobile robots in unknown cluttered environments. arXiv preprint 1401.6775 (2014)
2. Sarid, S., Shapiro, A., Gabriely, Y.: MRBUG: a competitive multi-robot path finding algorithm. In: Proceedings 2007 IEEE International Conference on Robotics and Automation, pp. 877–882 (2007)
3. Hart, P., Nilsson, N., Raphael, B.: A formal basis for the heuristic determination of minimum cost paths. IEEE Trans. Syst. Sci. Cybern. **4**, 100–107 (1968)
4. Zhao, L., Wang, W.: A new path search algorithm for providing paths among multiple origins and one single destination. Int. J. Comput. Sci. Appl. **3**, 29–33 (2014)
5. Botea, A., Müller, M., Schaeffer, J.: Near optimal hierarchical path-finding. J. Game Dev. **1**, 7–28 (2004)
6. Gholami, M.R., Delshad, D.E.: Presenting a model for the affordable choice of wiring route in the electrical and telecommunications networks in the residential areas based on the artificial intelligence A-STAR algorithm. J. Am. Sci. **9**, 189–191 (2013)

7. Pokorny, K.L., Ryan, E.V.: Multiple constraint satisfaction problems using the A-star (A*) search algorithm: classroom scheduling with preferences. J. Comput. Sci. Coll. **28**, 152–159 (2013)
8. Vincke, S.: Real-time pathfinding for multiple objects. J. Game Dev. **6**, 36–44 (1997)
9. Stout, B.: Smart moves: intelligent path-finding. J. Game Dev. **7**, 28–35 (1997)
10. Ferguson, D., Likhachev, M., Stentz, A.: A guide to heuristic-based path planning. In: Proceedings of the Workshop on Planning under Uncertainty for Autonomous Systems. International Conference on Automated Planning and Scheduling (ICAPS), pp. 9–18 (2005)

An Agent-Architecture for Automated Decision-Making on the Semantic Web

Richard Fox(✉) and Dustin Gulley

Department of Computer Science, Northern Kentucky University, Highland Heights, KY 41099, USA
foxr@nku.edu, gulleydu@mymail.nku.edu

Abstract. Semantic web research has largely concentrated on the representation of web-based knowledge and data, often in the form of ontologies. Only a minority of the research has explored the implementation of intelligent agents that can use the web-based knowledge. Even less research has generalized intelligent agent implementations into an architecture. Here, an implementation and architecture are described for a multi-agent decision-making system using available web-based data and information.

Keywords: Semantic web · Intelligent agent architecture · Decision-making

1 Introduction

It has been 17 years since Tim Berners-Lee published his introduction to the semantic web [1]. His vision was to implement intelligent agents which could communicate with each other and web-based knowledge sources to solve problems. In his seminal paper, he provided a planning example. In the interim, much semantic web research has explored ways of making the human-readable web information machine-readable by adding semantic information to the largely syntactic information and data found on the World Wide Web. Much of the research has explored the creation of ontologies although newer research has also explored linking data together, such as with the Linked Open Data project, as well as offering knowledge sources like WordNet.

Ontology construction, whether by hand or in an automated fashion, has made significant progress. It is estimated that over 2.5 billion web pages have been codified using one of the proscribed semantic web approaches, e.g., as OWL ontologies [2] or linked data [3]. Further, programs like IBM's Watson, CycCorp's Cyc, DBpedia and YAGO have utilized such knowledge to generate additional knowledge sources and/or make decisions [4–7]. However, only a small fraction of the semantic web research has targeted the intelligent agent aspect of Berners-Lee's vision.

Most intelligent agent research explores one (or more) of the following topics:

- discovery of agents, knowledge sources and web services;
- determining the reliability/trust of other agents and knowledge sources;
- communication protocols between agents and between agents and knowledge sources;
- issues with mobile agents and with autonomous agents;

R. Silhavy et al. (Eds.): CoMeSySo 2018, AISC 860, pp. 183–192, 2019.
https://doi.org/10.1007/978-3-030-00184-1_17

- theoretical issues;
- overarching agent architectures.

Few architectures for the semantic web have so far been proposed. Additionally, specific architectures to support decision-making problems are even rarer.

Research in this paper explores two related topics. First, a series of semantic web agents along with their knowledge sources is described for selecting restaurants based on user preferences. This form of decision-making is in many ways related to on-line travel agents such as Travelocity but tuned specifically for restaurant selection. Second, the implementation is abstracted to form a general-purpose architecture for decision-making.

This paper first presents related intelligent agent architectures. It then offers motivation for pursuing a semantic-web based agent architecture. It describes the implemented restaurant selection system in detail. The paper then provides a more generic view of the architecture behind the implementation. It finally concludes with a look at future work and expansion.

2 Related Work

One need only search semantic web topics to see that most semantic web research has revolved around ontologies, ontology-languages, and the use of linked data. A minority of the research has been dedicated to intelligent agents to serve the semantic web (there is plenty of agent-based research, but not necessarily of web-based agents). Even less research has targeted semantic web-based agent architectures. This paper's intent is to describe such an architecture specifically for decision-making.

Weiss [8] characterized agent architectures as logic-based, reactive, belief-desired-intention and layered. This is more of a classification of the type of agent, not the coordination of agent communication. A 2015 survey of agent platforms by Kravari and Bassiliades [9] notes that agent architectures may include many features including distributed agents, modeling and simulation-based agents, dynamic real-world environments, and event-driven agents. They also state that some agent architectures include end-to-end security via encryption or more minimally, authentication. This classification demonstrates again that there is less emphasis on architectures to support decision-making than on agent communication.

The MOMA methodology, to support business decision-making, requires that agents be able to communicate with ontologies, building their agents to interact with ontologies [10]. Missing from this approach is the ability to locate and communicate with other agents. Other researchers have concentrated on adding trust and negotiation between agents in their platforms [11, 12]. What these architectures lack is the emphasis on the types of agents needed to support decision-making endeavors specifically for the semantic web.

In 2008, Ketter et al. introduced an architecture to determine user preferences for decision-making specifically for business applications. In this research, the emphasis is on Advocate Agents that represent specific users and their preferences. These agents collaborate with each other to generate recommendations on given problems. Much of their architecture revolves around local knowledge sources rather than web-based knowledge and data [13].

This paper in some ways builds on Ketter's model, with a centralized User Agent, like Ketter's Advocate Agents, to represent the user's preferences. This agent locates and communicates via message passing with other agents. The approach here is to utilize a number of deliberative agents, each of which models some specific worldly information and each of which uses symbolic forms of reasoning. There are also aspects of reactive agents in the architecture in that an agent must react to changes in its environment such as when another agent is unavailable or returns an unexpected result.

3 Decision-Making on the Web

This research began with the following question: where can I buy truffles for my wife's birthday? To answer such a question, the person performed a Google search on candy stores in the area. Given the list of stores, the person narrowed down the list by looking at store locations, hours, cost, types of candy available, ratings supplied by customers, and whether the given store could specially make his request. Such decision-making processes are common among humans including for instance: where should we eat dinner tonight? what automobile dealership is reasonable and trustworthy? should we see a movie, play or concert tonight and if so, which one? where can I go to buy a new suit? where can I have my computer repaired?

The Internet plays a significant role in human decision-making to help answer such questions. Most people will use a search engine to locate places that can solve the given problem (e.g., names and locations of auto dealerships). From there, the human may explore other websites for driving directions, reviews, and so forth as well as the locations' websites for more information such as costs and available products. The user makes an informed decision based on the results of these searches and their own personal preferences, which might include restrictions on driving distance, desire for lower cost, emphasis on customer reviews, dietary restrictions (for restaurants) and size restrictions (for clothing stores).

The semantic web was proposed to solve these types of problems. To date, very few intelligent agents exist to support this type of general-purpose decision-making. Those that do exist are primarily travel-based (e.g., Travelocity) and proprietary. How then should such a decision-making system be constructed and are there pieces of such a system that can be generalized to support most forms of decision making? In the next two sections, such research is presented.

4 A Restaurant Recommendation System

Instead of purchasing candy, this research focused on selecting a restaurant based on personal preferences. User preferences might include some or all of the following.

- Maximum distance willing to travel.
- Parking options.

- Hours when the restaurant is open.
- Cost of the average dish (or cost of the cheapest dish).
- Type of food available.
- The number of dishes or diversity of food to choose from.
- Nutritional information on the food served.
- Reviews from previous patrons.
- Past experience with the restaurant.

Implementing such an automated decision-making system requires a central agent to represent the user's interest. The agent will need to know the user's preferences and be able to acquire information for restaurants within a given area. Locating restaurants can be accomplished with a search engine. Individual restaurant websites can be queried to obtain further information. A map engine (e.g., Google Maps) can be queried for distance and driving directions. Other web sources, like a restaurant reviews site, can be queried for further information.

The agent may require additional knowledge sources to translate returned data into data that the agent can use. For instance, a user may have specified a preference for Mexican food. A restaurant that advertises that it is Tex-Mex may or may not fit that preference. Ontologies can come into play to serve as knowledge translators. Other useful knowledge that the user agent may utilize includes how to gage the healthiness based on nutritional data, and whether parking can be found nearby or not.

Another useful piece of information that the agent might utilize is personal history with the restaurant. A user may prefer to try new restaurants; having visited a restaurant in the past month may be reason to downgrade a restaurant. Or alternatively, knowing that the user loved his or her experience with a previous restaurant might boost that restaurant's value.

The agent can benefit from a cache to store previously retrieved data from both the search engine and individual web sites. Although the agent should try to obtain the most recently posted information, if the site is not responsive the cache can be used to obtain older information.

The User Agent is the central agent, communicating with all other agents and maintaining user preferences. Communication is handled via message passing, sent to a Handler Agent, which is responsible for delivering messages to the appropriate agent. Messages are created using JSON with the properties shown in Table 1. Parts of the message format are based on the FIPA (www.fipa.org) ACL Message Structure Specification, particularly SenderAddress, ReceiverAddress, and ReceiverType, while ConversationId and Content are part of the specification.

The full architecture for restaurant decision-making is given in Fig. 1. The User Agent's first action is to send a general query to a search agent (labeled as Location Search Agent), to obtain a list of reasonably close restaurants. The Location Search Agent responds with restaurant names and URLs of their websites as well as locations and contact information. Given a restaurant's location and website, the User Agent can contact other agents to acquire more information. The Transportation Agent uses the address and returns driving directions, including distance, number of steps (roads, turns), time to drive there, and alternate distance for any alternate route suggested.

Table 1. Message format

Name	Description	Type
Performative	The domain of the agent	String
SenderAddress	Unique address of the sender	GUID
ReceiverAddress	Unique address of the receiver	GUID
ReceiverType	Type of receiving agent	String
ConversationId	A unique id for the conversation	GUID
ServiceConfiguration	A list of agents and jobs	Array
AgentJobs	A list of finished/unfinished jobs	Array
Content	Data sent among agents	JSON

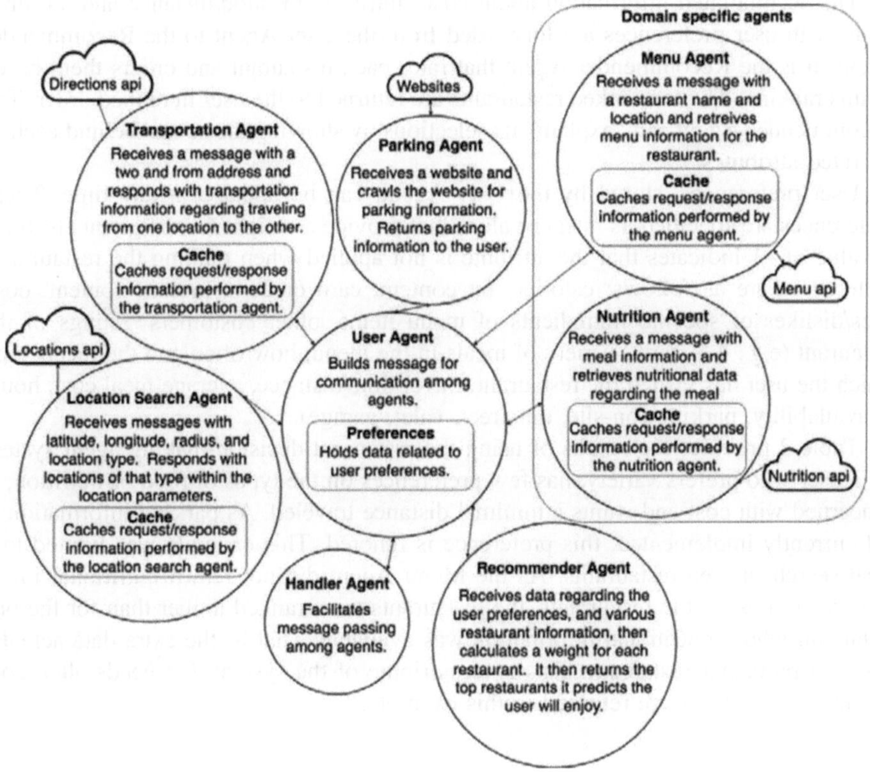

Fig. 1. Restaurant decision-making agent architecture

Parking Agent searches the restaurant's website for parking information using a follow-your-nose crawl [14] to identify specific information about the restaurant's parking lot. Should no such information be found, the Parking Agent can make inferences about parking based on the location of the restaurant. For instance, if there is no on-site parking and the restaurant is in the downtown area, then parking will require payment and may be limited. At the time of this writing, the Parking Agent has not been completed and so only provides minimal information.

A Menu Agent is called upon to respond with information about the restaurant's menu, which includes specific meals and costs. Although this information can also be obtained directly from the restaurant's website, there are web-based menu services available that provide such information for many restaurants, which include menu costs. A separate Nutritional Agent is called upon to look over available menu choices, then query a nutrition API which will respond with nutritional information (if available). This agent reasons over the nutritional data received from the API to score how nutritional the meals on the menu are. An example of the type of information this agent will examine is the amount of fat in a meal compared to the calories to determine if the meal is unhealthy.

The accumulated information about cost, nutrition, location/distance and so forth, along with user preferences are forwarded from the User Agent to the Recommender Agent. It is the Recommender Agent that rates each restaurant and orders them based on this ranking. The top ranked restaurants are returned to the user in ranked order. The Recommender Agent can "explain" its selections by showing the scores behind each of the rated attributes.

User preferences, stored by the User Agent, can be updated at any time. These preferences are specified as integer values that provide a weight for the given attribute. A value of –1 indicates that the attribute is not applied when ranking the restaurants. Preferences are as follows: calories, fat content, carb content, protein content, cost, likes/dislikes of specific ingredients of menu items, other customers' ratings of the restaurant (e.g., 4.5 stars), variety of meals in the menu, how often and the recency by which the user has visited the restaurant, travel time/distance, average meal cost, hours of availability, parking (on-site, on-street, valet, garage).

Table 2 provides the results of using the restaurant decision-making agent system for a user who prefers variety, has few preferences on the types of food or nutrition, is concerned with cost and wants a minimal distance traveled. As parking information is not currently implemented, this preference is ignored. This example was limited to a local search of area restaurants. As the Menu Agent did not return nutritional information for most of the restaurants, the restaurants were ranked higher than for the one restaurant whose nutritional information was available. That is, the extra data actually hurt that particular restaurant. This is an attribute of the system that needs alteration. Twenty restaurants were returned in this example.

5 A Decision-Making Architecture

Decision-making extends far beyond selecting restaurants. As mentioned in the introduction, people use the Internet to assist them in all types of selections including locating a nearby clothing store, finding an auto dealership or selecting a candy store. In order to promote a wider variety of decision-making, the system from Sect. 4 was adapted to one that can recommend auto dealerships. Domain-independent aspects of the system remain the same: Location Search Agent, Transportation Agent, Handler Agent, Parking Agent, Recommender Agent, along with the User Agent and its caches.

Table 2. Results from User 1's Preferences

Name	Cost	Rating	Duration	Distance	Steps	Menu size	Calories	Protein	Carbs	Fat	Score
Bob Evans	1	3.9	770	11340	6	0	1128	30	150	69	6623
Olive Garden	2	4.3	708	6883	7	465	380	30	31	19	347
FW Family Rest.0	4.5	937	11931	5	0	1128	30	150	69	6658	
BJ's	0	3.7	615	6464	4	0	1128	30	150	69	6231
Red Robin	2	4.1	672	6932	8	0	1128	30	150	69	6274
Rio Grande	0	4.2	816	10423	8	0	1128	30	150	69	6538
El Rio Grande	4.1	657	7115	7	0	1128	30	150	69	6274	
Smokin' BBQ	0	4.6	468	4948	5	0	1128	30	150	69	6106
Frisch's Big Boy1	3.5	725	9828	7	0	1128	30	150	69	6503	
Bob Evans	1	4.0	379	4198	3	0	1128	30	150	69 .	6050
Arby's	1	3.6	899	10549	6	0	1128	30	150	69	6567
Arby's	1	3.4	496	4870	5	0	1128	30	150	69	6111
Arby's	1	3.0	967	10789	7	0	1128	30	150	69	6583
Carrabba's	1	4.3	560	7270	5	0	1128	30	150	69	6298
Culver's	1	4.2	575	7254	7	0	1128	30	150	69	6298
White Castle	1	3.5	952	10700	9	0	1128	30	150	69	6575
White Castle	1	3.9	736	9057	6	0	1128	30	150	69	6447
White Castle	1	3.5	871	11267	8	0	1128	30	150	69	6623
White Castle	1	3.8	448	4912	3	0	1128	30	150	69	6111
White Castle	1	3.6	599	7133	7	0	1128	30	150	69	6287

The new domain requires its own domain-specific agents. These agents reason over a car's value by consulting bluebook and Carfax data, and reviews of customers for the given auto dealership. Additionally, an agent is required to acquire information on local car sales outside of auto dealerships.

Figure 2 illustrates the revised architecture. Notice in Fig. 2 the inclusion of a new component, a domain-specific expert. Within the automobile domain, this expert is expected to answer questions and make inferences to assist the Recommender Agent that are specific to purchasing an automobile. As an example, a car may be labeled as a Coupe, but the user may not want a 2-door car. This domain-specific agent will have to infer that since it is a Coupe, it is a 2-door car. Another example is to identify safety factors of the given car for a user who prefers a car with a high safety rating.

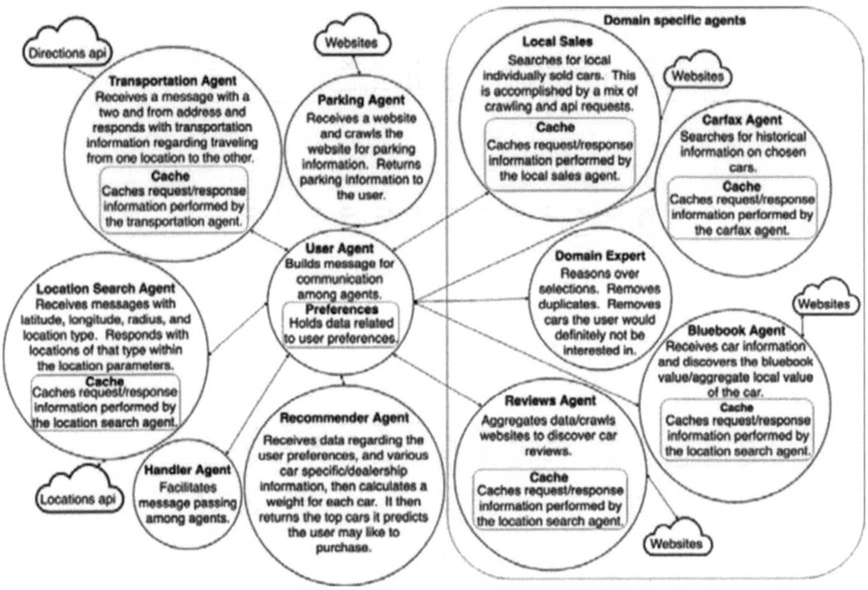

Fig. 2. Car decision-making agent architecture

The domain expert should either have the necessary domain knowledge internally or be able to obtain it (for instance by searching Consumer Reports or by communicating with one or more Car ontologies). The domain expert then is both an agent that can reason over data but also one that can communicate with other agents, dealership websites, and domain ontologies.

As the domain changes, so must the types of attributes that describe user preferences. With each domain, the Recommender Agent will have a different way of ranking the selected locations. For instance, rather than nutritional information, auto dealership reputation and comparison between the advertised price of the car and its bluebook value will be taken into account.

To generalize from these two examples, it is clear that the domain-independent architecture stays the same between different decision-making domains, aside from the preferences. Thus, to adapt this architecture into one of a specific domain, one must identify the attributes unique to that domain and create Agents capable of acquiring that information from one or more web-based knowledge sources. A clothing store decision-making system, for instance, might require an Agent to reason over whether the store does alterations while another Agent might reason over the store's return policy and a third might reason over whether a given store carries the name brands and/or sizes of interest to the user.

6 Conclusions

The web has become an integral part of human decision making. People regularly use search engines to obtain listings of choices to select between, whether the choices are restaurants, auto dealerships, clothing stores or events. The semantic web has been envisioned as a means for automating this decision-making process so that the user is less involved. However, research has lacked in how to make this work. Many specific instances of decision making agents have been made available, but there has been less effort in creating a general-purpose agent-based architecture. This paper has provided an example of such an architecture and demonstrated it with an implementation on selecting restaurants.

The architecture proposed here is still incomplete. The Parking Agent, for instance, was not implemented (primarily because most restaurants do not include parking data on their web sites). The proposed domain-independent architecture has not yet been tested. Although the paper presents the changes needed to adapt the architecture into a second domain, the Bluebook, Carfax Local Sales and Reviews Agents are not yet implemented. Nor has there been an attempt to build a Domain Expert Agent for the automotive domain.

Work continues to expand the system. Future work may include a peer-to-peer architecture with more emphasis on agent discovery, research into other types of agents such as reasoning and deductive agents which may serve as the domain expert, a way to configure the recommender agent based on preference configurations, and agents with more emphasis on crawling which can accumulate meaningful data which may be queried when needed. The goal is to complete the second domain, auto dealership, and a third domain to continue the proof of concept proposed here.

References

1. Berners-Lee, T., Hendler, J., Lassila, O.: The semantic web - a new form of web content that is meaningful to computers will unleash a revolution of new possibilities. Sci. Am. **284**(5), 34–43 (2001)
2. Dou, D., Wang, H., Liu, H.: Semantic data mining: a survey of ontology-based approaches. In: 2015 IEEE International Conference on Semantic Computing (ICSC) (2015)

3. Kontokostas, D., Westphal, P., Auer, S., Hellmann, S., Lehmann, J., Cornelissen, R., Zaveri, A.: Test-driven evaluation of linked data quality. In: Proceedings of the 23rd International Conference on World Wide Web, pp. 747–758. ACM, New York (2014)
4. Lally, A., Bagchi, S., Barborak, M., Buchanan, D., Chu-Carroll, J., Ferrucci, D., Glass, M., Kalyanpur, A., Mueller, E., Murdock, J., Patwardham, S.: WatsonPaths: scenario-based question answering and inference over unstructured information. AI Mag. **38**(2), 59 (2017)
5. Pierce, C., Booth, D., Ogbuji, C., Deaton, C., Blackstone, E., Lenat, D.: SemanticDB: a semantic web infrastructure for clinical research and quality reporting. Curr. Bioinf. **7**(3) (2012)
6. Lehmann, J., Isele, R., Jakob, M., Jentzsch, A., Kontokostas, D., Mendes, P., Hellmann, S., Morsey, M., Van Kleef, P., Auer, S., Bizer, C.: DBpedia–a large-scale, multilingual knowledge base extracted from Wikipedia. Semant. Web **6**(2), 167–195 (2015)
7. Rebele, T., Suchanek, F., Hoffart, J., Biega, J., Kuzey, E., Weikum, G.: YAGO: a multilingual knowledge base from wikipedia, wordnet, and geonames. In: International Semantic Web Conference, pp. 177–185 (2016)
8. Weiss, G.: Multiagent Systems, MIT Press (2013)
9. Kravari, K., Bassiliades, N.: A survey of agent platforms. J. Artif. Soc. Soc. Simul. **18**(1), 11 (2015)
10. Ying, W., Ray, P., Lewis, L.: A methodology for creating ontology-based multi-agent systems with an experiment in financial application development. In: System Sciences (HICSS), pp. 3397–3406 (2013)
11. Cao, M., Luo, X., Luo, X., Dai, X.: Automated negotiation for e-commerce decision making: a goal deliberated agent architecture for multi-strategy selection. Decis. Support Syst. **73**, 1–14 (2015)
12. Messina, F., Pappalardo, G., Rosaci, D., Sarne, G.: A trust-based, multi-agent architecture supporting inter-cloud VM migration in IaaS federations. In: International Conference on Internet and Distributed Computing Systems, pp. 74–83 (2014)
13. Ketter, W., Batchu, A., Berosik, G., Mccreary, D.: A semantic web architecture for advocate agents to determine preferences and facilitate decision making. In: ACM International Conference Proceeding Series (2008)
14. Liyang, Y.: A Developers Guide to the Semantic Web. Springer (2015)

An Efficient Hardware Realization of DCT Based Color Image Mosaicing System on FPGA

H. Jayalaxmi[1(✉)] and S. Ramachandran[2]

[1] Department of ECE, Acharya Institute of Technology, Bengaluru, India
jayalaxmiresearchwork@gmail.com
[2] Department of ECE, SJBIT, Bangalore, India

Abstract. Researchers of Surveillance systems in today's world are exploring various ways to capture continuous image sequences. Capturing these image sequences in a single plane is highly challenging. As a consequence, camera arrays have sprung into action to capture the image sequences in a single plane. This concept of combining the captured images by an array of cameras is referred to as mosaicing. The image mosaicing generally encounters the problem of sacrificing the image quality because of improper mode of image capturing. In order to tackle this problem, Cross correlation (Ccr) method can be implemented. However, it faces quality problems again under brightness variations. This paper introduces a Discrete Cosine Transform (DCT) based image mosaicing method that works effectively against the brightness issues suitable for FPGA implementation. This proposal comprises functional blocks for DCT, inverse DCT, image registration and multipliers and dividers. The image quality assessing parameters like PSNR, MSE, device utilization and timing analysis were considered for performance analysis of the proposed scheme by using Xilinx 14.7 ISE and simulated by using Modelsim6.3 f. The proposed image mosaicing system gives excellent results in terms of image quality (PSNR of over 35 dB) and fast execution time of 23 ms per image of size 1600 × 1200 pixels.

Keywords: CCr · DCT · Device utilization · Execution time
Image mosaicing · Image registration · MSE · PSNR · Timing analysis

1 Introduction

Modern aerial surveillance systems utilizing camera arrays can capture several square miles of ground activity at high resolution from a single plane [1]. A camera array uses multiple cameras to capture images synchronously with partial overlap between cameras' fields of view [2]. This allows a wide area to be monitored continuously and without any blind spots. Instead of observing the different images separately, a single image can be generated by combining all the individual images in absence of large image capturing camera [3]. This kind of combined image will have all the required information from all the images collected from the camera array which is commonly known as mosaic [4, 5]. The image analyst can easily study the mosiaced image having

R. Silhavy et al. (Eds.): CoMeSySo 2018, AISC 860, pp. 193–203, 2019.
https://doi.org/10.1007/978-3-030-00184-1_18

different angle perception and with redundant scene data [6]. For further image processing or data exploration from the mosaiced image, it is desirable to settle the issues like image misalignment with pixels overlapping [7].

The principle concept of the image mosaicing is a synthetic composition obtained from an image sequence and which can be generated by knowing the geometrical relationship among the images [8]. Mainly, coordinate transformations can be considered as the geometric relations which relates the coordinates of the images. The implementation of proper transformations through warping operation and merging of the warped images overlapping regions can help to construct the single image with bigger size covering all the visible areas of the different images [9]. This process of merging the different images as single image is the motivation factor for image mosaicing. The image mosaicing involves stages like feature extraction, registration, stitching and bending [10, 11]. The stage image registration is meant to align the set of images geometrics. The image set may have multiple numbers of images captured of single scene at different time interval from different viewpoints or sensors. The registration process aims to bring the geometric correspondence among the images so that the images are transformed and analyzed within the common reference frame [12]. This kind of registration process is practically implemented in different areas like remote sensing, computer vision and medical imaging. The image registration can be categorized [13] into:

(a) *Correlation Technique (CrT),* where pixel values of the image is directly used in the algorithm.
(b) *Fast Fourier Transform (FFT)* technique, where frequency domain is used in the algorithm.
(c) *Feature based techniques*, where edges and corners of the images are used as low level features in the algorithm.
(d) *Graph theory approaches*, where, the high level features of the images are used in the algorithm.

The concept of image mosaicing has acquired lot of attention in recent years. The applicability of image mosaicing is mainly found in image processing, texture mapping, multimedia etc. Nowadays, lots of image mosaicing devices are in existence in real world which offer users to generate their own multimedia content and panorama [14]. The image mosaicing involves with the issues of image quality variation due to non-significant method of image capturing. The implementation of Cross correlation (Ccr) is simple but faces problem under brightness variation. Also, the normalized form of Ccr may lead to misaligned images within the mosaic [15]. This paper presents a computationally efficient DCT based image mosaicing system, guarding against the issues like illumination, blurring etc.

The paper is organized in the following manner. Section 2 gives the review of most recent research works. The main problem focused on this paper is described in Sect. 3. Section 4 explains the proposed system while results of the proposed system are discussed in Sect. 5. Finally the Sect. 6 concludes the significance of the method.

2 Background

The tiles decorate an original image to imitating the original visual data rendered into another mosaic-like image. The created mosaic image from PC is another research interest in recent days. Different mosaics can be made for an image contingent upon the decision of the tile data set and the forced requirements for situating, disfigurements, and so on. Mosaic images will be images made by establishing together little shaded tiles. Also, these are the antiquated cases of discrete images of primitive type. A photo/photograph is isolated into little areas and every one of which is supplanted with other image which matches the objective image or recreate the image by appropriately painting the tiles [16].

The classification of image mosaic can be done as crystallization, ancient, photo, and puzzle image mosaic. The initial two sorts of mosaics break down a source image into various shading, size and rotation tiles, recreating the image by legitimately painting the tiles. Thus, these can be assembled together under the section of tile mosaics [17].The last two sorts are acquired by fitting images from a database to cover a relegated source image. Similarly, they might be gathered together under the division of multi-picture mosaics. This scientific categorization ought not to be planned as an inflexible one. Numerous mosaic systems may fit in excess of a solitary class and it is likely that other new kinds of mosaics are available in future [18].

A general system for image mosaicing shown in (Fig. 1) and archive image mosaicing can be performed by calculation for the system [19] and is given as: (1) selection of images from the disk of captured image. (2) Arrangement of these images in amends request and introduction by utilizing the highlights separated. (3) Construction of the mosaic image. (4) Performing distortion strategies to enhance the nature of mosaic image. (5) Finally, obtain the high resolution mosaic image.

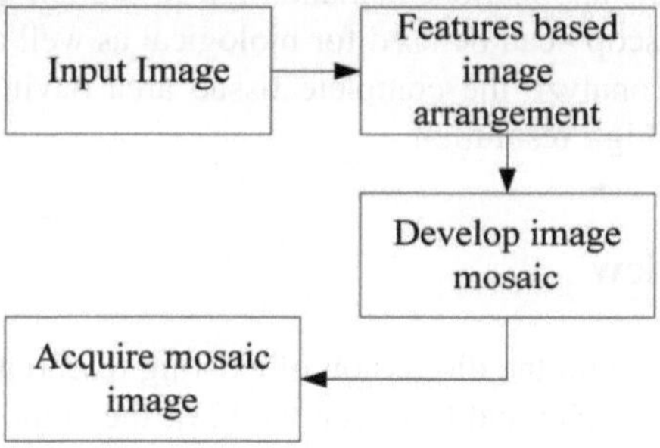

Fig. 1. General model for image mosaicing

Image registration alludes to the geometric arrangement of images. The set may comprise of at least two advanced images taken of a solitary scene at various circumstances, from various sensors, or from various perspectives. The objective of registration is to set up geometric correspondence between the images with the goal that they might be changed, thought about and examined in a reference outline [10, 11].

Image registration is the assignment of coordinating at least two images. It has been a focal issue for an assortment of issues in image handling, for example, question acknowledgment, observing satellite images, coordinating stereo images for reproducing profundity, coordinating biomedical images for analysis and so on. Registration is additionally the focal undertaking of image mosaicing techniques. Deliberately adjusted and prerecorded camera parameters might be utilized to fulfill the programmed registration requirements. Client association additionally is a dependable source for physically enlisting images (e.g., by picking, comparing focuses and utilizing important changes on screen with visual input) [12, 13, 20].

The automated way of image registration can be utilized for image mosaicing and are classified as follows [21, 22]:

(a) **Feature based Approaches: -** These approaches rely on the image features detection in accurate manner. The similar image features helps to perform good image alignment but not work in absence of distinctive features.
(b) **Exhaustively Searching Approach: -** This approach is best one for all the necessary motion parameters of image. By making use of hierarchical processing, the computational speed can be improved. Also, the approach has the ability of performing parallel processing to achieve further performance improvement.
(c) **Frequency Domain based Approach: -** This helps to identify the rotation/scale and displacement of image which are computationally efficient but sensitive towards the noise. These approaches need overlap extent to acquire the important part of image.
(d) **Iterative Approach: -** The adjustment of camera-motion parameters may lead to the local minimums until a desired initial estimation is provided. This initial estimation can be performed by utilizing a coarse global search or by implementation of frequency domain approach.

Currently, most of the applications demand high resolution images for information. For example: the microscopy can be used for biological as well as medical application. Thus, it is required to analyze the complete tissue area having several tens of millimeters dimensions at high resolution.

3 Literature Review

This section commences with the discussion of existing researches in image mosaicing techniques. In the work of Zhi and Cooperstock [23], the authors have considered the parallax and object motion issues in causing artifact results in image registration. To overcome these issues, a dynamic mosaicing approach was presented by incorporating background/foreground segmentation. This approach helps to achieve good quality image mosaicing.

Shao et al. [24] have considered the fluorescence image that stitches two asymmetrical images for image mosaicing. It incorporates an optimization approach which evaluates the blending coefficients for image mosaicing. The authors claim that this work gives significant results than other methods.

Prados et al. [25] have introduced strategical solutions to overcome the issues of underwater giga-mosaics that help to perform image pre-processing, image enhancement and bending steps to improve the visual quality for photo mosaics. The outcomes of the research were validated by large optical survey of deep sea and achieve high quality of image registration.

Rahman et al. [26] have proposed the Object based mosaicing (OMc) and achieved visually enhanced roof shapes within the scene, optimized processing time, hot spot detection etc.

Further, the work of Yigitsoy and Navab [27] presented an image registration to stitch sub-images by using the perceptual grouping mechanism. The effectiveness of this research was analyzed over the medical and synthetic images and it is found that the stitching of sub-images is possible even at existence of physical gap in the image.

Similarly, Molina and Zhu [28] have addressed the issues in the image registration with low frame rates. Here, set of panoramic mosaics are constructed for representation and outcomes were found effective with batch error correction.

The research work towards automated mosaicing is found in Lurie et al. [29], which exploits the known rigid connection among the white light and optical coherence (OC) tomography imaging system and reduces the computational complexity.

The recent works of Bellavia and Colombo [30] have presented a compositional architecture to classify the color correction mechanisms as per their two main computational segments. The experimental analysis was performed by perceptual image quality metrics for image stitching.

The work of Jain et al. [31] introduced a random corner mechanism based image mosaicing which assembles the multiple overlapping images with automatic image mosaicing ability resulting in enhanced accuracy, robustness and efficiency.

- The problem of generating an image mosaic from the image sequences is considered by an unmanned aerial vehicle is addressed in the work of Hadrovic et al. [32]. Here, the feature matching based approach is used to perform both image stitching and mosaicing process in real time scenario.

A bidirectional algorithm for image mosaicing was introduced in Dhokley et al. [33] that involve implementation of direct intensity based mechanism for matching of pixel value.

An interesting work of Chen et al. [34] utilizes the microscopic image mosaicing. In this work, the registered images were blended as composite image by using weighted smooth algorithm. This work yielded effective results in generating panoramic image.

The review of the existing researches reveals that only very few work were focused on image quality enhancement in image mosaicing and that only few authors have considered FPGA implementation for image mosaicing. No other researcher has reported image mosaicing using DCT which promises better mosaiced image than any other methods reported so far. This paper is a work filling this gap by designing DCT based image mosaicing system to improve image quality than before.

4 Proposed System

This paper introduces the computationally efficient image mosaicing system by using DCT based approach. The implemented methodology for the system design is presented in Fig. 2. The system is designed using MATLAB and Verilog RTL suitable for FPGA implementation, where MATLAB is used to perform image processing operations and Verilog for processing in real time. The system considers sample of two images for mosaicing. A preprocessing can be applied over these images to remove the existing noise factors in the images from the text files generated by MATLAB. Discrete Cosine Transformation (DCT) is applied for the color images and given to the multiplier and divider module as shown. The output of this multiplier and divider is subjected to inverse DCT (IDCT) processor and phase correlation is estimated. By utilizing the estimated phase correlation, image registration is performed for mosaicing. Finally, the system performance analysis is performed by considering the parameters like PSNR, MSE, device utilization and timing analysis.

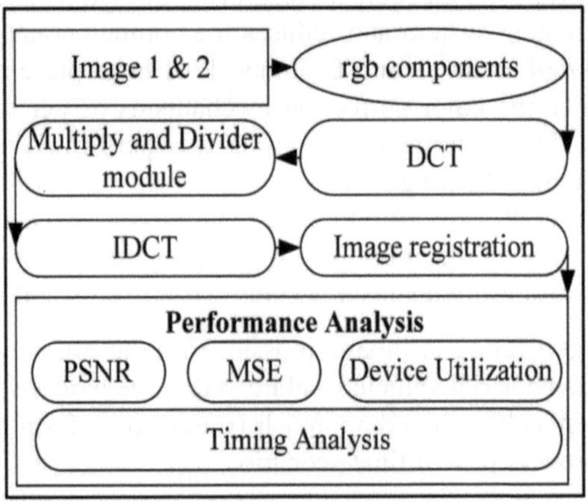

Fig. 2. Methodology implemented for system design

The design steps of the system is as follows. Here, a text file is generated for two colour images 1 and 2 in RGB format which helps to perform image mosaicing. Further, these text files in RGB format are fed to the top module of DCT based image mosaicing. Thereafter DCT is processed over two images where the images will be parted into different frequencies. Then it will be forwarded to multiplier and divider module. Later, IDCT engine is used to calculate the phase correlation value. From that image registration is done, and registration values are generated. Image registration values are considered as the final mosaiced image.

The cross correlation is a simple technique which helps to overcome the issues associated with image brightness, but after the generation of mosaic image, it leads to image misalignment. Thus, the phase correlation approach plays a major role in solving these brightness issues and also solves the blurring and color variation problems.

The Phase correction value addresses the solution to overlapping images with better mosaic image output than without it. The DCT approach is used to convert the image information into frequency coefficients. Mainly, the DCT approach is helpful in image and sound compression. The applicability of DCT is not limited to the JPEG formats which can be used for MPEG audio and digital VCR. In short, the DCT helps to pack complex image signals in as few coefficients as possible leading to compression. The DCT performs near lossless compression by adapting the principle of noisy components removal without changing the perception of the image.

5 Results and Discussions

This section presents the discussion and analysis of the proposed DCT based color image mosaicing system. The System was designed using RTL CompliantVerilog coding. The RTL Code was simulated using ModelSim 6.3 f and validation of the design was performed by using Xilinx 14.7 ISE. The System design was targeted on Artix-7 FPGA board using device 7A100T-3 CSG324. The successful execution of the system brings the following outcomes.

Figure 3 presents the DCT based image mosaicing for Lena image of dimension 1600 × 1200 pixels. Finally, features of both the images were matched and the complete image mosaicing performed.

Fig. 3. DCT based image mosaicing of Lina image of size 1600 × 1200 pixels

Figure 4 presents the RTL Schematic circuit diagram of the DCT based image mosaicing system coded in RTL Verilog. The RTL schema involves two multiplier-divider modules and five each of DCT and IDCT modules and having three models of image registration.

The performance analysis of the proposed system model is performed using criteria such as PSNR, MSE, device utilization and timing analysis. Tables 1, 2, 3 and 4

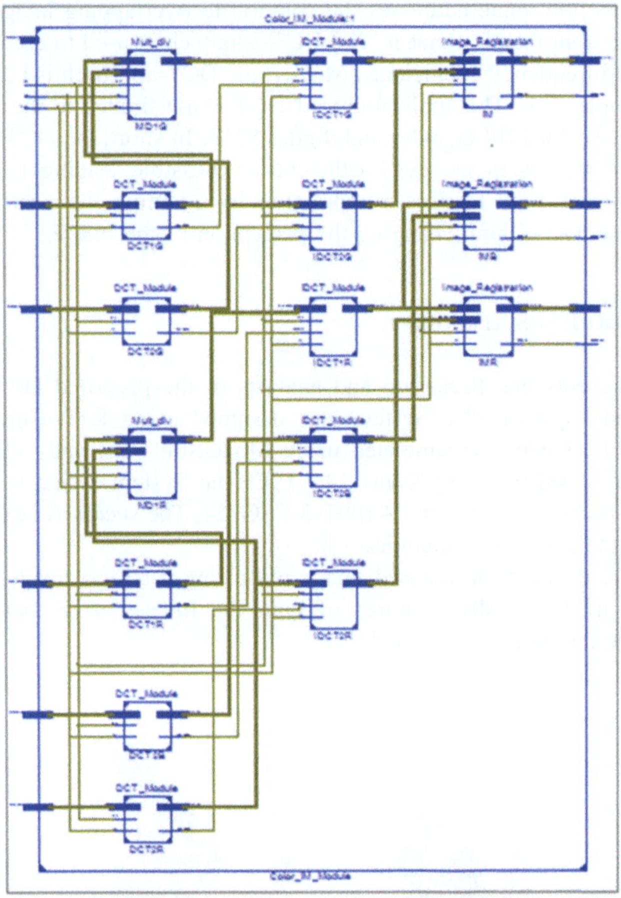

Fig. 4. RTL schematic diagram of DCT based image mosaicing obtained using Xilinx tool

presents the results of PSNR, MSE, and device utilization and timing analysis respectively for image of size 1600 × 1200 pixels. The PSNR obtained is better than 35 dB(indistinguishable from original image) for most of the images of size 1600 × 1200 pixels. Further, if the image size is increased, it gives even better image qualities.

Table 1. PSNR (image quality)

Image (1600 × 1200 pixels)/Parameter	PSNR (dB)
Lena.jpg	37.0

The device utilization of the proposed system is given in Table 3 for image of size 1600 × 1200 pixels and is targeted on Artix-7 FPGA board.

Table 2. MSE

Image (1600 × 1200 pixels)/Parameter	MSE
Lena.jpg	12.7

Table 3. Device utilization

Device utilization	Available	Proposed system
Number of slice registers	126800	66988
Number of slice LUTs	63400	94612
Number of fully used LUT-FF pairs	97127	64473
Number of bonded IOBs	210	70
Number of BUFG/BUFGCTRLs	32	2
Number of DSP48E1s	240	110

Table 4. Timing analysis of DCT based image mosaicing system designed

Timing utilization	Proposed system
Minimum period (ns)	14.95
Maximum frequency (MHz)	66.9
Setup time (ns)	1.747
Hold time (ns)	0.64

Similarly, timing analysis of the proposed system is given in Table 4.

The processing time obtained for image of size: 1600 × 1200 pixels is 23 ms for image mosaicing using the DCT method.

6 Conclusion

This paper presented a computationally efficient image mosaicing system using DCT based approach against brightness issues in image mosaicing. The design of the system was carried out using Verilog coding. The validation of the designed model was performed by using Xilinx 14.7 ISE and simulated by using Modelsim6.3f. The entire system design was targeted over Artix-7 FPGA board of device 7A100T-3 CSG324. The performance analysis of the proposed system was carried out using PSNR, MSE, device utilization and timing analysis. From the analysis, it is found that the proposed scheme offers high image quality and faster execution for mosaicing.

The proposed research concept can be further implemented for the applications like Video Codec, High resolution, high definition video streaming.

References

1. Dustin, M.C.: Monitoring parks with inexpensive UAVs: cost benefits analysis for monitoring and maintaining parks facilities. Dissertion, University of Southern California (2015)
2. Wilburn, B., et al.: High performance imaging using large camera arrays. ACM Trans. Graph. (TOG) **24**(3), 765–776 (2005)
3. Duarte, M.F., et al.: Single-pixel imaging via compressive sampling. IEEE Signal Process. Mag. **25**(2), 83–91 (2008)
4. Szeliski, R.: Computer Vision: Algorithms and Applications. Springer Science & Business Media (2010)
5. Sawhney, H.S., et al.: Method and apparatus for synthesizing new video and/or still imagery from a collection of real video and/or still imagery. U.S. Patent No. 7,085,409, 1 August 2006
6. Aigrain, P., Zhang, H., Petkovic, D.: Content-based representation and retrieval of visual media: a state-of-the-art review. Multimedia Tools Appl. **3**(3), 179–202 (1996)
7. Bachmann, C.M., Ainsworth, T.L., Fusina, R.A.: Exploiting manifold geometry in hyperspectral imagery. IEEE Trans. Geosci. Remote Sens. **43**(3), 441–454 (2005)
8. Shum, H., Kang, S.B.: Review of image-based rendering techniques. In: International Society for Optics and Photonics, Visual Communications and Image Processing, vol. 4067 (2000)
9. Geng, J.: Structured-light 3D surface imaging: a tutorial. Adv. Opt. Photonics **3**(2), 128–160 (2011)
10. Liang, J., Doermann, D., Li, H.: Camera-based analysis of text and documents: a survey. IJDAR **7**(2-3), 84–104 (2005)
11. Vercauteren, T.: Image registration and mosaicing for dynamic in vivo fibered confocal microscopy. Dissertion, École Nationale Supérieure des Mines de Paris (2008)
12. Zhang, Z., et al.: A robust technique for matching two uncalibrated images through the recovery of the unknown epipolar geometry. Artif. Intell. **78**(1–2), 87–119 (1995)
13. Zitova, B., Flusser, J.: Image registration methods: a survey. Image Vis. Comput. **21**(11), 977–1000 (2003)
14. Shum, H.-Y., Kang, S.B., Chan, S.-C.: Survey of image-based representations and compression techniques. IEEE Trans. Circ. Syst. Video Technol. **13**(11), 1020–1037 (2003)
15. Jung, K., Kim, K.I., Jain, A.K.: Text information extraction in images and video: a survey. Pattern Recogn. **37**(5), 977–997 (2004)
16. Kyprianidis, J.E., et al.: State of the "Art": a taxonomy of artistic stylization techniques for images and video. IEEE Trans. Vis. Comput. Graph. **19**(5), 866–885 (2013)
17. Battiato, S., et al.: Digital mosaic frameworks - an overview. In: Computer Graphics Forum, vol. 26, no. 4. Blackwell Publishing Ltd. (2007)
18. Battiato, S., et al.: A survey of digital mosaic techniques. In: Eurographics Italian Chapter Conference (2006)
19. Jacob, J.C., et al.: Montage: a grid portal and software toolkit for science-grade astronomical image mosaicking. Int. J. Comput. Sci. Eng. **4**(2), 73–87 (2009)
20. Pisanich, G., et al.: Initial efforts toward mission-representative imaging surveys from aerial explorers. In: Real-Time Imaging VIII, International Society for Optics and Photonics, vol. 5297 (2004)
21. Fan, X.: Automatic registration of multi-modal airborne imagery. Rochester Institute of Technology (2011)

22. Pluim, J.P.W., Maintz, J.B.A., Viergever, M.A.: Mutual-information-based registration of medical images: a survey. IEEE Trans. Med. Imaging **22**(8), 986–1004 (2003)
23. Zhi, Q., Cooperstock, J.R.: Toward dynamic image mosaic generation with robustness to parallax. IEEE Trans. Image Process. **21**(1), 366–378 (2012)
24. Shao, H.C., Hwang, W.L., Chen, Y.C.: Optimal multiresolution blending of confocal microscope images. IEEE Trans. Biomed. Eng. **59**(2), 531–541 (2012)
25. Prados, R., Garcia, R., Gracias, N., Escartin, J., Neumann, L.: A novel blending technique for underwater gigamosaicing. IEEE J. Oceanic Eng. **37**(4), 626–644 (2012)
26. Rahman, M.M., et al.: Geographic Object-Based Mosaicing (OBM) of high-resolution thermal airborne imagery (TABI-1800) to improve the interpretation of urban image objects. IEEE Geosc. Remote Sens. Lett. **10**(4), 918–922 (2013)
27. Yigitsoy, M., Navab, N.: Structure propagation for image registration. IEEE Trans. Med. Imaging **32**(9), 1657–1670 (2013)
28. Molina, E., Zhu, Z.: Persistent aerial video registration and fast multi-view mosaicing. IEEE Trans. Image Process. **23**(5), 2184–2192 (2014)
29. Lurie, K.L., Angst, R., Ellerbee, A.K.: Automated mosaicing of feature-poor optical coherence tomography volumes with an integrated white light imaging system. IEEE Trans. Biomed. Eng. **61**(7), 2141–2153 (2014)
30. Bellavia, F., Colombo, C.: Dissecting and reassembling color correction algorithms for image stitching. IEEE Trans. Image Process. **27**(2), 735–748 (2018)
31. Jain, D.K., Saxena, G., Singh, V.K.: Image mosaicing using corner techniques. In: 2012 International Conference on Communication Systems and Network Technologies, Rajkot, pp. 79–84 (2012)
32. Hadrović, E., Osmanković, D., Velagić, J.: Aerial image mosaicing approach based on feature matching. In: 2017 International Symposium ELMAR, Zadar, pp. 177–180 (2017)
33. Dhokley, W., Zamirhasan, C., Zahid, K., Asif, S.: Image mosaicing using bidirectional algorithm. In: 2017 International Conference on Innovations in Information, Embedded and Communication Systems (ICIIECS), Coimbatore, pp. 1–6 (2017)
34. Chen, X., Liu, H., Zhou, M., Zhang, F., Li, Q.: Medical image mosaic based on low-overlapping regions. In: 2017 10th International Congress on Image and Signal Processing, BioMedical Engineering and Informatics (CISP-BMEI), Shanghai, pp. 1–5 (2017)

Semantic History: Ontology-Based Modeling of Users' Web Browsing Behaviors for Improved Web Page Revisitation

Imam ud Din, Shah Khusro(✉), Irfan Ullah, and Azhar Rauf

Department of Computer Science, University of Peshawar, Peshawar, Pakistan
meetimam2@gmail.com,
{khusro, cs.irfan, azhar.rauf}@uop.edu.pk

Abstract. Web Browsers are software solutions that facilitate users in browsing the Web. However, the huge size of the Web makes it difficult to find relevant resources, resulting in information and cognitive overload. To mitigate this overload, researchers have attempted to find ways for re-visitation of web pages that are deemed useful and more likely to be revisited. Also, web browsers have several built-in tools including history, bookmarks, backward & forward buttons, Uniform Resource Locator (URL) auto-completion, and so on. This research focuses on web browser history, which maintains details of visited web pages with their associated metadata to enable users in finding and re-finding (revisitation) web pages without encountering the information and cognitive overload. In addition to the built-in history tools in web browsers, several third-party tools in the form of toolbars, extensions, and add-ons are available. However, these solutions exploit no or limited web page-level semantics and fail to provide full revisitation support to the users. It is, therefore, necessary to fill this semantic gap by exploiting web page-level semantics, which is the aim of this paper. We contribute "Browser History Ontology," and use it in our developed Chrome-based browser extension, namely "Semantic History." Experimental results show that our proposed solution provides better re-visitation support to the users by semantically organizing the web browser history.

Keywords: Re-visitation · Web browser · Web browser history Ontology · Web page

1 Introduction

With the exponential growth of the Web, retrieving and refining relevant resources has become challenging [1, 2] even for experienced users and using sophisticated search engine technologies. After finding relevant resources, additional efforts are required to remember their URLs (so that they may be revisited if needed), which is not feasible especially when the visited web pages are too many. To handle his problem, a web

R. Silhavy et al. (Eds.): CoMeSySo 2018, AISC 860, pp. 204–215, 2019.
https://doi.org/10.1007/978-3-030-00184-1_19

browser contains several built-in tools, extensions, add-ons, etc. Also, several third-party applications including Google Toolbar[1], back & forward buttons, URL auto-completion, bookmarks, etc., are available for improving web page revisitation [3]. Moreover, several notable add-ons including WebMynd[2], Delicious[3], Hooeey[4], etc. are available to help users in re-finding interesting information [4]. Users also use search engines [5], and their memory for finding and returning to web-based information [6]. The state-of-the-art literature contains several latest history-based revisitation solutions including MyWebSteps [1] and dynamic bookmarks [6].

However, to the best of our knowledge, none of the existing history-based revisitation solutions exploits all the necessary aspects about revisitation [7] including page-level, history-related semantics, and therefore, unable to fully support users in revisitation. Unlike the existing approaches, surveyed in [7], we extend the state-of-the-art by creating "Browser History Ontology" to capture all the required semantics including clicking, scrolling, bookmarking, time spent on a web page, text highlighting, etc., of the corresponding visited web pages and use it in our developed browser extension. This way, the history of the visited web pages could be kept well-organized and semantically retrievable to make revisitation an easy and user-friendly task. Section 2 covers related work. Section 3 covers the methodology for developing "Browser History Ontology" and "Semantic History," our Google Chrome-based extension. Section 4 presents results. Section 5 concludes our discussion and presents some future work. References are given at the end of the paper.

2 Related Work

Almost all modern web browsers come with some history tools and techniques to allow revisitation. Researchers have proposed several approaches towards history-based revisitation. These include: (i) step-wise path-following, permitting users to backtrack their paths; (ii) directed jumps to a homepage; (iii) directed jumps to "previously-visited" web pages; (iv) presenting history in linear form; (v) searching history lists [8]; (vi) branching history display [9, 10]; (vii) history views within session; (viii) personalized URLs lists [8]; and (ix) new systems integrating history URLs' personalized lists [11].

According to [1], a user adopts one of the three methods for revisiting a web page, including an explicit web history; automatically recorded web browsing history, and using search engines [1]. Most commonly, browsing history is presented as a list of textual lines containing web page title, URL, time, session, etc., which users scan to revisit a specific page. Several browsers adopt this method for bookmarking, URL-auto-completion, and history lists in addition to research tools including Search Bar [12], Re: Search Engine [13] and so on [1].

[1] http://www.toolbar.google.com.

[2] http://www.webmynd.com.

[3] http://www.delicious.com.

[4] http://www.hooeey.com.

Do and Ruddl [1] identified ten reasons why users are unable to revisit them. They designed MyWebSteps [1], a novel web history tool that attempts to mitigate these issues and implemented it as Mozilla Firefox browser add-on.

Momento system [14] unifies content and contextual aids in revisiting web pages, where a web page's context by exploiting the preceding or following pages visited in the same session and extracting topic-phrases from (the full content of) these visited pages using the topics list of Wikipedia [14]. A more related work is presented in [15] that extracts terms from segments of the web pages that currently displayed to the user, which are assigned probabilistic values considering the browsing behaviors (highlighting and dwell time), the subject headings of the page, term frequency-inverse document frequency (TF-IDF), which reflects the impression and interest of the user in the page. The extracted contextual information includes access time, location and concurrent activities, which are automatically inferred from the running programs.

The authors in [16] argue that the existing browser tools are unable to efficiently visualize the visited items and connect them in generating personalized re-visitation recommendations [16]. Also, they are unable to record how much time a user spends on a specific web page. They proposed a conceptual framework showing that how Linked Data can mitigate this issue and improve user experience and quality of recommendations. By developing a history tool, they demonstrated the potential of Linked Data in generating quality recommendations and better browser history experience [17]. They applied a clustering algorithm to keywords (meta tags) of visited web pages together with exploiting other metadata including start time, URLs, titles, and visit counts in generating interactive visualizations so that user see the temporal patterns and contents of the visited pages, which will help them find the desired page to be revisited.

In conclusion, several browser history tools are available; they are unable to efficiently visualize the visited pages and relate them in offering a more personalized revisitation experience to the user [16]. This research paper summarizes our work in progress regarding the development of semantics-aware revisitation solution, namely Semantic History, and demonstrates that semantically and ontologically modeling web pages does improve the revisitation of web pages.

3 Methodology

This Section discusses the proposed solution, which is in the form of a web browser extension developed explicitly for Google Chrome, one of the famous and widely used web browser. However, it can be adapted and modified for any other web browser. The extension uses ontology to keep track of all the activities of the user so that semantic search capabilities could be provided during web page revisitation. Section 3.1 discusses the proposed solution; it's working, and briefly describes the tools used for implementation. Section 3.2 explains how the Browser History Ontology (BHO) was developed and evaluated. Section 3.3 presents the evaluation metrics namely precision, recall, and f-measure to compare the performance of Semantic History to that of the state-of-the-art Chrome History tool, as the extension was developed and tested for Google Chrome.

3.1 The Proposed Semantic History Extension

Figure 1 depicts the proposed Semantic History web browser extension, which is a layered architecture, consisting of three layers namely the user interface, application logic layer, and data layer. The user interface is responsible for enabling/disabling the extension as well as for search and browsing the visited web pages during page revisitation. The application logic layer contains the business intelligence responsible for collecting the required information about visited web pages, database storage, information extraction and ontology population, and inferencing behind search & browsing during page revisitation. For this purpose, it employs different modules/components including collection engine module, semantics management module, and search & browsing module.

To enable the browser to semantically retrieve the required web page from the massive list of visited web pages, it is necessary to identify the user preferences and interests, which in this case, should be captured implicitly from the browsing activities, the visited web pages, and their metadata, as users are usually not willing to spend time on sharing their interests with the browser explicitly by rating, tagging, and reviewing each visited web page. The extension, through its *collection engine*, captures and stores the required associated metadata about visited web pages inconspicuously and implicitly so that user actions are not interrupted during browsing. The metadata, captured by the extension, includes web page title, URL, date visited, session, scrolling, time spent on the web page, number of times a page is visited, and whether the page is bookmarked/saved/printed. All this information is first stored in the database and then semantically modeled using the *semantics management module* in the form of Resource Description Framework (RDF) triples to populate the BHO ontology. This module is also responsible for making necessary inferencing and relating different informational entities to retrieve the most contextually relevant and required web page matching the search query during revisitation. The revisitation is facilitated by the *search & browsing module* which allows access to the already visited pages when requested by the user using search query on the extension user interface.

A proof-of-concept version of the proposed ontology-based extension was developed for Google Chrome web browser, which can be modified and implemented for any other browser. For the implementation of the extension, we used Adobe Dreamweaver CC as IDE, PHP, and JavaScript as programming and scripting language, Chrome API for interacting with the Chrome browser, and SQLite for database development. For developing BHO ontology, we used Web Ontology Language (OWL) as ontology development language and Protégé 4.3 as ontology development tool.

3.2 Browser History Ontology

In developing BHO, we used ontology development methodology discussed in [18]. The details of the BHO are given in the following subsections.

Domain and Scope. BHO is designed with the primary aim of facilitating users in revisitation by semantically modeling user activities by covering all the related concepts (classes), object and datatype properties. To meet the domain requirements, the scope

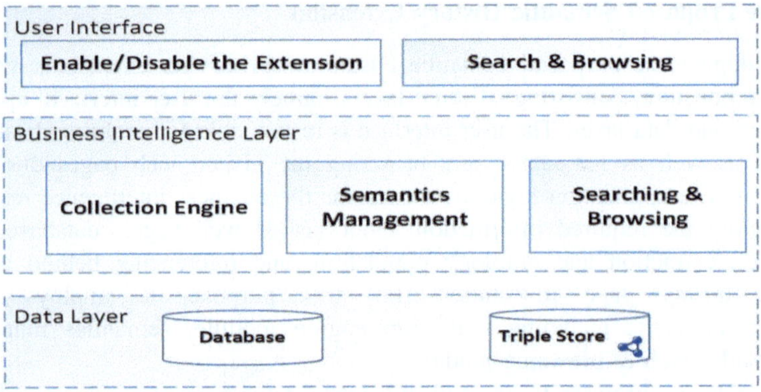

Fig. 1. The architecture of the Semantic History

of the ontology is covered by identifying competency questions (CQs), which are written in natural language; capture all the functional requirements of the application, and help in evaluating the final ontology [18]. The collected CQs were categorized into groups to make them more understandable and validated to find missing or conflicting CQs. In total, 62 CQs were identified in this process. Table 1 shows an excerpt of these CQs with possible answers and divided into several groups.

Ontology Reuse. Before building an ontology, it is first necessary to check whether the available ontologies could be used [18]. To the best of our knowledge and besides the availability of several ontologies, no ontology exists that meet our browser-specific requirements. Therefore, we focused on developing BHO from scratch.

Domain Lexicon. The domain lexicon contains the terminology to be used for representing different classes, and related object and datatype properties [18], which are identified from the validated requirements of the ontology, asking domain experts, and a multiplicity of other resources including the Web (16%), existing ontologies (52%), technical reports (11%), and books (14%) contributed to the DL of BHO. The lexicon can either be manually generated or using some automated tools including, e.g., OntoLearn [19]. In total 317 most specific and meaningful terms were collected from a large initial set of lexical words. An excerpt of the lexicon is shown in Table 2.

Class Hierarchy, Classes, Slots, Axioms, and Individuals. The hierarchical relationships among classes, i.e., the class hierarchy can be achieved in several ways including top-down, bottom-up, or both [20], and their use depends on the target domain. For BHO, we adopted a top-down approach, which starts with the definition of most general concepts in the domain and ends at specific subsequent concepts. An excerpt from the class hierarchy is shown in Fig. 2.

The classes are concepts, representing sets of identical objects (entities) that serve the role of business object, agents, or processes in a specific domain. These can be places, people, things, roles, events, organizational units, etc., which are related to each other based on the semantic relationship between them. The slots or properties establish the necessary relations among classes, objects, and literals in a domain-range

Table 1. The list of competency questions (CQs)

S. No.	Competency question	Answer
CQ1	What is the title of the Web page?	Any string value
CQ2	What is the URL of the Web page?	Any URL like www.google.com
CQ3	Is a Web page saved?	Yes or No
CQ4	What type of menu is used in a Web page?	Horizontal or Vertical etc.
CQ5	What is the name of a form element?	Any string value like a password
CQ6	What is the frame size?	Any value in pixel or percentage
CQ7	What is the frame source?	Any URL
CQ8	What is the style of text?	Bold, Italic, etc.
CQ9	What is the type of URL used in the hyperlink?	Absolute or Relative, etc.

Table 2. An excerpt of the BHO domain lexicon

Web page	Title	Header	Footer	Hyperlink	Label text	Element name
Border	Chat room	Comment	Web feed	Sidebar	Frame type	Frame scrolling
Frame	Menu	Bookmark	Text	Metadata	URL text	Header image

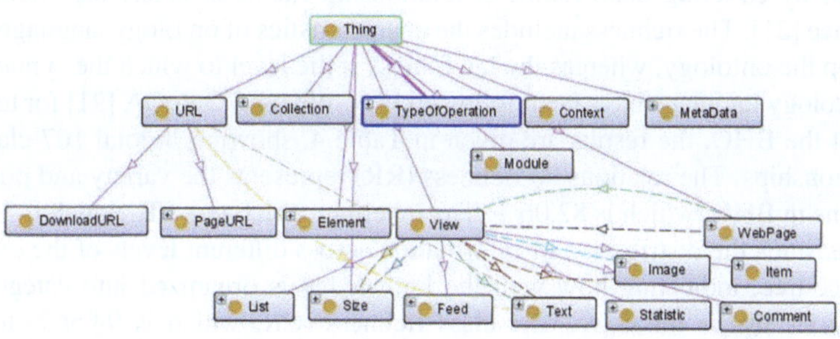

Fig. 2. An excerpt of the BHO class hierarchy in Protégé 4.3

relationship. In the lexicon, classes are identified from the nouns, whereas properties are recognized from the verbs. A property can be either object property or datatype property. An object property represents the exchange of information among classes and relates class instances. A datatype property establishes the relationship among instances of classes and literals. Table 3 depicts an excerpt of the classes and object properties and datatype properties of BHO.

The slots in BHO have different facets describing the allowed values, value type, and the cardinality constraints, etc. For example, the value of the size slot (as in "the size of a web page") is a float. A slot may have single or multiple values, e.g., the

Table 3. An excerpt of the classes, object and datatype properties of BHO

Classes				
Web page	Header	Footer	Menu	Logo
Sidebar	Breadcrumb	Bookmark	Image	Audio
Form	Button	Textbox	Textarea	Frame
Object properties				
hasBreadCrumb	hasMenu	hasTab	hasURL	AccordionText
FooterImage	HeaderImage	HyperlinkURL	HyperlinkTarget	TabMap
Datatype properties				
hasTitle	hasText	hasVisitTime	isSaved	TextColor
hasName	hasValue	MenuType	isChecked	TitleText

"downloads" (as in "a user downloads a file") may have multiple values, where the values are instances of the class "WebPage." The final step of the methodology [18] is to create individuals of the classes. This includes selecting class, creating its instances, and assigning or stuffing slot values.

Quality Testing and Evaluation. The quality testing helps identify all the possible errors and inconsistencies in the ontology. Three types of quality tests are conducted. These include syntactic quality testing, semantic quality testing, and pragmatic quality testing.

Syntactic Quality. It measures how well an ontology is written in terms richness and lawfulness by covering class richness, relationship richness, inheritance richness and tree balance [21]. The richness includes the characteristics of ontology language selected to develop the ontology, whereas the lawfulness is the level to which the syntactic rules of the ontology language have been followed [21]. We used OntoQA [21] for testing the quality of the BHO, the results are given in Table 4, showing in total 107 classes and 198 relationships. The relationship richness (RR) represents the variety and positioning of relations in BHO, which is 82.06%. The inheritance richness (IR, which is 1.62% for BHO) measures the distribution of information across different levels of the ontology's inheritance tree, indicating how well the knowledge is organized into categories and subcategories in the ontology. The class richness (CR, which is 94.56% for BHO) measures how the instances are distributed across classes, the number of classes that have instances is compared to the total number of classes, giving a general idea of how many instances are related to the classes defined in the schema [21].

Semantic Quality. It measures the meaning of terms used in the ontology regarding consistency and interoperability. Consistency evaluates whether terms have a consistent meaning in ontology by finding any contradiction and correct use of modeling constructs. Interoperability guesses the meaning of terms in ontology and is acquired by checking the terms in the ontology in another autonomous semantic source [22]. Consistency of the ontology can be checked by using reasoner (e.g. Hermit, Fact++,

Table 4. Syntactic quality testing of BHO using OntoQA

Classes	Relationships	RR	IR	Tree balance	CR
107	198	82.06%	1.62%	0.82	94.56%

RacerPro, and Pellet). A reasoner works by taking a manually created class hierarchy (called asserted ontology) to automatically compute an inferred class hierarchy (called inferred ontology) using the descriptions of classes and relationships [23]. Figure 3 depicts a snippet of the BHO asserted and inferred class hierarchies after reasoning using Fact++ in Protégé 4.3, showing that the ontology is semantically consistent.

Pragmatic Quality. Testing: Pragmatic quality testing measures the usefulness of ontology for its users regardless of syntax and semantics [22]. It is measured in terms of comprehensiveness, accuracy and relevancy. A comprehensive ontology is the one that meets all the requirements and constraints of the problem domain and can be measured in terms of domain coverage, goals, and degree of response to competency questions [24]. Accuracy decides that the maintained knowledge in ontology corresponds to the domain knowledge of the experts, and therefore, requires that the classes, individuals and properties are correctly defined. Accuracy can be measured using recall and precision rates [23]. However, in our case, we measured the accuracy of the ontology by defining use-cases based on CQs and creating and running SPARQL queries to query the ontology in Protégé 4.3. Relevance concerns with the coverage assert that whether ontology meets a user's needs [23]. The relevance of BHO was tested by considering CQs to retrieve specific answers from the ontology using SPARQL queries.

3.3 Evaluation Metrics

There are several evaluation metrics used for the assessing the performance of an information retrieval solution. Among these, precision, recall, and f-measure are the widely adopted measures, and therefore, we used these for comparing Semantic History against the traditional Chrome-based history management tool. For a search, the precision is the ratio of the number of relevant items retrieved over the total number of items retrieved. The recall is the ratio of the number of relevant items retrieved over the total number of relevant items. Let T_p = number of relevant documents which are retrieved (called true positive); F_p = number of irrelevant documents which are retrieved (called false positive); and F_n = number of relevant documents which are not retrieved (called false negative). According to [25] Eqs. (1) and (2) give the precision (P), recall (R) respectively. If a system is good at recall, it doesn't mean that the system is good at precision. That's why we also need F score which is the (harmonic) mean of recall and precision to evaluate a system. F can be calculated using Eq. (3).

$$P = \frac{T_p}{T_p + F_p} \tag{1}$$

$$R = \frac{T_p}{T_p + F_n} \tag{2}$$

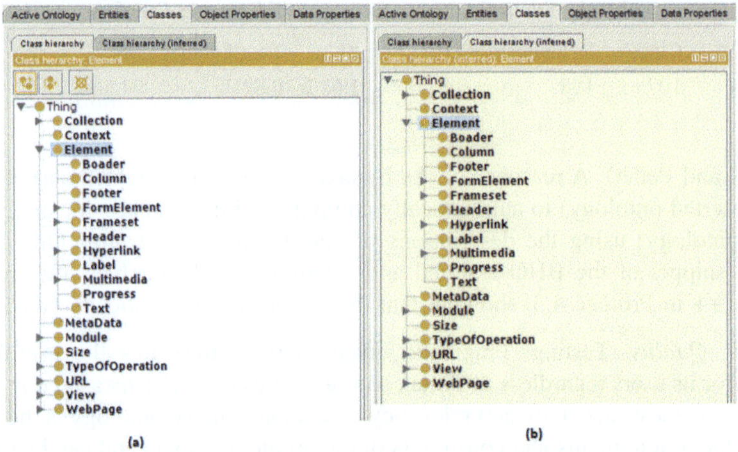

Fig. 3. Asserted (a) and inferred (b) BHO class hierarchies after reasoning

$$F = 2.\frac{P.R}{P+R} \tag{3}$$

Where 2 is a constant that scales F to 1, when $P = R = 1$, i.e., the f-measure increases when both precision and recall increase, F = 100% if P = 100%, R = 100%.

4 Experimental Settings and Results

A list of 20 different keywords/phrases, which are representative of their corresponding topics, was generated. Each of these keywords was used in querying Google web search engine and a list of 20 top relevant pages (total: 500 web pages visited) were manually recorded as well as were browsed so that the default Chrome history tool as well as our proposed tool could capture the required metadata and store them in their database and populate their OWL file, respectively. In the next step, each of the keywords/phrases was queried against the Chrome's default history tool and recorded *number of relevant pages retrieved*, *number of irrelevant web pages retrieved* and *total number of web pages retrieved* respectively, retrieved by it. This information is required in measuring the precision, recall and f-measure values, which helps in assessing the performance of the system under consideration. These are discussed in the next paragraph. Similarly, each of these keywords/phrases was used in querying the Semantic History tool to record *number of relevant pages retrieved*, *number of irrelevant web pages retrieved* and *total number of web pages retrieved* respectively, retrieved by it.

Using Eqs. (1), (2), and (3), the precision, recall, and f-values were respectively computed per competitive system per keyword and then average on all results of the 20 records. Figures 4, 5, and 6 plot the respective values of P, R, and F for both systems to visualize the difference between these systems. In Fig. 3, the precision values are

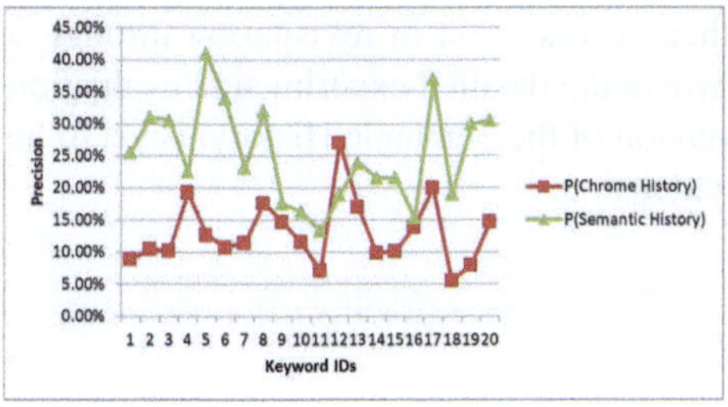

Fig. 4. Precision for traditional Chrome-based History vs. Semantic History

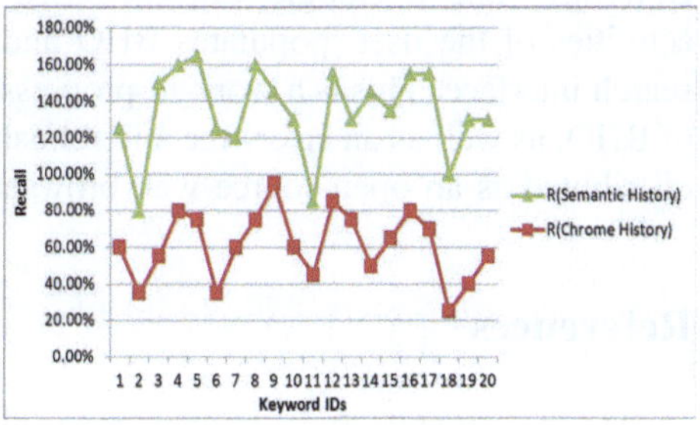

Fig. 5. Recall for traditional vs. ontology-based history systems

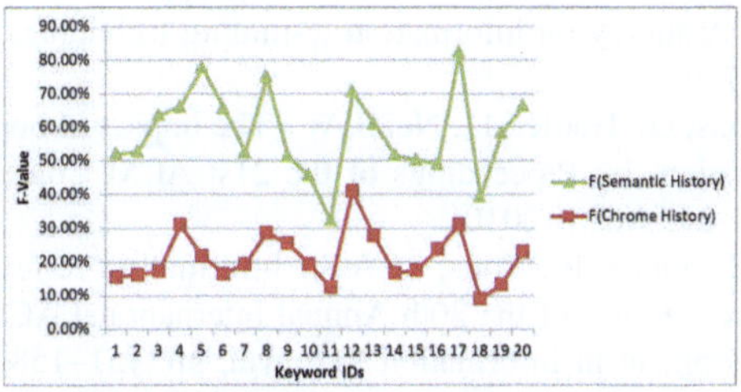

Fig. 6. F-measure for traditional vs. ontology-based history systems

compared for both systems, which at some points overlap or the precision of the traditional approach get better than the proposed one. This is due to the fact that some keywords appear to be the most representative keywords for the browsed pages. In other cases, precision of the proposed system is higher because of the exploitation of semantics, which the traditional system cannot capture. This is a clear demonstration that making web browser history semantics-aware does improve the revisitation performance. This is further supported by Figs. 5 and 6 in terms of recall and f-measure.

5 Conclusion and Future Work

Modern web browsers come with a set of built-in revisitation tools including history list and bookmarks to aid the people in searching the interesting web pages that they have already visited. These tools are limited in facilitating users in the revisitation of previously visited but important web pages. One prominent reason behind their failure is the lack of the full exploitation of semantics of the visited pages. This work is an initial attempt to bridge this semantic gap with the help of a comprehensive browser history ontology to making re-visitation better using Semantic Web technologies. The study is a demonstration that Semantic Web technologies have the potential to improve the performance of the existing re-visitation approaches, especially a Web browser's history tool. To this goal, we have contributed and briefly explained our "Browser History

Ontology" and a Google Chrome-based extension that records implicitly different activities of the user, populates BHO and then serves users in revisitation through a search interface. This is a work-in-progress, where the detailed experimental evaluation of BHO, as well as an error-free and robust version of the Semantic History, is yet to be distributed as an open-source web browser extension.

References

1. Do, T.V., Ruddle, R.A.: MyWebSteps: aiding revisiting with a visual web history. Interact. Comput. **29**(4), 530–551 (2017). https://doi.org/10.1093/iwc/iww038
2. Sadeghi, S., Blanco, R., Mika, P., Sanderson, M., Scholer, F., Vallet, D.: Predicting re-finding activity and difficulty. In: Hanbury, A., Kazai, G., Rauber, A., Fuhr, N. (eds.) Advances in Information Retrieval: 37th European Conference on IR Research, ECIR 2015, Vienna, Austria, 29 March–2 April 2015, Proceedings, pp. 715–727. Springer, Cham (2015)
3. Deng, T., Feng, L.: A survey on information re-finding techniques. Int. J. Web Inf. Syst. **7** (4), 313–332 (2011)
4. Kawase, R., Papadakis, G., Herder, E., Nejdl, W.: The impact of bookmarks and annotations on refinding information. In: Proceedings of the 21st ACM Conference on Hypertext and Hypermedia, pp. 29–34. ACM (2010)
5. Teevan, J., Adar, E., Jones, R., Potts, M.A.S.: Information re-retrieval: repeat queries in Yahoo's logs. In: Proceedings of the 30th Annual International ACM SIGIR Conference on Research and Development in Information Retrieval, pp. 151–158. ACM (2007)
6. Bruce, H., Jones, W., Dumais, S.: Keeping and re-finding information on the web: what do people do and what do they need? Proc. Assoc. Inf. Sci. Technol. **41**(1), 129–137 (2004)
7. Papadakis, G., Kawase, R., Herder, E., Nejdl, W.: Methods for web revisitation prediction: survey and experimentation. User Model. User-Adapt. Interact. **25**(4), 331–369 (2015). https://doi.org/10.1007/s11257-015-9161-7
8. Tauscher, L., Greenberg, S.: How people revisit web pages: empirical findings and implications for the design of history systems. Int. J. Hum Comput Stud. **47**(1), 97–137 (1997)
9. Cockburn, A., Jones, S.: Which way now? Analysing and easing inadequacies in WWW navigation. Int. J. Hum Comput Stud. **45**(1), 105–129 (1996)
10. Ayers, E.Z., Stasko, J.T.: Using graphic history in browsing the World Wide Web. In. Georgia Institute of Technology (1995)
11. Brown, M.H., Shillner, R.A.: DeckScape: an experimental web browser. Comput. Netw. ISDN Syst. **27**(6), 1097–1104 (1995)
12. Morris, D., Morris, M.R., Venolia, G.: SearchBar: a search-centric web history for task resumption and information re-finding. In: Proceedings of the SIGCHI Conference on Human Factors in Computing Systems, pp. 1207–1216. ACM (2008)
13. Teevan, J.: The re:search engine: simultaneous support for finding and re-finding. In: Proceedings of the 20th Annual ACM Symposium on User Interface Software and Technology, Newport, Rhode Island, USA
14. Kulkarni, C.E., Raju, S., Udupa, R.: Memento: unifying content and context to aid webpage re-visitation. In: The Adjunct Proceedings of the 23nd Annual ACM Symposium on User Interface Software and Technology, New York, USA
15. Jin, L., Feng, L., Liu, G., Wang, C.: Personal web revisitation by context and content keywords with relevance feedback. IEEE Trans. Knowl. Data Eng. **29**(7), 1508–1521 (2017). https://doi.org/10.1109/TKDE.2017.2672747

16. Kandala, H., Tripathy, B.K., Manoj Kumar, K.: A framework to collect and visualize user's browser history for better user experience and personalized recommendations. In: Satapathy, S.C., Joshi, A. (eds.) Information and Communication Technology for Intelligent Systems (ICTIS 2017) -, vol. 1, pp. 218–224. Springer, Cham (2018)
17. Du, W.: Personal Web Library: Organizing and Visualizing Web Browsing History. Purdu University (2017)
18. Noy, N.F., McGuinness, D.L.: Ontology development 101: a guide to creating your first ontology. In: Stanford Knowledge Systems Laboratory Technical Report KSL-01-05 and Stanford Medical Informatics Technical Report SMI-2001-0880, Stanford, CA (2001)
19. Navigli, R., Velardi, P.: Learning domain ontologies from document warehouses and dedicated web sites. Comput. Linguist. **30**(2), 151–179 (2004)
20. Uschold, M., Gruninger, M.: Ontologies: principles, methods and applications. Knowl. Eng. Rev. **11**(2), 93–136 (1996)
21. Tartir, S., Arpinar, I.B., Moore, M., Sheth, A.P., Aleman-Meza, B.: OntoQA: metric-based ontology quality analysis. In: The IEEE ICDM Workshop on Knowledge Acquisition from Distributed, Autonomous, Semantically Heterogeneous Data and Knowledge Sources, Houston, TX, 27 November 2005
22. Burton-Jones, A., Storey, V.C., Sugumaran, V., Ahluwalia, P.: A semiotic metrics suite for assessing the quality of ontologies. Data Knowl. Eng. **55**(1), 84–102 (2005). https://doi.org/10.1016/j.datak.2004.11.010
23. Ali, S., Khusro, S.: POEM: practical ontology engineering model for semantic web ontologies. Cogent Eng. **3**(1), 1193959 (2016). https://doi.org/10.1080/23311916.2016.1193959
24. Maynard, D., Peters, W., Li, Y.: Metrics for evaluation of ontology-based information extraction. In: International World Wide Web Conference, pp. 1–8. Edinburgh, UK (2006)
25. Joshi, R.: Accuracy, precision, recall & f1 score: interpretation of performance measures: how to evaluate the performance of a model in Azure ML and understanding "confusion metrics". In: Exsilio Solutions, 9 September 2016

A Recovery Technique for the Fog-Computing-Based Information and Control Systems

Eduard Melnik[1], Anna Klimenko[2](✉), and Vladislav Klimenko[2]

[1] Southern Scientific Center of the Russian Academy of Science, Rostov-on-Don, Russia
[2] Scientific Research Institute of Multiprocessor Computer Systems of Southern Federal University, Taganrog, Russia
anna_klimenko@mail.ru

Abstract. The current paper deals with the dependability issue of the fog-computing-based information and control systems (ICS). Comparing with the traditional, "cloud" ICS architectures, fog-computing paradigm brings the possibility to distribute the workload not only among the central computational units (CU), but to shift it to the fog-layer and edge-devices of the ICS infrastructure. This allows to reconfigure ICS in case of CU failure such as to maximize the CU reliability function value by workload shift to the edge of the network. In this paper the new recovery technique for the fog-computing-based ICSs is presented and discussed. The technique is based on the simplified models of the workload distribution in the fog-computing-based systems. Besides, some important conclusions are made and selected simulation results are presented.

Keywords: Control systems · Fog computing
Information and control systems · Recovery · Reliability · Fault-tolerance

1 Introduction

Contemporary information and control systems (ICS) are used in a huge amount of human activities, including power stations, spacecraft, aircraft, oil and gas production, robotics, autonomous collaborative robotics, etc. Starting at the point of simple embedded systems, present days ICSs involve new concepts and technologies of distributed computing. For example, the well-known Industry 4.0 concept engages the Industrial Internet of Things technology [1, 2], the latter using widely fog- and edge computing paradigms and so on.

There is a plenty of studies considering ICSs based on fog-computing concept: some of them emphasize the new ICS architectures for usage in particular subject areas [3–5], others consider security of fog-computing-based systems, etc., while the dependability issue has been studied quite poorly [6].

In this paper the dependability issue of fog-computing-based ICS is under concern. Dependability is a complex characteristic of a system, and can be considered in terms of reliability, maintainability, availability, etc. [7]. Yet the dependability is connected closely to the term "fault-tolerance", which can be implemented by the reconfiguration

R. Silhavy et al. (Eds.): CoMeSySo 2018, AISC 860, pp. 216–227, 2019.
https://doi.org/10.1007/978-3-030-00184-1_20

as a fault-recovery method [7]. The latter is related to the system redundancy as only a system featuring redundancy can be fault-tolerant.

As a continuation of our previous work, we consider a class of ICS with performance redundancy [8, 9], in which recovery performs as a procedure of monitoring and control tasks re-distribution among the computational units (CU) of a system. This particular procedure is called "reconfiguration" in the context of the current paper. Within the class of systems selected, the term "configuration" is a cornerstone: it means the distribution of the monitoring and control tasks among the functional CUs. The configuration forming (CF) problem was presented, described and solved partially in our previous works [10, 11], yet the CF problem was formalized for ICSs with "traditional", centralized architecture, while the application of fog-computing concept to the ICS design and development leads to some new system features, which make it different from previously described ones.

The general difference of the fog-computing-based ICS from the ICS with traditional architecture is the ability to shift the computational workload to the fog-layer and to the edge of the network, localizing the data processing much nearer to the data sources. In our previous works the CF problem was formalized for central CUs only, while the fog-computing concept allows to use fog-layer and endpoint devices for workload re-distribution.

So, in this paper we present a new recovery technique of the fog-computing-based control systems. It allows to perform the reconfiguration procedure in the way to improve the reliability function of the central computational devices.

The following sections of this paper include:

- the brief review of system reconfiguration procedure for ICS with performance redundancy;
- the ICS architectures description;
- the simplified model of fog-computing-based ICS workload distribution;
- the recovery technique for the fog-computing-based ICSs;
- the simulation results and conclusions.

2 Reconfiguration Procedure of ICS with Performance Redundancy

Performance redundancy [12] presupposes that there are no major and stand-by elements in a system. When the failure occurs, tasks from the faulted nodes are moved to the operational ones, so a set of tasks can be spread within more than one node according to the configurations prepared.

Taking into consideration the peculiarities of performance redundancy method, the reconfiguration procedure differs from the one with featuring of structural redundancy. The general scheme of reconfiguration is shown on Figs. 1, 2 and 3.

So, each state of the system after the recovery is determined by possible configurations, which are the solution of the CF problem mentioned above. Configurations can be developed either on design stage of the ICS or "on-the fly" during the reconfiguration itself. The latter presupposes the approximate decisions usage due to the high

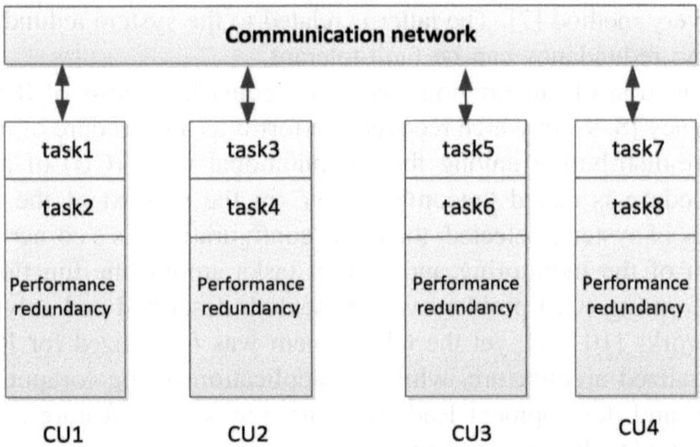

Fig. 1. The system with performance redundancy before failure.

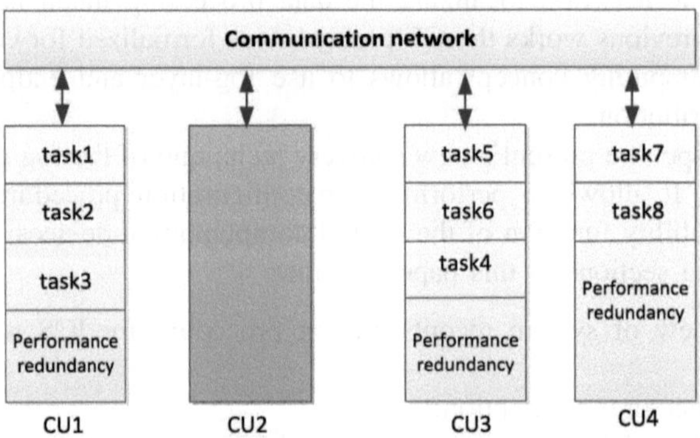

Fig. 2. The system with one faulted node.

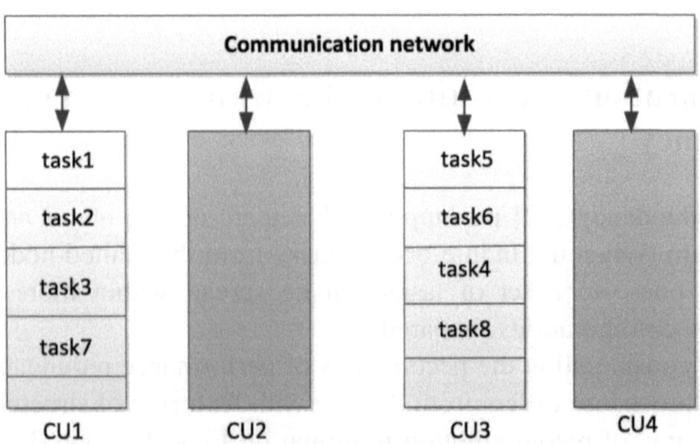

Fig. 3. The system with two faulted nodes.

computational complexity of the CF problem. Yet some experimental research [13] shows that solutions of a good quality can be found by means of simulated annealing algorithms with quenching temperature scheme [14].

As the CF problem is considered in this paper, it is expedient to recall its generic model (which is described in details in [11]).

There are N tasks with computational complexities g_i, M CUs with equal performance m_j, $U = \{u_{ij}\}$ – the percentage of j CU performance allocated for the i task, T' – planned completion time for the N tasks, $F = \{f_k\}$, $k \in \{1, \ldots M\}$, – the set of simultaneously failed CUs.

Through the resource allocation every task links to the CU, and it can be described by the following tuple:

$a_i = \langle j, u_{ij}, t_i \rangle$, where j – the CU identifier, u_{ij} – the allocated resource ratio, t_i – the time of task i start.

So, the set $A = \{a_i\}$ determines the configuration of ICS before failure, the set $A' = \{a'_i\}$ determines the configuration of ICS after the reconfiguration. In fact, A' is the solution of configuration forming problem, and a'_i – the tuples which describe the new task assignments.

The subtraction operator for sets A and A' must be determined:

$$a_i - a'_i = \begin{cases} 0, & \text{if } j \text{ is equal to } j'; \\ 1, & \text{otherwise.} \end{cases} \quad (1)$$

The objective functions are as follows:

$$F_1 = \sum_{i=1}^{N} (a_i - a'_i) \rightarrow MIN \quad (2)$$

$$F_2 = |A| - |A'| \rightarrow MIN \quad (3)$$

$$F_3 = \sum_{k=1}^{K} u'_{kj} - \sum_{l=1}^{L} u'_{lq} \rightarrow MIN, \ \forall j, q \quad (4)$$

Where F_1 is the number of tasks, which are relocated from the operational nodes; F_2 is the number of non-critical tasks eliminated from the CUs; F_3 is the workload dispersion through the CUs. The solution of CF problem is the task distribution among the CUs.

According to [15], failure rate of an element depends on its temperature:

$$\lambda = \lambda_0 \cdot 2^{\Delta T/10} \quad (5)$$

where ΔT is the temperature increase.

As temperature depends on workload, the Eq. (5) can be presented as follows:

$$\lambda = \lambda_0 \cdot 2^{kD/10} \quad (6)$$

Where k is the ratio between the workload D and the temperature increase.

The reliability function is described by the following equation:

$$P(t) = e^{-\lambda t} \tag{7}$$

where t is the time of element functioning.

The solution of CF problem – a new configuration A' - affects the CU workload, and, as a consequence, affects the CU reliability function, as is shown in Eqs. (5–7).

3 Traditional and Fog-Computing-Based ICS Architecture

Traditional, cloud-computing-based, or centralized ICS architecture can be presented as is shown in Fig. 4. Here the data from the sensors are processed by the computational environment (CE), with/or without operators involvement.

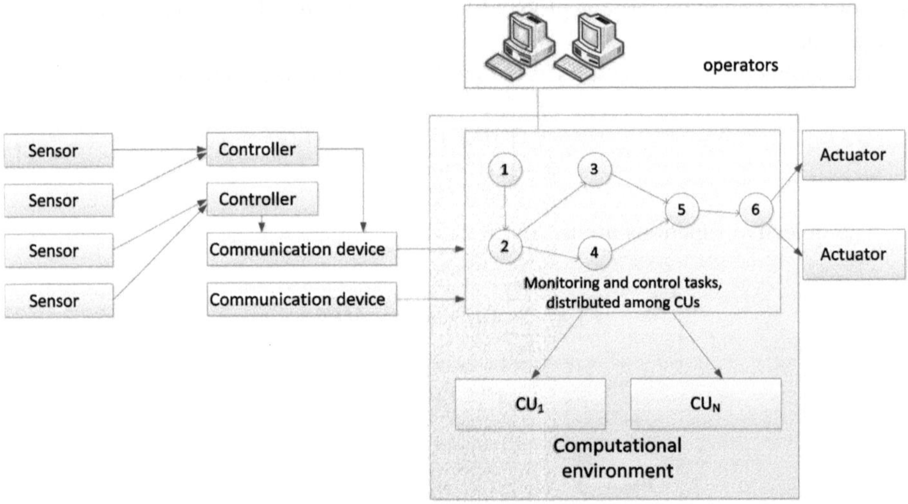

Fig. 4. Traditional ICS architecture

Within the IIoT concept the amount of data to be transferred and processed is huge, and it was the reason to admit the insufficiency of the cloud-computing paradigm [16].

Fog-computing, as was mentioned earlier, presupposes the computational workload shift to the fog-layer, which consists of controllers, communication devices, gateways, etc. According to Cisco White Papers [16], data are aggregated in the fog-nodes, where the decision must be made about where these data have to be sent and processed. Here the difference between terms the "fog-computing" and "edge-computing" must be accentuated: edge-computing presupposes the data to be processed on the endpoint devices (e.g., sensors), and no fog-nodes participate in decision-making about the place of the data processing.

A simplified scheme of fog-computing-based ICS is shown in Fig. 5.

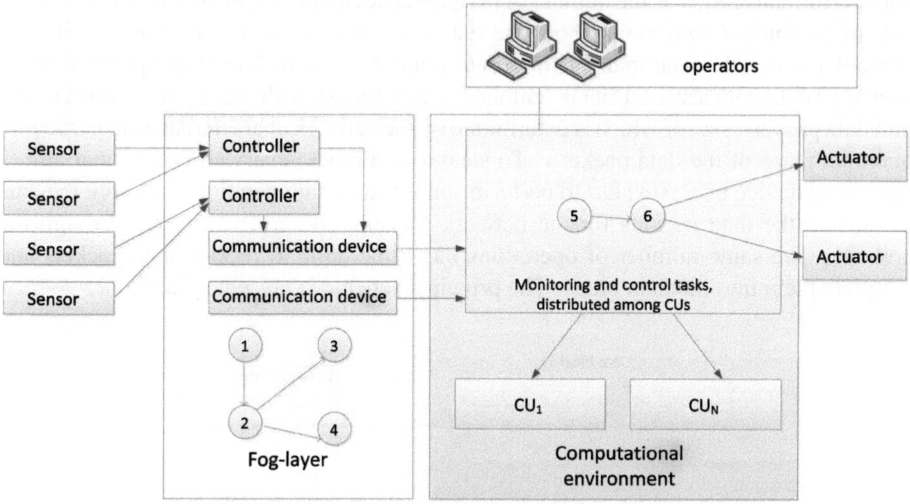

Fig. 5. Fog-computing-based ICS architecture

So, as data processing is shifted to the fog-layer, the workload distribution is shifted too and it can be used in a recovery procedure. Yet as reliability function of a device depends on its computational workload (Eqs. 5–7), some formal models should be built for the further research.

4 Fog-Computing-Based ICS Workload Distribution Model

Firstly, to develop a model of the fog-computing-based ICS workload distribution, we will simplify the basic model of the fog-computing-based ICS (Fig. 6).

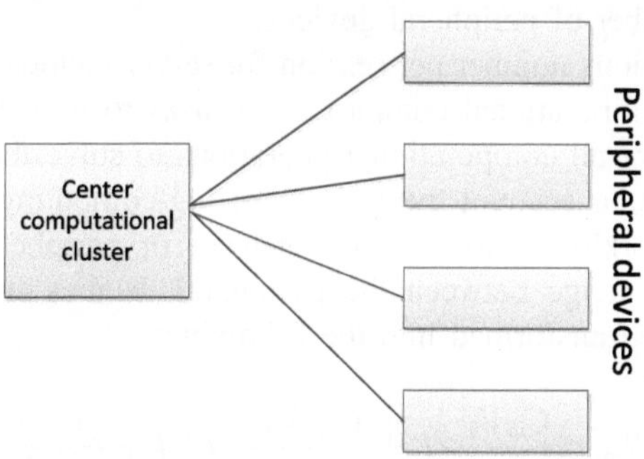

Fig. 6. Simplified fog-computing-based scheme

Assume that the computational complexity of all tasks have to be solved by the ICS is L [ops], f is the fraction of L which is solved by center computational cluster (or "server") with performance R_s, N is the number of peripheral devices with performance R_p. Also we will make some assumptions about the data exchange between server and peripheral devices: through the time span T, which determines the session length, peripheral devices exchange data with server. Data is sent via C_{in} data blocks with size V_{in} and must be split into data packets, size of which is determined by the network characteristics, in particular, maximum size of the data packet v. To send a data block, every computational unit (a peripheral device or server) has to perform some additional operations l_s to split the data block into the data packets for the network transmission and send them through the network. The same number of operations has to be done to receive data packets and integrate them into the data block. The principal scheme is presented in Fig. 7.

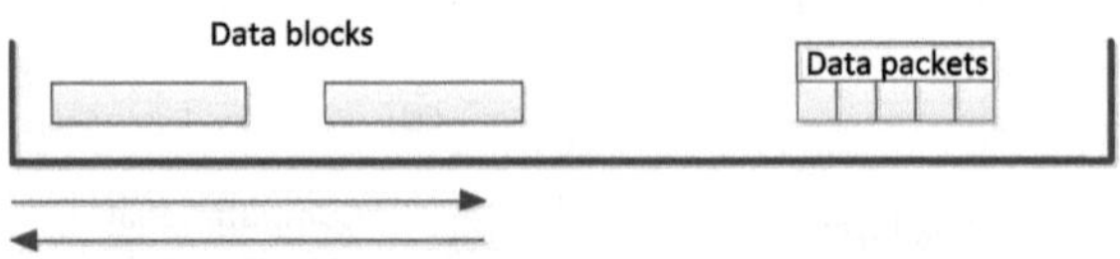

Fig. 7. The scheme of data exchange

The models of central server and peripheral device overall operations number can be synthesized, assuming that the peripheral workload can be distributed uniformly among all peripheral devices:

$$WL_p = \frac{C_{in} V_{in}}{v} \cdot l_s + \frac{C_{out} V_{out}}{v} \cdot l_s + \frac{(1-f)L}{N} \tag{8}$$

$$WL_s = (\frac{C_{in} V_{in}}{v} \cdot l_s + \frac{C_{out} V_{out}}{v} \cdot l_s) \cdot N + f \cdot L \tag{9}$$

The overall operations number per session for peripheral device includes all data transmission operations and the computational complexity of tasks per peripheral device: (1-server workload fraction)*overall computational operations to solve the operational tasks/number of peripheral devices.

The overall operations number per session for server includes all data transmission operations and the computational complexity of tasks to be solved on server: server workload fraction* overall computational operations to solve the operational tasks.

We have to take into account the following: fog-computing concept presupposes the computational workload shift to the fog-layer or peripheral devices and so to decrease the data exchange between the peripheral devices and the server. So, the Eqs. (8 and 9) can be transformed into the following:

$$WL_p = \frac{C_{in} V_{in}}{v} \cdot l_s + \frac{C_{out} V_{out}}{v} \cdot l_s \cdot f + \frac{(1-f)L}{N} \tag{10}$$

$$WL_s = (\frac{C_{in} V_{in}}{v} \cdot l_s + \frac{C_{out} V_{out}}{v} \cdot l_s) \cdot f \cdot N + f \cdot L \tag{11}$$

We assume that with decreasing of server fraction f the overall amount of data transmitted decreases too both for server and for the peripheral devices. The Eqs. (10 and 11) determine the overall operations number, which have to be performed in session time.

Now we have to transform the Eqs. (10 and 11) to estimations of the units workload. Having performances R_s and R_p, session length T and operations number WL_p and WL_s, the workload in fractions for peripheral device and server are:

$$L_p = \frac{WL_p}{T \cdot R_p} \tag{12}$$

$$L_s = \frac{WL_s}{T \cdot R_s} \tag{13}$$

Taking into account Eqs. (5–7), the reliability functions for peripheral device $P_p(t)$ and server $P_s(t)$ can be formulated as follows:

$$P_p(t) = e^{-\lambda_p t} \tag{14}$$

where $\lambda = \lambda_0 \cdot 2^{kL_p/10}$, k is the ratio between the workload and the temperature increase;

$$P_s(t) = e^{-\lambda_s t} \tag{15}$$

where $\lambda = \lambda_0 \cdot 2^{kL_s/10}$.

So, the reliability function equations are got for the center and peripheral computational devices within the fog-computing-based system. The workload distribution between server and peripheral devices impacts the reliability function value, so as the time moment t. As it will be shown in the following section, these two parameters are extremely important for the further task distribution.

5 A Recovery Technique for the Fog-Computing-Based ICS

To begin with, the following statement must be proved.

Assume Z is a CF problem solution, which determines the task distribution among the CUs, and $Z(t_1)$ and $Z(t_2)$ are the same task distributions in time moments t_1 and t_2 accordingly. Assume $P(Z(t_1))$ to be the reliability function value of $Z(t_1)$ and $P(Z(t_2))$ to be the reliability function value of $Z(t_2)$ configuration. Given $t_2 > t_1$ the configuration $Z(t_1)$ is not equal to the configuration $Z(t_2)$ in terms of reliability function value.

Indeed, assume the statement above is not true, and $P(Z(t_1))$ is equal to $P(Z(t_2))$.

Then, $P(Z(t_1)) = e^{-\lambda t_1} = P_z(Z(t_2)) = e^{-\lambda t_2}$.

It is possible only if $t_1 = t_2$ in conditions of the same task distribution, but contradicts with condition $t_1 < t_2$.Q.E.D.

Here the important conclusion must be made: if $t_1 < t_2$ the equality of $P(Z(t_1))$ and $P(Z(t_2))$ is reachable by the workload correction.

This conclusion is extremely useful for the fog-computing-based ICS because of possibility to shift the workload from the central computational server to the fog-layer (or to the end-point devices). So, the configurations must be formed with involving of additional informative parameters, like L_p, L_s (Eqs. 8 and 9), server fraction f and time moment of reconfiguration t_0.

Within this paper the recovery technique for the fog-computing-based ICSs is presented. This technique enhances the reconfiguration procedure, adopting it to the fog-computing concept.

So, in case of reconfiguration in time moment t_0 the following steps have to be done:

1. for the CE device:
   - estimation of L_s, which gives the desirable level of $P_s(t_0)$;
   - estimation of L_s', which is a result of reconfiguration, paying attention to the faulted node;
   - if $L_s' > L_s$, the set of tasks must be formed, which gives L_s. Others are transferred to the fog-layer; else tasks are distributed among the CUs according the heuristic rule of $P_s(t)$ maximization;
   - if some tasks are received from the fog-layer, they are distributed among the CUs according the heuristic rule of $P_s(t)$ maximization.
2. For the fog-layer devices:
   - estimation of L_p, which gives the desirable level of $P_p(t_0)$;
   - estimation of L_p', which is a result of reconfiguration, paying attention to the faulted node;
   - if $L_p' > L_p$, the set of tasks must be formed, which gives L_p. Others are transferred to the fog-layer; else tasks are distributed among the fog-nodes according the heuristic rule of $P_p(t)$ maximization;
   - if some tasks are received from the fog-layer, they are distributed among the fog-nodes according the heuristic rule of $P_p(t)$ maximization.

6 Simulation Results

We simulated the reliability function values for the following conditions: (t_0 = 1, 200,400,1000; $R_s = 10000$; $R_p = 1000$; $L = 107$; $v = 200$; $l_s = 50$; $C_{in} = C_{out} = 100$; $V_{in} = V_{out} = 2000$). Server fraction changes from 0 to 1.

The following figures show the $P_s(t_0)$ dependency on the server workload fraction f.

It is seen that with the reconfiguration time moment t_0 the server fraction f must be decreased to keep the reliability function value, for example, more than 0,8.

As to the reliability function of the peripheral device, with the same simulation conditions the reliability function changes with the workload distribution f and t_0 (Fig. 9).

Comparing the plots in Figs. 8 and 9 we can see that the workload distribution in the fog-computing-based ICS is actually the interplay between the reliability of the central server and the ones of the peripheral computational devices. For example, for t0 = 200 only 0.1 of general operations amount can be processed on server to keep the reliability function value more than 0.8. Such workload distribution is rather acceptable for the peripheral devices (reliability function value >0.8), yet the situation changes when t0 = 400: server fraction must be <0.1 to keep the reliability function value > 0.8, while for the peripheral device server fraction must be >0.2 to keep the same reliability function value.

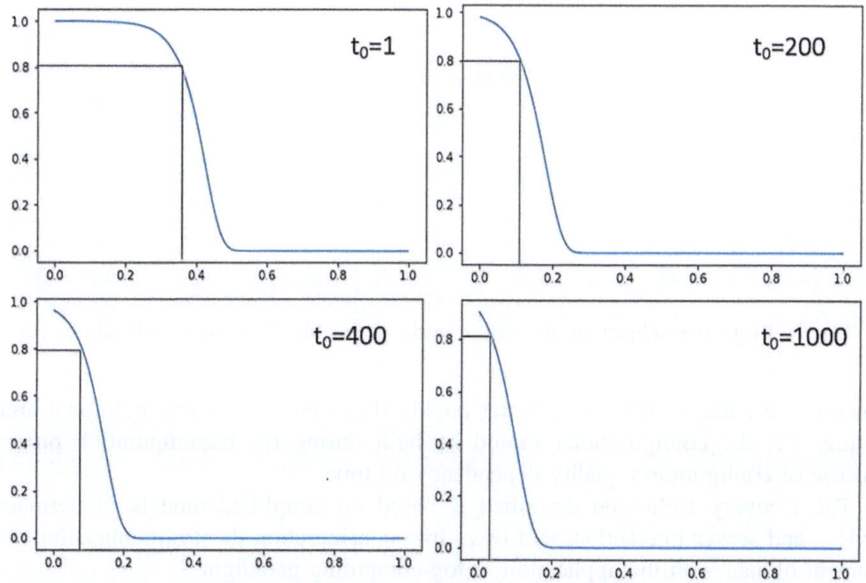

Fig. 8. $P_s(t_0)$ dependency on the server workload fraction *f* for the central devices.

These simulated results confirm our statement about the need to form configurations for the fog-computing-based ICSs considering the reconfiguration time t_0. This leads to one more important conclusion: in fog-computing-based ICSs it is pointless to develop all possible configurations on the system design stage. Prepared configuration can be good enough, when the reconfiguration time moment occurs at the beginning of system functioning time, and the same configuration can give unacceptable result when time has passed, while the fog-computing concept gives the promising possibility to shift some workload to the fog-layer and to get configuration of a higher quality at least for the CE devices.

7 Conclusions

In this paper the new fog-computing-based ICS recovery technique is presented. The fog-computing paradigm brings the possibility to shift the workload from the server, or central computational environment, to the fog-layer of the system. This allows to

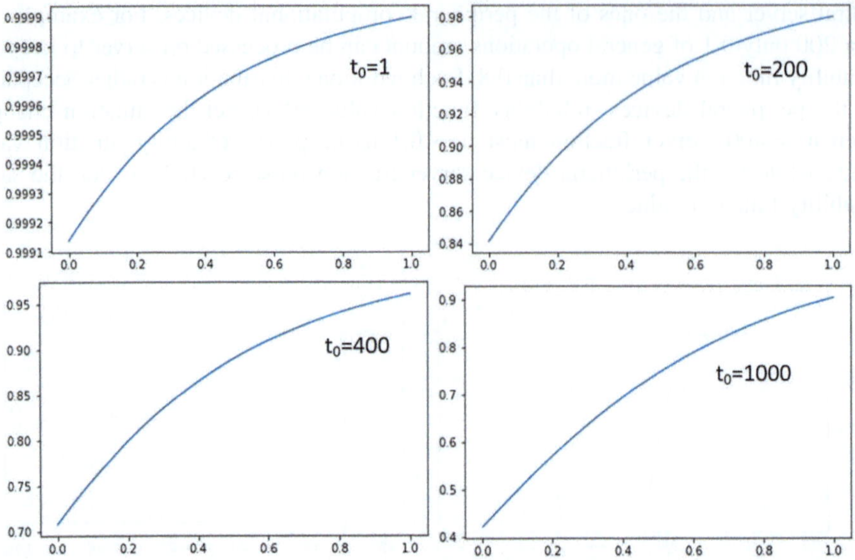

Fig. 9. Ps(t0) dependency on the server workload fraction f for the peripheral devices

generate the configurations of a better quality then ones within the traditional architecture. Yet the configurations should be built during the reconfiguration process, because of configuration quality dependency on time.

The recovery technique described is based on simplified models of peripheral devices and server interaction, and takes into consideration the communication environment offload with the application of fog-computing paradigm.

Then, forming the task distribution, we consider the time of recovery occurred t_0 and the fraction f of general workload to be processed on a server. Considering these parameters, the configurations can be built on the basis of the recovery procedure.

Simulation results prove the need to consider the time moment of reconfiguration to form the configurations of a high quality.

Acknowledgement. The paper has been prepared within the GZ SSC RAS N GR project 01201354238 and RFBR projects 17-08-01605 A, 18-29-03229 mk, 18-05-80092 and SO SSC RAS N GR project 01201354238, the program of RAS presidium fundamental research I.30 "Theory and technologies of multilevel decentralized group management in conflict and cooperation conditions" (project № AAAA-A18-118011290099-9).

References

1. Rüßmann, M., et al.: Industry 4.0. The Boston Consulting Group, p. 20 (2015). https://doi.org/10.1007/s12599-014-0334-4
2. Lasi, H., et al.: Industry 4.0. Bus. Inf. Syst. Eng. **6**(4), 239–242 (2014). https://doi.org/10.1007/s12599-014-0334-4

3. Aazam, M., Huh, E.N.: E-HAMC: leveraging fog computing for emergency alert service. In: 2015 IEEE International Conference on Pervasive Computing and Communication Workshops, PerCom Workshops 2015, pp. 518–523 (2015). https://doi.org/10.1109/percomw.2015.7134091
4. Al Faruque, M.A., Vatanparvar, K.: Energy management-as-a-service over fog computing platform. IEEE Internet Things J. **3**(2), 161–169 (2016). https://doi.org/10.1109/JIOT.2015.2471260
5. Chen, R.Y.: Fog computing-based intelligent inference performance evaluation system integrated internet of thing in food cold chain. In: 2015 12th International Conference on Fuzzy Systems and Knowledge Discovery, FSKD 2015, pp. 879–886 (2016). https://doi.org/10.1109/fskd.2015.7382059
6. Inaltekin, H., Gorlatova, M., Mung, C.: Virtualized control over fog: interplay between reliability and latency. CoRR, abs/1712.0 (2017). https://arxiv.org/pdf/1712.00100.pdf
7. Avizienis, A., Laprie, J.C., Randell, B.: Fundamental concepts of dependability. Technical report seriesuniversity of newcastle upon tyne computing science, 1145, pp. 7–12 (2001). 10.1124.6074
8. Klimenko, A.B., Ivanov, D., Melnik, E.V.: The configuration generation problem for the informational and control systems with the performance redundancy. In: 2016 2nd International Conference on Industrial Engineering, Applications and Manufacturing, ICIEAM 2016 – Proceedings (2016). https://doi.org/10.1109/icieam.2016.7910901
9. Korovin, I., Melnik, E., Klimenko, A.: A recovery method for the robotic decentralized control system with performance redundancy. In: Lecture Notes in Computer Science, Including Subseries Lecture Notes in Artificial Intelligence and Lecture Notes in Bioinformatics (2016). https://doi.org/10.1007/978-3-319-43955-6_2
10. Melnik, E., Korobkin, V., Klimenko, A.: System reconfiguration using multiagent cooperative principles. Adv. Intell. Syst. Comput. (2016). https://doi.org/10.1007/978-3-319-33816-3_38
11. Melnik, E., Klimenko, A., Korobkin, V.: Reconfigurable distributed information and control system multiagent management approach. Adv. Intell. Syst. Comput. (2018). https://doi.org/10.1007/978-3-319-68324-9_10
12. Melnik, E.V., et al.: A novel approach to fault tolerant information and control system design. In: 2016 5th International Conference on Informatics, Electronics and Vision, ICIEV 2016 (2016). https://doi.org/10.1109/iciev.2016.7760182
13. Klimenko, A.B., Klimenko, V.V., Melnik, E.V.: The parallel simulated annealing-based reconfiguration speedup algorithm for the real time distributed control system fault-tolerance providing. In: 9th International Conference on Application of Information and Communication Technologies, AICT 2015 – Proceedings (2015). https://doi.org/10.1109/icaict.2015.7338562
14. Practice, T., Fu, M.C.: Optimization for simulation: theory vs. practice. INFORMS J. Comput. **14**(3), 79–84 (2002). https://doi.org/10.1287/educ.1080.0050
15. Strogonov, A.: Dolgovechnost Integralnih schem I proizvodstvenniye metody ee prognozirovaniya. ChipNews **6**, 44–49 (2002)
16. CISCO: Fog Computing and the Internet of Things: Extend the Cloud to Where the Things Are (2015). https://www.cisco.com/c/dam/en_us/solutions/trends/iot/docs/computing-overview.pdf

An Examination of Message Ferry Movement for Delay Tolerant Networks

Takahiro Koita(✉) and Minami Yoneda

Doshisha University, Kyoto, Japan
tkoita@gmail.com

Abstract. After a disaster has occurred, it is extremely difficult to confirm the safety of victims in an environment in which the communication infrastructure has been destroyed. For this reason, Delay Tolerant Networking (DTN), which makes communication possible even in environments which do not have adequate communication infrastructure, is being actively researched. Moreover, a relaying technology called message ferrying, which utilizes Unmanned Aerial Vehicles (UAVs), has been attracting attention as a means of mediating between sites that are not connected for communication during a disaster, thus facilitating the exchange of data between them. In order to improve the performance of message ferrying, the efficient movement of UAVs is necessary to ensure that safety information of more victims can be recovered in a limited time during disaster. In this study, evaluation experiments utilizing actual UAVs are conducted, UAV movement conditions are examined in light of the results of the experiments, and the ideal movement conditions of UAVs are described, with the overall objective of clarifying the necessary conditions for efficient movement of UAVs.

Keywords: Delay tolerant networks · Ad-hoc communication
Unmanned aerial vehicles

1 Introduction

The confirmation of individuals' safety during disasters has been raised as a current problem in information communications. For example, in the aftermath of large-scale earthquakes, in affected areas there are often many sites that, from a networking perspective, become isolated as a result of the blocking of transportation routes by landslides and the destruction of communication infrastructure, making communication with those outside these areas difficult. In the Great East Japan Earthquake, the communication infrastructure sustained an unprecedented level of catastrophic damage, with power outages, collapsing and flooding of mobile phone base stations and communication buildings, disconnection of underground cables, collapsing of power poles, etc. As a result, at the peak of the damage, there are communication carriers that lost two-thirds of their capabilities in power outages and the like [1]. In addition, up to 95% of the calls on mobile phones are restricted. Information about the safety of individuals is necessary in order to provide prompt relief to disaster victims, but collecting and sharing information in such an environment is extremely difficult. There is, for this

R. Silhavy et al. (Eds.): CoMeSySo 2018, AISC 860, pp. 228–239, 2019.
https://doi.org/10.1007/978-3-030-00184-1_21

reason, a need for communication methods that can be utilized in environments in which the communication infrastructure has been destroyed in a disaster.

Recently, Delay Tolerant Networking (DTN) has been actively researched as an information communication technology to use during disasters. DTN is an extended network architecture in order to provide end-to-end information even when a communication environment is not available, which is not possible with conventional TCP/IP technology. Moreover, a routing technology called message ferrying has been attracting attention as a means of mediating between sites that are not connected for communication during a disaster, thus facilitating the exchange of data between them. Message ferrying, in which nodes called ferry nodes relay messages, is able to increase the probability that communication can be achieved even if the nodes move around locally.

In order to improve the performance of message ferrying, it is important to make the movement of Unmanned Aerial Vehicles (UAVs) that utilize ferry nodes efficient. Due to UAV battery limitations, ferry nodes must move around in order to exchange messages with a larger number of other nodes in a short time. Currently, not only DTN, but also networks utilizing UAVs are being actively researched. However, the movement efficiency of UAVs is still an issue [2]. It is thus necessary to conceive of movement methods that facilitate communication with a large number of terminals in a limited amount of time.

This study considers the efficient movement of UAVs with the purpose of improving message ferrying performance. Efficient movement refers to the ability to communicate with more terminals in a limited flying time, while also maintaining reasonable reliability. Currently, research into the movement of UAVs is being actively carried out, but obstacles in actual environments are not sufficiently taken into account, which makes it difficult for existing research results to be reflected as is. For this reason, it is necessary to investigate the ideal UAV movement conditions with a focus on communication between UAVs and terminals in actual environments. An objective of this study is to clarify, after considering the operations necessary for UAVs and terminals to communicate, how close UAVs should get to terminals, how fast they should move, and how long they should remain. By making the movement of UAVs more efficient, the performance of message ferrying can be improved, which enables information to be communicated during disasters.

2 Delay Tolerant Network

2.1 Overview

With the current advances in the expansion and speeding up of mobile communication networks, TCP/IP is unable to sufficiently support the diversification of ad hoc communication. DTN technology has emerged in order to extend conventional TCP/IP technology and to reevaluate the capabilities of relay nodes and end nodes. DTN designates networking that withstands a variety of obstacles, including delays, variability, and discontinuities, but here the focus will be on a network architecture that attempts to more efficiently utilize insufficient network resources, improve performance, and decrease actual costs by sacrificing real-time characteristics.

One of the most significant features of DTN is that, since moving nodes relay messages to one another, communication can be carried out even in environments in which the communication infrastructure is inadequate, such as during disasters. At present, the epidemic routing method is widely used as a DTN routing method [3]. An overview of this method is shown in Fig. 1. As shown in the figure, in this method, when a terminal that holds a message becomes able to communicate with other terminals, it replicates the message in its possession and sends it to all of the terminals with which it can communicate. Provided that there are sufficient network resources, this method exhibits excellent performance and is capable of sending messages to terminals over long distances.

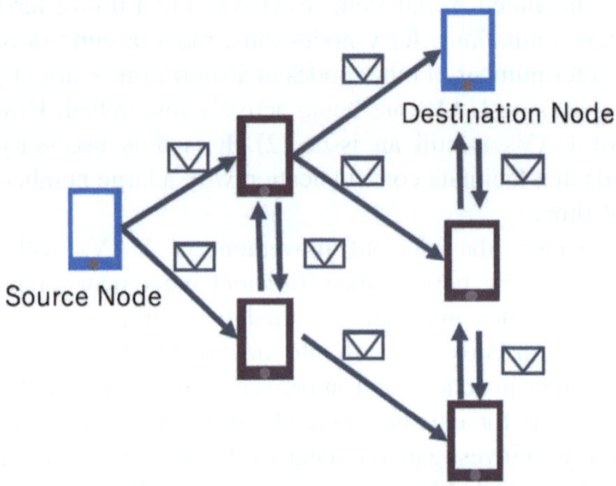

Fig. 1. Overview of epidemic routing method

In order to improve DTN network performance, it is necessary for messages to be relayed efficiently between terminals. Since it is impossible to know in advance the timing at which communication with other terminals will become possible, it is necessary to determine the timing of the interruption and reopening of communication and the possibility of messages from sender nodes to arrive at receiver nodes. In addition, it can be expected that in a DTN environment, the connectivity between nodes may be intermittent, or the network resources that can be used may become extremely scarce. In light of this, it is important to effectively take advantage of opportunities for nodes to become able to communicate with one another in order to improve communication performance.

2.2 Message Ferrying

During disasters, victims will in most cases gather in shelters and other such facilities, and the range of movement of these victims' computers and terminals will not expand significantly. Moreover, there will be areas in which, as a result of the collapse of roads and buildings, the transportation network will be fragmented, thus limiting the range of movement. If, as a result, nodes move around locally or decrease in number, then, with the epidemic routing method introduced in previous section, communication efficiency will degrade, and the probability that messages will fail to arrive at receiver nodes will increase. In addition, it can also be pointed out that since copies of messages are sent to all terminals with which communication is possible, network resources are significantly consumed.

Recently, message ferrying has been attracting attention as a means of enabling the transmission of messages even if node movement is local or if nodes are few in number. With message ferrying, in order to send and receive messages among terminals that cannot communicate directly, such as those separated by long distances, ferry nodes that move along paths prescribed in advance serve to transport messages.

For example, as shown in Fig. 2, when a ferry node enters the communication range, a terminal sends messages to the ferry node. The ferry node that has received these messages then moves to the area in which the destination node is located, and sends the message to a terminal with which it is able to communicate. Subsequently, terminals within that area then come into contact with one another, and ultimately the message is transported to the destination node.

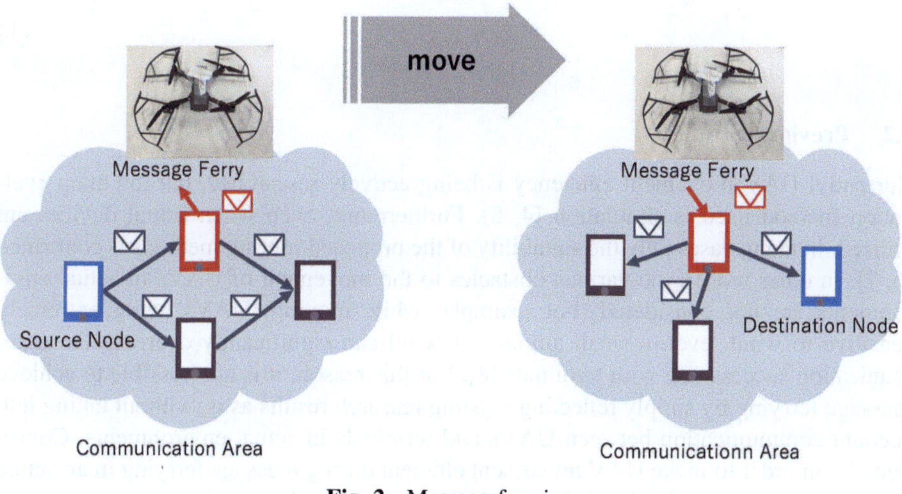

Fig. 2. Message ferrying

In this way, communication between areas that are isolated from each other network-wise is enabled by ferry nodes that serve as intermediaries in transporting messages. Moreover, the number of packets that are copied in order to relay messages can be decreased, thus serving to reduce network resource consumption.

In light of this, we consider message ferrying to be the most effective routing method for communicating information in areas that have been isolated during the type of disasters assumed here. In this study, message ferrying is implemented using UAVs, with the purpose of improving information communication performance during disasters.

3 Problem

In this chapter, existing research and current issues related to UAV movement efficiency, along with the approach taken by this research, will be described.

3.1 UAV Movement Efficiency

First, let's define movement efficiency. When confirming the safety of individuals during disasters, in order to utilize message ferrying, all terminals must be able to send and receive messages. For this reason, the ideal UAV movement implies covering a wider range while communicating reliably with terminals with which the UAV is in contact. Meanwhile, due to limitations of the UAV battery, the time spent sending and receiving messages and moving between regions is short. As a result, from the perspective of a UAV involved in message ferrying, it is necessary to move in such a way that communication with more terminals is possible in a short time, while also maintaining at least a certain level of reliability. In this study, movement efficiency E is defined by the Eq. (1), with P being the reliability (i.e., the communication success rate), T (seconds) being the moving time, and N (units) being the number of units capable of communication. Movement efficiency can be increased by increasing the values for N and P, and by decreasing the value for T.

$$E = N \times P \div T \tag{1}$$

3.2 Previous Studies

Currently, UAV movement efficiency is being actively researched, but the main evaluation method used is simulation [4, 5]. Furthermore, even when actual devices are utilized, in many cases only the suitability of the proposed routing methods is confirmed [6, 7]. In other words, the various obstacles to the movement of UAVs in actual environments are not considered. For example, while moving, UAVs being extremely sensitive to wind, even a small amount of wind can significantly decrease the communication success rate with terminals [8]. For this reason, it is not possible to achieve message ferrying by simply reflecting existing research results as is, without taking into account communication between UAVs and terminals in actual environments. Consequently, in order to make UAV movement efficient during message ferrying in an actual environment, communication between UAVs and terminals is an issue to consider.

3.3 Our Approach

Focusing on the issue described above, we conducted experiments simulating the actual environment and evaluated the conditions for communication between UAVs and terminals. In this study, the operations necessary for communication between a UAV and a single terminal are broadly divided into the following two types.

(1) The UAV moves toward or away from the terminal;
(2) The UAV exchanges data with the terminal.

For each of these operations, we will concretely describe the communication conditions that increase the efficiency of UAV movement (Fig. 3). Regarding (1), it is necessary to find out how slowly a UAV must move toward (or away from) a terminal, as well as how close to (or far from) the terminal the UAV must move, in order for the UAV to reliably send and receive data from the terminal. Next, regarding (2), it is necessary to find out how long the UAV should stay within the communication range of the terminal in order for the UAV to reliably send and receive data from the terminal. In light of this, UAV moving speed, distance from terminals, and staying time within the communication range are evaluated in the next chapter.

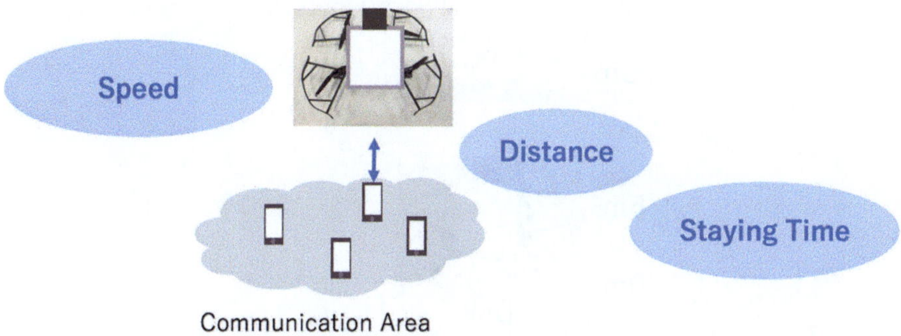

Fig. 3. Communication conditions

By determining the conditions required for communication between UAVs and terminals, the ideal UAV movement method in actual environments will become clear, and this can be useful in making UAV movement efficient.

4 Evaluation Experiments

We will evaluate the appropriate UAV moving speed, distance from the terminals, and staying time within the communication range for message ferrying, and will describe and analyze the observed results.

4.1 Evaluation Experiment Related to Distance (Experiment 1)

In this experiment, we evaluate the effects of the distance between the UAV and the terminal on the communication success rate. Relative to a 0 m location, the UAV and the terminal are placed at heights of 5 m and 25 m, respectively, along a straight line.

The location of the UAV is then varied along the same straight line, and the communication success rate is measured (Fig. 4). The size of the transmission data from the terminal is 100 bytes per second. In addition, the UAV is operated in the two ways described below, and the variations due to the difference in operation are observed.

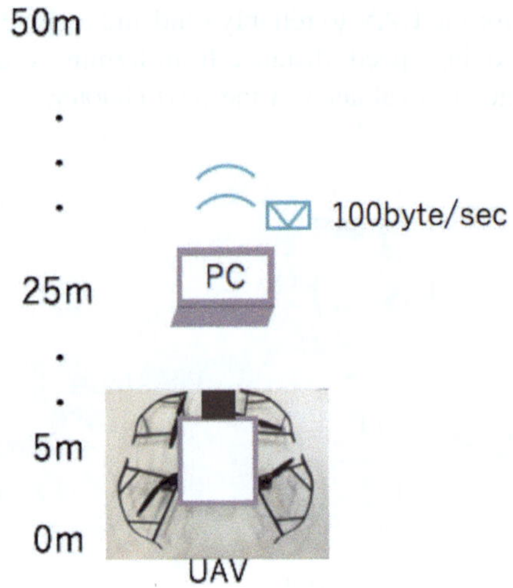

Fig. 4. Positional relationship between UAV and terminal during experiment

(1) The UAV stops every 5 m and receives data from the terminal in this resting state.
(2) The UAV receives data from the terminal as he keeps moving at a speed of 1 m/s.

(1) and (2) are each measured three times. The average reception signal strength (RSSI) and communication success rate at each UAV position are shown in Figs. 5 and 6, respectively.

According to the results, the communication success rate improves as the distance between the UAV and the terminal decrease. On the whole, the communication success rate decreases more in (2), with the UAV moving at a constant speed, than in (1), with the UAV communicating while at rest. This confirms that the UAV's vibration resulting from its movement affects the communication success rate. In addition, in the case of the UAV moving with resting as in (1), when comparing the UAV being at 20–25 m moving closer to the terminal, with the UAV being at 25–30 m moving farther from the terminal, the communication success rate is higher at the 20–25 m position. We are thus able to determine that the orientation of the wireless LAN has an effect on the communication success rate. Based on the fact that, in the environment of this experiment, a communication success rate of 80% or higher is achieved in the resting state in (1), and 60% or higher in the moving state in (2), we consider that stable communication is possible when the distance between the UAV and terminal is 5 m or less. Consequently, the communication range for ferry nodes is defined to be within 5 m for these experiments.

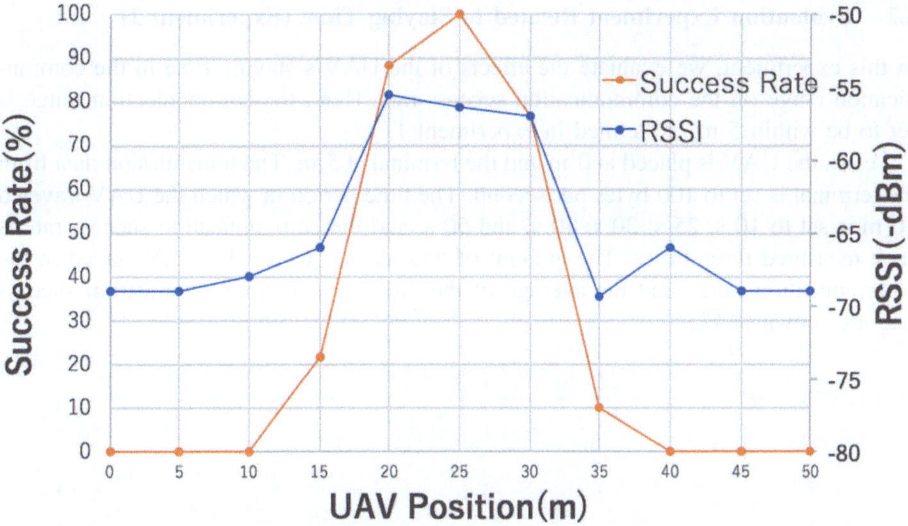

Fig. 5. UAV position and communication success rate (Rest at 5 m intervals)

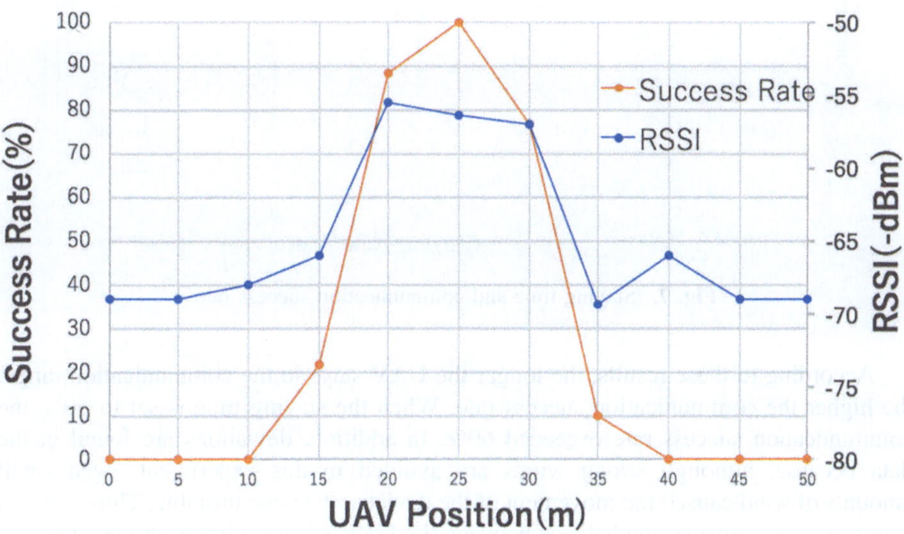

Fig. 6. UAV position and communication success rate (moving at constant speed)

4.2 Evaluation Experiment Related to Staying Time (Experiment 2)

In this experiment, we evaluate the effects of the UAV's staying time in the communication range on the communication success rate. Here, the communication range is set to be within 5 m, as defined in experiment 1.

First, the UAV is placed at 0 m and the terminal at 5 m. The transmission data from the terminal is set to 100 bytes per second. The time period in which the UAV traveled 10 m is set to 10 s, 25 s, 30 s, 40 s, and 50 s, and the communication success rate is each measured three times. The amount of time during which the UAV stayed in the communication range and the average of the three measured communication success rate are shown in Fig. 7.

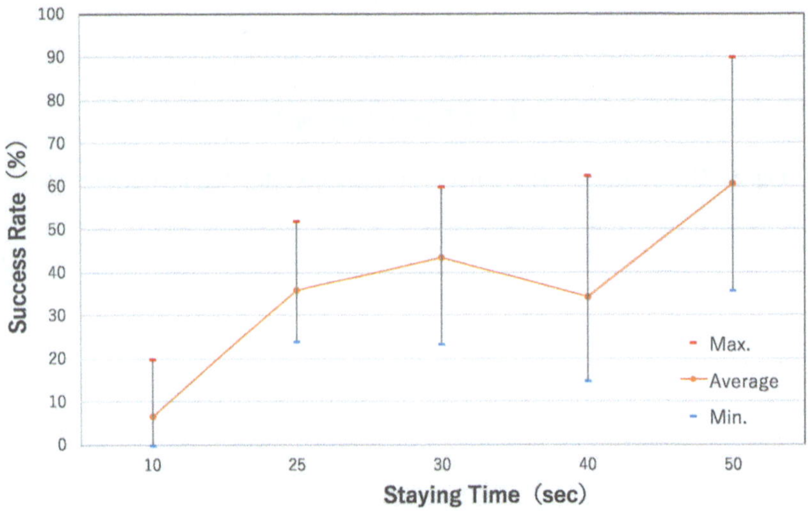

Fig. 7. Staying time and communication success rate

According to these results, the longer the UAV stays in the communication range, the higher the communication success rate. When the staying time is set to 50 s, the communication success rate exceeded 60%. In addition, deviations are found in the data because, although strong winds are avoided in this experiment, even small amounts of wind caused the movement of the device to become unstable. Thus, while it is necessary to consider the balance between the UAV's battery limits and the need for real-time communication, the UAV's staying time within the communication range must be of at least 50 s.

4.3 Evaluation Experiment Related to Moving Speed (Experiment 3)

In this experiment, we evaluate the effects of the UAV's staying time in the communication range on the communication success rate. As in experiment 2, the communication range is set to be within 5 m, as defined in Sect. 4.3.

As in experiment 2, the UAV is placed at 0 m and the terminal at 5 m, and as in experiments 1 and 2, the transmission data from the terminal is set to 100 bytes per second. Using the two methods below, the UAV is moved 10 m over a total time period of 50 s, measurements are each taken three times, and the relationship between UAV moving speed and communication success rate is evaluated.

(1) Constant speed mode, with the UAV moving 50 m in 50 s
(2) Variable speed mode, with the UAV moving to the 23 m position in 2 s, resting for 46 s, and moving to the 50 m position in the remaining 2 s

The results of the experiment are shown in Fig. 8. According to the results, the communication success rate is higher for constant speed mode, reaching a level of 60%. We consider this to be due to the fact that, in variable speed mode, the body of the device vibrates during speed changes, causing its movement to become unstable. Thus, the UAV must have a constant moving speed, with a stable movement.

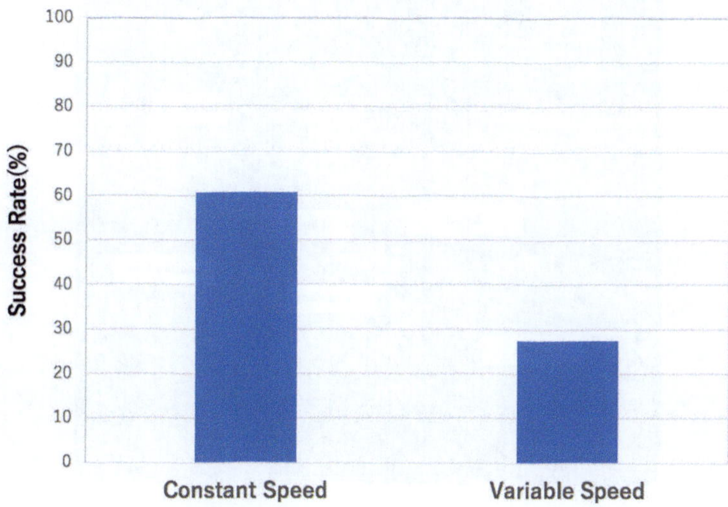

Fig. 8. Moving speed and communication success rate

4.4 Discussion

Based on the above results, a summary of the UAV moving speed, distance from terminals, and staying time within the communication range that increase the efficiency of the UAV movement is shown in Table 1. By satisfying the conditions in Table 1, a communication success rate of at least 60% is achieved in the environment in which experiments 1, 2, and 3 are performed. In addition, according to the results of experiments 1, 2 and 3, if the communication range of the UAV is a circle with a radius of 5 m, the staying time within the communication range is 50 s, and the moving speed of the UAV is constant, then the ideal moving speed of the UAV in the environment of these experiments is 0.2 m/s.

Table 1. Ideal movement conditions of UAV

Distance from terminals	Within 5 m
Staying time within communication range	At least 50 s
Moving speed	0.2 m/sec

With these results, we will now consider the safety confirmation of individuals on the Kyotanabe campus of Doshisha University. Let's assume that approximately 9,000 students are gathered in the main buildings on the campus shown in Fig. 9, and suppose that we collect the safety information emitted by the terminals of these students. Since, in existing research, simply circulating among the access points has been considered to be the most efficient moving route [5], we assumed that the UAV would move around the route in Fig. 9.

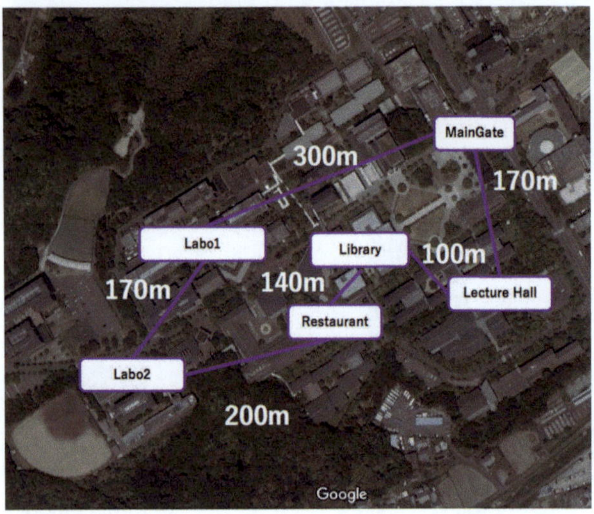

Fig. 9. Distance between major buildings of Doshisha University

One trip around the route in Fig. 9 is 1,080 m. The flying time of the UAV used in this experiment being 27 min, and the moving speed of the UAV 0.2 m/s, the distance that a single UAV can travel is 324 m. Consequently, calculation results indicate that in order to travel 1,080 m, at least 4 UAVs are necessary, and that when 4 UAVs are used, the time necessary to make one round-trip around the route will be 1323 s (22 min, 3 s). This is an appropriate outcome for confirming individuals' safety, allowing us to consider message ferrying feasible.

Moreover, when applying the above estimation results to Eq. 1, as defined in Sect. 3.1, the movement efficiency E can now be calculated with the number of terminals N = 9000, the communication success rate P = 0.6, and the moving time T = 1323 and UAV moving routes can be compared numerically. In this way, the effect of various obstacles in an actual environment can be ascertained, and the ideal movement conditions under which UAVs can communicate with terminals can be found, which can be useful when investigating efficient UAV moving routes.

5 Conclusion

The objective of this study is to make UAV's movement more efficient to achieve DTN with message ferrying during disasters. To accomplish this, we conducted experiments simulating the actual environment to evaluate the necessary conditions for the communication between UAVs and terminals. As a result, we showed that the efficiency of the UAV movement can be improved by defining the conditions of the distance between UAVs and terminals, the UAV's staying time within the communication range, and UAV movement time. In addition, the device's vibration and noise have a significant effect on the communication success rate, and pose serious problems when trying to improve message ferrying performance. Considering the above conditions, we believe that message ferrying that provides fairly stable information communication can be achieved.

References

1. Murakami, K.: The great East Japan earthquake: development of a safety information system, and the problems therein: toward a future discussion. NHK Publishing, Broadcast Research and Study, June Issue, pp. 18–33 (2011)
2. Kim, J., Origasa, K., Kanda, K.: Environment data monitoring and sensor module development using autonomous flight of unmanned aerial vehicles. Agric. Inf. Res. **26**(2), 34–43 (2017)
3. Tsuru, M., Uchida, M., Takine, T., Nagata, A., Matsuda, T., Minami, H., Yamamura, S.: DTN technology: status and prospects. Commun. Soc. Mag. 16, 57–68 (2011)
4. Jawhar, I., Mohamed, N., Al-Jaroodi, J., Agrawal, D.P., Zhang, S.: Communication and networking of UAV-based systems: classification and associated architectures. J. Netw. Comput. Appl. **84**, 93–108 (2017)
5. Tsuji, H., Minami, H.: Unmanned Aerial Vehicle (UAV) Flight Plan Law for UAV-Utilizing MANET. Institute of Electronics, Information and Communication Engineering, pp. 337–342 (2014)
6. Bellur, B.R., Lewis, M.G., Templin, F.L.: An Ad-hoc network for teams of autonomous vehicles. The United States Office of Naval Research
7. Danilov, C., Henderson, T.R., Goff, T., Kim, J.H., Macker, J., Weston, J., Neogi, N., Ortiz, A., Uhlig, D.: Experiment and field demonstration of a 802.11-based ground-UAV mobile ad-hoc network. In: Military Communications Conference, IEEE (2009)
8. Uchida, N., Kawamura, N., Ishida, T., Shibata, Y.: Proposal of autonomous flight wireless nodes with delay tolerant networks for disaster use. In: Proceedings of the 8th International Conference on Innovative Mobile and Internet Services in Ubiquitous, pp. 146–151 (2014)
9. Mavic Pro – dji. https://www.dji.com/mavic
10. WLI-UC-GNME-BUFFALO. http://buffalo.jp/product/wireless-lan/client/wli-uc-gnme/

Comparison of ERP Systems with Blockchain Platform

Boris Sokolov[1,2] and Anton Kolosov[3](✉)

[1] St. Petersburg Institute for Informatics and Automation of the Russian Academy of Sciences, St. Petersburg, Russia
sokolov_boris@inbox.ru
[2] ITMO University, St. Petersburg, Russia
international@mail.ifmo.ru
[3] National Research University Higher School of Economics St. Petersburg Branch, St. Petersburg, Russia
akolosov@hse.ru

Abstract. This research paper is devoted to a comparison of ERP systems and Blockchain platform. Despite it is too early talk about the widespread of blockchain platform, people want to know how this technology can be used for their works with compare to standard program product. One of this product is ERP system. Today experts have a different opinion about applicability and opportunities of blockchain technology. This research paper presents a definition of the main terms, benefits, the sphere of applicability, advantages and disadvantages of ERP systems and blockchain technology. In addition, this survey describes how ERP systems and Blockchain platform can be work in the different area. Results show the ambiguity of allying blockchain platform in the current moment. ERP systems also have some features reflected in the conclusions of the research paper.

Keywords: Blockchain · ERP system · Smart-contract

1 Introduction

Now each global company has software for automatization and optimization of the business process. Different programs can automatize different process. It can either be a single process or group of processes. For automatization marketing and sales procedure, it can be CRM-systems, for supplier relationship management – SRM-systems. However, if companies need to automatize many different processes they with a higher degree of probability will use ERP systems. ERP (enterprise resource planning) systems are the most applicable business-management software [7]. As enterprise data and processes based into ERP systems, businesses are able to align separate departments and improve workflow, resulting in significant bottom-line savings [9]. ERP systems provide visibility, analytics, and efficiency across every aspect of a business. Using the latest technologies, ERP systems facilitate the flow of real-time information across departments, so businesses can make data-driven decisions and manage performance. ERP systems can automate core of business processes and help ensure regulatory

R. Silhavy et al. (Eds.): CoMeSySo 2018, AISC 860, pp. 240–247, 2019.
https://doi.org/10.1007/978-3-030-00184-1_22

compliance, reduce risk, fast-track reporting – and so much more [10]. ERP systems include automatization of management, production, sales, procurement and a lot of another process. Eventually, ERP systems have many advantages but also have some problems include excessive cost, a lengthy period of creating documents, the operation of the systems need lawyer support, documents need a special signature, staff needs local space of work for have access in the system and other. The most of researchers describe how can correct these problems within ERP systems (Hamilton, Samara, O'Sullivan, Rico, Goldensohn and others) but they do not describe can be possible use different programs or not.

One of the alternative technologies which can be used for automatization business process is blockchain. This technology allows automating the process connect with a digital transaction of information in the areas of government, health, science, insurance, logistics and some other direction [5]. Blockchain technology allows for decentralization digital process. It is essentially a database managed by a peer-to-peer network of computers. Blockchain facilitates the transfer and payment of digital money without the need for a trusted entity, such as a bank. The data it holds is public and immutable. Blockchain can ensure that the terms of programmable autonomous contracts, known as smart contracts, are met. It can be used to secure identity and many other situations where transparency and security are lacking. Many authors describe blockchain as an innovation (Kelly, Diedrich, Alferes, Bertossi, Governatori, Fodor and others) but they do not describe how to apply it for ERP systems and they do not know possible it or not possible.

The research question for this research paper is "can companies use the blockchain platform instead of ERP systems or not". The thesis statement based on the assumption that blockchain platform is an alternative to classical ERP systems and in the future company will use it in the area of ERP. The research focuses on key features of ERP systems. Purposes of research include analyzing the area of application ERP systems and blockchain platforms, their advantages, disadvantages, and features and comparing ERP systems and blockchain platforms how an alternative to each other.

The structure of the research includes literature review, methods, and results of the research.

2 ERP Systems and Blockchain Technology. What Is It and How It Can Be Used?

2.1 Definition of the Main Terms of ERP Systems and Blockchain Technology

To answer the main research question, it is necessary to consider the concepts of the research. The acronym ERP stands for enterprise resource planning. It refers to the systems and software packages used by organizations to manage day-to-day business activities, such as accounting, procurement, project management, and manufacturing. ERP systems tie together and define a plethora of business processes and enable the flow of data between them. By collecting an organizations shared transactional data from multiple sources, ERP systems eliminate data duplication and provide data integrity with a "single source of truth." [8]. One of the most popular producers of ERP systems is the SAP, Microsoft, and Oracle [2]. Today, ERP systems are critical for managing thousands of businesses of all sizes and in all industries. To these companies, ERP is as indispensable as the electricity that keeps the lights on. ERP systems are designed around a common, defined data structure (schema) that usually has a common database. ERP systems provide access to enterprise data from multiple activities using common constructs and definitions and common user experiences. A key ERP principle is the central collection of data for wide distribution. Instead of several standalone databases with an endless inventory of disconnected spreadsheets, ERP systems bring order to the chaos so that all users (from the CEO to accounts payable clerks) create, store, and use the same data derived through common processes. With a secure and centralized data repository, everyone in the organization can be confident that the datum is correct, up to date, and complete. Data integrity is assured for every task performed throughout the organization, from a quarterly financial statement to a single outstanding receivables report, without deploying error-prone spreadsheets [8].

Blockchain technology, in general, it is a decentralized trustless system using for transacted digital cash (bitcoin) via the Internet. This technology appeared in the 2008 year. In 2015 Etherium blockchain platform was created where the technology of smart-contract was implemented. It is a digitally signed, computable agreement between two or more parties. A virtual third party (a software agent) can execute and enforce at least some of the terms of such agreements. The goal of contract scripting languages is to build what already exists, but it will be more interactive and more automated in comparison with origin. Blockchain-based transactions are only the starting point for what could become quite elaborate, multiparty rule-based environments. If to compare traditional contracts and smart-contracts it can be seen the difference: the average time of creation for tradition contracts is 1–3 days, for smart-contract – minutes. At the stage of making a traditional contract lawyer support is needed, for smart-contract it is not necessary. When a traditional contract is made a remittance can be only manual if it is a smart-contract a remittance is automatic. Classical contracts need a physical presence for signing it, for the smart-contract only virtual sign is needed. by means of the smart-contracts, each user can reach out and grab knowledge and other digital or digitally represented assets [5].

2.2 Benefits and Sphere of Applicability of ERP Systems and Blockchain Technology

To understand the purpose of use ERP systems it is necessary to know key benefits and sphere of applicability. Examples of specific business benefits published by SAP, Microsoft and Oracle include: improved business insight (from real-time information generated by reports); lower operational costs (through defined and more streamlined business processes); improved efficiency (through a common user experience across many business functions and managed business processes); reduced risk (through improved data integrity and financial controls) [9]; reduced level of inventory (through improved planning and control); improved production efficiency (due to minimization of deficiency of stocks and interruptions in deliveries); reduced materials cost (through improved procurement and payment protocols); reduced labour cost (through better allocation of staff and reduced overtime); increased sales revenue (driven by better managed customer relationships); increased gross margin percentage; reduced administrative costs; reduced regulatory compliance costs; [3] reduced operational costs (through automate warehouse processes); improved forecasting accuracy (to boost product availability and minimise inventory costs) and increased the consistency of data (across organisation and deployments) [8].

The sphere of applicability ERP system include: Retail, insurance, oil & gas, utilities, public sector, media, plant maintenance, R&D engineering, finance, commodity management, HCM [10], Warehouse Management, Transportation Management, Demand Forecasting, Product Change Management, Serial Number Tracking Enhancements for Service and Warranty Scenarios, Trade Allowance Management, Vendor Rebate Management, Broker and Royalty Contract Management, eProcurement, CRM Integration, Project Client Integration and Visual Scheduling, Modern Point of Sale (POS), eCommerce and Social, Retail Industry Excellence, Retail Enterprise Manageability, Call Center Sales, Retail Essentials, Order Entry for High-volume (Consumer) Transactions, Human Resources, Budget Planning, Management Reporter, Master Data Management (MDM), Business Insights and Multi-Threading Compiler (according Microsoft web site).

Blockchain technology also has some benefits. For supply chain management, the blockchain technology offers the benefits of traceability and cost-effectiveness. Blockchain can be used to track the movement of goods, their origin, quantity and so forth. Blockchain can use for providing quality assurance, accounting, contractual transactions, voting, securities and commodities trading on the stock exchange and fast, secure and cheap transferring of funds across the globe [4].

Blockchain technology has three versions. Each version has a different sphere of applicability. Blockchain 1.0 was created how technology for cryptocurrency transaction. The deployment of cryptocurrencies in applications related to cash, such as currency transfer, remittance, and digital payment systems. The technology of Blockchain 2.0 related to contracts. The entire slate of economic, market, and financial applications using the blockchain. This technology more extensive than simple cash transactions, it includes stocks, bonds, futures, loans, mortgages, titles, smart property, and smart contracts. Blockchain 3.0 appeared how blockchain applications beyond currency, finance, and markets - particularly in the areas of government, health, science, literacy, culture, and art [5].

Now ERP systems have more benefit and more applicability than blockchain technology but blockchain continues developing and expands the opportunities.

2.3 Advantages and Disadvantages of ERP Systems and Blockchain Technology

Companies that develop ERP systems (Microsoft, Oracle, SAP) allocate many advantages of these systems. For some of them it is allowed easier global integration (barriers of currency exchange rates, language and culture can be bridged automatically); updates only need to be done once to be implemented company-wide; provides real-time information, reducing the possibility of redundancy errors; may create a more efficient work environment for employees; vendors have past knowledge and expertise on how to best build and implement a system; user interface is completely customizable allowing end users to dictate the operational structure of the product.

ERP systems also have disadvantages. It connects with locked into relationship by contract and manageability with vendor (a contract can hold a company to the vendor until it expires, and it can be unprofitable to switch vendors if switching costs are too high); inflexibility (vendor packages may not fit a company's business model well and customization can be expensive); return on investment may take too long to be profitable; implementations have a risk of project failure [6].

Advantages of blockchain technology include autonomy (only one user need for making the agreement also knocks out the danger of manipulation by a third party), trust (documents are encrypted in a shared ledger), backup (On the blockchain, each and every one of users have other backup, documents are duplicated many times over), safety (cryptography, the encryption of websites, keeps documents safe), speed (smart contracts use software code to automate tasks, thereby shaving hours off a range of business processes), savings (smart contracts save money since they knock out the presence of an intermediary) and accuracy (automated avoid the errors that come from manually filling out heaps of forms).

Disadvantages of blockchain technology relate to novelty and lack of regulation. Companies which will use blockchain platform can detect problems such as developers do not know how to fix code if bugs get in; governments have not unified opinion about regulating blockchain and smart-contracts (Some government forbid it other try to regulate); also, governments do not know how to take tax from smart contract transactions; developers do not know what happens if users send the wrong code, or, send the right code, but service is not rendered (If this were the traditional contract, user could rescind it in court, but the blockchain is a different situation) [11]; current maximum limit of 7 transactions per second (the VISA credit card processing network routinely handles 2,000 transactions per second and can accommodate peak volumes of 10,000 transactions per second). Some of the main issues of blockchain technology connect with increasing the block size, addressing blockchain bloat, countering vulnerability to 51 percent mining attacks, and implementing hard forks (changes that are not backwards compatible) to the code [5].

Today ERP systems are the most usable program for automatization. It includes many spheres of applicability and a lot of features. The blockchain is a new technology with many problems but also with many advantages. If developers can fix the most of problem and expand opportunities, it will be so competitive.

3 Methods

As the research paper is mostly descriptive so the main methods of the research based on the analyses and systematization of information. For purpose of analyses need systematization information about features of ERP systems and Blockchain technology. The next step in the analysis it is described how these features work into ERP system and blockchain technology. At the end of analyses need to describe problems and opportunities which users have when they use ERP systems and products realize on blockchain platform.

The most features of ERP systems and blockchain technology are different but also have few features which match. Treat the match features such as: using in the finance area, using for goals of supply chain management, and using for online transaction. Each of these features correlated with other. However, companies develop ERP systems and program products based on blockchain technology realize these features so different. literature review described a lot of advantages of ERP system but, in case of using some features correlated with other features, ERP system has some problems compared with blockchain technology. In ERP system each feature is separate and work as the individual module and therefore if companies want to work with different modules they have to buy each module. Companies can buy only one module or package of modules. If users need buy another module after installation the first package, they will have some problem and additional cost for installation. Companies are installing additional module must pay for services for the implementation of the module and clear or buy resources on the server of the company. Blockchain technology realizes on the cloud platform, therefore, when a company buy some ads for their systems they do not have some additional pay for services and installation also modules realize on blockchain platform cheaper and simply integrated into the work process.

If says about transfer information in ERP systems also can see some problem. Each module of ERP system can transfer information to other, but it takes a long time and works so slow. Besides, connection inside the module can be unstable and transfer the incorrect information. If companies will use blockchain technology all features will realize in one module and connection inside this module will so fast and stable. Also, if companies buy program product realize on blockchain technology it will be cheaper than if they buy ERP system. In addition to price and speed, work with this product will be more efficiency and safety.

Despite all the advantages of Blockchain technology in work with features of systems, data analysis of literature review shows that ERP systems now better in another area. It can be used in a lot of sphere of activity and companies will not have problems with government regulation. Also, companies which develop ERP systems fix the most program problems and their systems work (not so fast and more expensive but work without critical problem).

Developers, which realize your products on blockchain technology, now need some time and some resources to solve this problem and do blockchain technology more accomplished.

4 Results

Responding to a research question "can companies use the blockchain platform instead of ERP systems or not", now it is not possible because program platform based on blockchain technology very young and have many problems with their work (it is program problems and problems with government regulation) but in the feature (if companies who develop program product on blockchain technology fix the most of their problem) it can be possible.

The potential opportunities for the blockchain technology not a limitation of purpose for using instead of ERP system. They extend into governmental, humanitarian, scientific and social domains. Some group already harness the technological capacity of the blockchain for the solution of real problems. For example, blockchain technology can be used for decision making in decentralized cloud functions that previously needed manual management by jurisdictionally bound organizations. In addition, the blockchain can be apply as the public records repository for the registry of all documents, events, identities, and assets. In the future all property could become smart property, every asset will be encoded in the blockchain with a unique identifier such that the asset can be tracked, controlled, and exchanged on the blockchain platform. This means that all manner of tangible assets (real estate, cars, gold ingots and other) and digital assets (intellectual property, data) could be registered and transacted on the blockchain.

Smart contracts as a part of blockchain technology also will help people in different cases. It includes exchange money, property, shares, goods or any other items while avoiding the services of an intermediary. Smart contracts not only define the rules and penalties around different agreements it also automatically enforce those obligations. Smart contracts could be streamlines ordinarily business operations (while before waiting for approvals) and search of the solution of internal and external problems. It also cuts out discrepancies that typically occur with independent processing that may lead to settlement delays and additional cost.

5 Discussion and Conclusions

If companies want to make blockchain and smart-contract equal to ERP many roles inside the business ecosystem must be transformed. Staff (managers, lawyers, sellers, purchasers, accountants and other) must learn how to write computable code and how to interpret it or rely on expert witnesses that will suggest valid solution. Developers must bring into the program based on blockchain technology the edgiest techniques from various disciplines. Besides, developers need to determine what are the best techniques and awareness of the leading knowledge [1].

Smart-contracts might help with the problem of regulations are among the least flexible and automated elements of the business ecosystem. Use of smart-contract can be a starting point to solve that problem, but it is no simple solution. The base of business technologies, processes, and procedures reflects many assumptions that must be revised for smart-contract implementation. In addition, the list of improvements for smart-contracts includes: improve the legal and regulatory environment, systematize the business ecosystem, focus on a more applicable sphere of business and develop best practices for next implementation.

Blockchain platform and smart-contracts can become viable, but one of the major issues is that many other ways to codify agreements already exist and they also can be used.

Acknowledgement. The research carried out on this subject was partially supported by RFBS grants (##17-29-07073-ofi-m, section 2), state assignments of the Ministry of education and science No. 2.3135.2017/4.6, section 1,5), within the budget theme #0073–2018–0003, section 3,4) , International project ERASMUS +, Capacity building in higher education, No. 73751-EPP-1-2016-1-DE-EPPKA2-CBHE-JP.

References

1. Morrison, A.: How Smart Contracts Automate Digital Business. PWC, US Blogs (2016)
2. Pang, C., Anderson, R.P., Montgomery, N., Kostoulas, J., Eschinger, C.: Market share analysis: ERP software, worldwide, 2016. Gartner, Inc. (2017)
3. Reali, K.: Calculating the benefits of implementing ERP software. INDUSA Technical Corp. (2015)
4. Agrawal, J.: 8 Benefits of Blockchain to Industries Beyond Cryptocurrency. Entrepreneur (2018)
5. Swan, M.: Blockchain: Blueprint for a New Economy. O'Reilly Media Inc. (2015)
6. Monk, E.F., Wagner, B.J.: Concepts in Enterprise Resource Planning, 3rd edn. Course Technology Cengage Learning, Boston (2009)
7. Statista. https://www.statista.com/topics/1823/business-software/
8. Microsoft. https://www.microsoft.com/en-gb/dynamics365/
9. Oracle. https://www.oracle.com/uk/applications/erp/what-is-erp.html
10. SAP. https://www.sap.com/products/what-is-erp.html
11. Blockgeeks. https://blockgeeks.com/guides/smart-contracts/

Image Content Protection Using Hybrid Approach of Blocking and Embedding Algorithm in the Context of Social Network

M. Naveen[1(✉)] and G. N. K. Suresh Babu[2]

[1] Visvesvaraya Technological University (VTU), Belgaum, Karnataka, India
patadnaveen@gmail.com
[2] Department of Computer Science and Engineering,
Acharya Institute of Technology, Bengaluru, India

Abstract. The security of an informative data plays a crucial role in current world. The reversible hiding mechanisms for data are able to perform the embedding of secret data in an image in secure way so that the receiver can receive image data without any data loss in secure manner. But, the existing data hiding mechanisms are able to embed the data only after the image encryption. This may cause error during the extraction and recover of image. The security of image content shared on collaborative (social) networks is the de-facto research today to minimize losses. This paper presents a hybrid mechanism of blocking and embedding for image data protection in the context of collaborative network. The outcomes of the system give the improved execution time for optimized transformation (29.3 s) than normal transformation (139.17 s). Also, it is observed that the PSNR value of 53.15 and 53.13 is observed for two set of image data which represents the improved image quality after recovery.

Keywords: Image security · Block operation · Data embedding
PSNR · Execution time

1 Introduction

The technological advancement in current digital imaging techniques (DIT) and availability of high performance providing computation platform (HPCP) provisions many critical as well as useful applications for human society in various domains including healthcare, surveillance, security, monitoring, tracking, identification, recognition, classification, etc. [1–5]. In all these applications the core input is either image or image sequences (video as frames per second). In case of surveillance system, people with malicious intention try to compromise the captured images content so that they are not caught by the forensic team [6, 7]. The rapid adoption of e-systems for governance processes leads the uses of document images. For any nation it is a top most priority today to have very advance system of securing such image data's to minimize the consequences and economic losses [8, 9].

R. Silhavy et al. (Eds.): CoMeSySo 2018, AISC 860, pp. 248–258, 2019.
https://doi.org/10.1007/978-3-030-00184-1_23

The collaborative network (Coll-N) is used widely to propagate information. The Coll-N is used in both open and close context, where corporate stack holders and general public participate to share textual, visual data (image/video) [9]. There have been various approaches found in literatures towards methods for image security which includes (1) approaches to handle confidentiality, (2) approaches to handle image data integrity, (3) approaches to handle image authentication [8, 9]. The various methods such as image encryption, steganography, watermarking, etc. are the popular approaches. Section 2 describes in detail as a review of literature for related work.

This paper proposes an efficient cryptographic protocol for protecting images which need to be shared and communicated images in the social network paradigm by mechanism of blocking and embedding.

2 Literature Review

The conventional Fourier transformation such as fast Fourier transformation (FFT) is used initially for the transformation of image data to get significant co-efficient, which is computationally overhead due to incremental linearity in the key space. Author Tao et al. the image encryption is achieved by summation of different orders inverse discrete fraction Fourier transform (FFRT) by interpolation of some images [10]. The attacks such as plain text attack, statistical analysis is as well as differential attack etc. are effective on FFT based image encryption processes [11]. In order to achieve higher level of security Ahmed et al. have proposed two-layer architecture uses technique of chaotic Baker map (CBM) and random phase encoding (RPE) [12]. But most of the digital images nowadays have to travel through the communication channel so images to be encrypted before compression, therefore a joint algorithm development approach are needed for both encryption and compression together. One of the states of art work by Zhou et al. [13] proposes an prediction error clustering based mechanism for compression and random permutation (RP) for encryption. The performance matrices in their work is been considered are rate and PSNR [13].

With the improvement into digital imaging technique from black and white to grey images and further color images came into existence and is widely used. Traditional cryptographic encryption methods such as advanced encryption standard (AES) is not directly applicable to color images so a method of color image encryption is proposed by Li and Liu [14], where Henonmap(HM) and Chebyshev map(CM) is used to generate keys into AES algorithm mapped for image encryption. Towads investigating the effectiveness of algorithm key sensitivity and statistical characteristic analysis is performed and found method to be effective and secured as compared to the state of art methods of that time [14].

In order to bring the effectiveness in the image security over a watermarking scheme, a Chaos-Fractal coding algorithm is being introduced in Keyvanpour and Farnoosh [15]. This coding algorithm determines the set of contractive transformation for each blocks of image. The simulation outputs are suggested with better improvement in the capacity and data hiding. This research idea has proven effective for various watermarking schemes without any alteration in the embedding algorithm [15].

Further to have lossless image data embedding for security aspects is found in Malakooti and Khederzdeh [16] by using randomized key generation and DCT mechanism. In this, the important information is embedded with a cover image and achieves image security with quality. The embedding algorithm exhibits various security levels to hide important data [16].

The Steganalysis is also an effective mechanism to detect the hidden informations in the stego media which can helps to provide the security against the disastrous events. The research work towards such mechanism is found in Fatnassi et al. [17] which talks about the Steganalys is and Steganography. This research evolves with the review of varioys Steganalysis algorithm which helps to analyze the stego media image characteristics against security attacks [17]. The security enhancement by using both the image encryption standards and watermarking techniques is presented in Al-haj and Abdel-Nabi [18]. This system adopts the hybrid algorithm which utilizes the combined approach (encryption and watermarking techniques) for security enhancement of medical images. Here, spatial or encrypted or hybrid domain for integrity and authenticity of the transmitted image. The proposed approach adopts the bit planes concept to combine the 8 bit planes of plain image and encrypted image and formed 16 bit single image. The final analysis of the system suggests that the system is computationally efficient, less complex and offers high embedding capacity [18].

Similarly the other researches like Zhang [19] have presented the reversible data hiding concept for security enhancement by adopting the encryption, embedding, extraction and recovery phases and achieved computationally efficient results [19]. An decompression algorithm is implemented to embed the image data by Zhang et al. for data (reversible) hiding for better security [20]. For similar purpose Zhang utilized the iterative flow and its outcomes were suggested for its implementation in various data values for security enhancement [21].

From the survey analysis of the various researches it is found that most of the researches were implemented the basic encryption mechanisms and achieved the security to some extent and embedding mechanisms to hide the important data in another image. With a closer view, it is observed that most of the researches were not efficient towards data compression. Also, the blocking operation for embedding was not efficient and the techniques were presented is lags with higher computational complexity. Also, very rare works were focused on the image quality after transmitted over the collaborative network. Thus, this paper aims to presents a computationally efficient and image quality enhanced hybrid mechanism of blocking and embedding for image data in the context of collaborative network. The following section gives the system modelling and analysis of the results associated with the model.

3 Proposed System

In order to achieve secured data transmission as image, a hybrid mechanism of blocking and embedding is introduced in the context of collaborative network. The hybrid mechanism involves image data embedding (encryption) and extraction of data. The process involved in image encryption is shown Fig. 1, where an image to be shared (I_s) and image to be travelled (I_f) are considered for encryption. Both these images I_s and I_f are subjected to the alpha (α) quantile of standard deviation (Sd) whose value ranges from 20 to 90. The value of α helps to perform the blocking operation. Later, reverse image transformation by using the parameters like lamda (λ), overflow pixel value (O*f*) and underflow pixel value (U*f*). The value of λ, O*f* and U*f*is ranges with even number, maximum 260 and less than zero (0) respectively. Later, transformation approach is applied and the image data transmitted over the collaborative network. For the transmitted image an secrete key is provided to embedd I_s in I_f. This process gives an encrypted image (Ie).

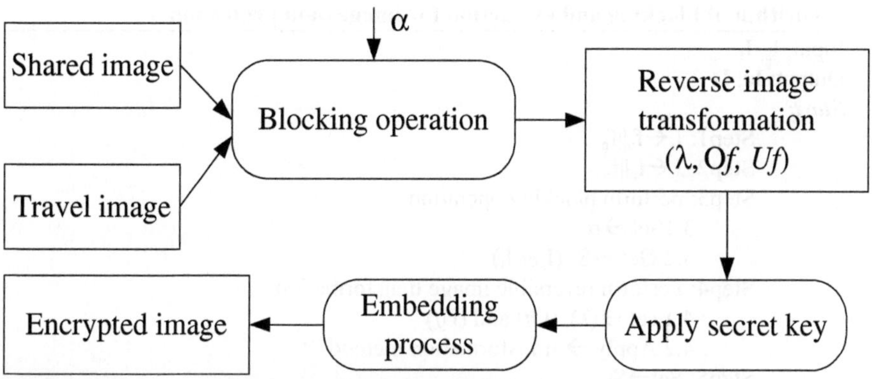

Fig. 1. Image encryption (Embedding) process

The following Fig. 2 states the steps involved in the extraction of information from encrypted image (Ie) and get recovered image. Here, the Ie image will be subjected with the secret key which is provided during the embedding of I_s and I_f. This helps to extract information's of the Ie and then recover I_s.

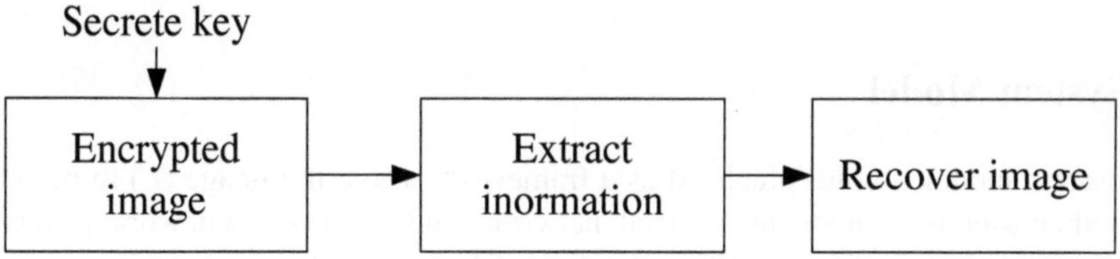

Fig. 2. Extraction of embedded image

The algorithm employed is initiates with the selecting the image to be shared (I_s) and image to be travelled (I_t) from the desired file path (f_p) and filename (f_n) from respective file location. Then, the image blocking operation is performed by providing the value of alpha (α) quintile of standard deviation (S_d) whose value ranges from 20 to 90. This helps to get the value Sds for image to be shared (I_s) and image to be travelled (I_t) for red, green and yellow components of the image. Further, the reversible image transmission is performed by setting values of lamda (λ), overflow (Of) pixel value and underflow (Uf) pixel value and then applying the transformation method. Time taken for transformation is evaluated and then a secret key (S_k) is applied for the image over collaborative network and then perfom the embedding of both I_s in I_f. The corresponding PSNR value is calculated for encrypted image. Then the encrypted image is saved in particular folder.

For extraction the saved encrypted image (Ie) is selected and then applied the secret key (S_k) which is given during embedding. Once the key is accessed, the informations are extracted from the image. Finally, the image is recovered (I_R) from Ie.

Algorithm of blocking and extraction for image data protection

Input: I_s, I_t
Output: Ie, I_R
Start:
Step1: $I_s \leftarrow f_n || f_p$
Step2: $I_t \leftarrow f_n || f_p$
Step3: perform blocking operation
3.1Set $\rightarrow \alpha$
3.2 Get $\rightarrow S_d$ (I_s& I_f)
Step4: Perform reversible image transformation
4.1 set $\rightarrow$(λ), (Of) and (Uf)
4.2 Apply $\rightarrow$ transformation method
Step5: Set$\rightarrow S_k$
Step6: Embedd (I_s& I_t)
Step7: Obtain $\rightarrow$Ie
Step8: Select$\rightarrow$Ie
Step9: Enter $\rightarrow S_k$
Step10: Extract$\rightarrow$ information
Step11: $I_R \rightarrow$recover (Ie)
End

4 System Model

The system model is conceptualized as a framework where the image (I_s) to be shared to another user or a node in a social network context. The framework provisions sampling and quantization for the digitization of I_s. In order to protect the content of I_s another image is digitized which will travel across the network and this image is represented by I_t.

4.1 Blocking Operation

For blocking operation, the independent variable parameter is alpha quantile of standard deviation (S_d) within the threshold value of lower limit 20 and upper limit 90. The block size (B_s) value is initialized. The operations of blocking are devised from gray and color image structure.

Process for Gray Image

$$[I_{SB}\ I_{tB}] \leftarrow \underset{B\rightarrow C}{f}\left({}^{I_s}\!/\!{}_{I_t} : B_s\right)$$

$$[\sigma_{ISB}\ \sigma_{ItB}] \leftarrow f_\sigma(I_{SB}, I_{tB} : w)$$

$$([V_{ISB}, V_{ITB}], [\text{Index}]) \leftarrow f_\sigma(\sigma_{ISB}, \sigma_{ItB})$$

Both I_s and I_t blocks are rearranged distinguishably with pads of '0' into columns with Bs. The standard deviations σI_{SB}, σI_{tB} of both I_{sB} and I_{tB} is computed with a weight vector 'w'. The values (V_{ISB}, V_{ITB}) and the corresponding indexes are computed by sorting array elements of I_{sB} and $I_{tB.}$ The Fig. 3 illustrates the pattern of S_d of block size (B_s = 4) vs. sorted block number for both I_S and $I_{t.}$

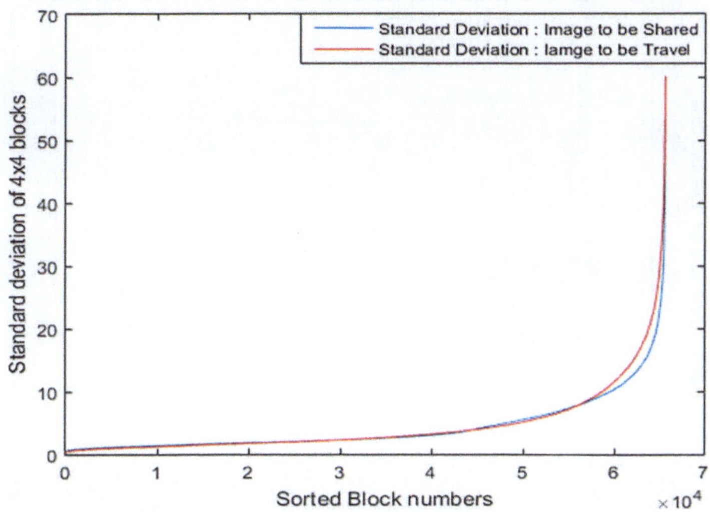

Fig. 3. Sd of block size (B_s = 4) vs sorted block number for both I_S and I_t

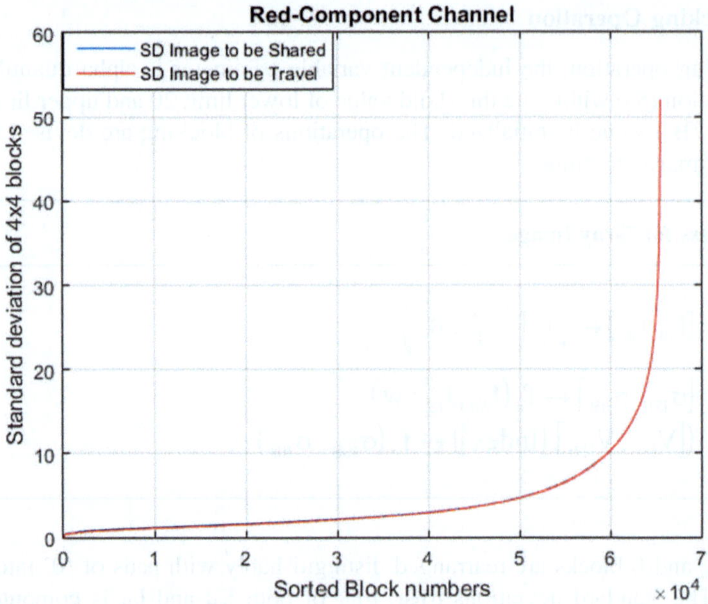

Fig. 4. Sd for images to be shared and travel vs Sorted block numbers (For Red-Component channel)

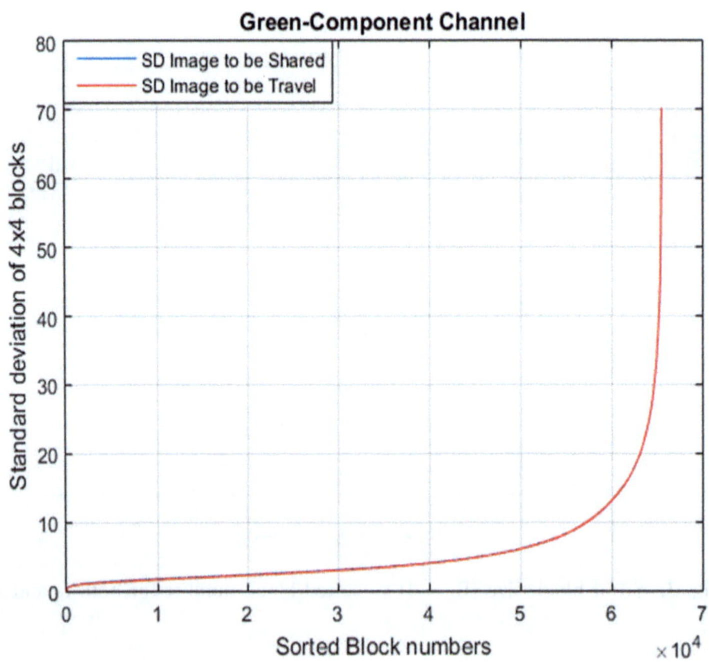

Fig. 5. Sd for images to be shared and travel vs Sorted block numbers (For Green-Component channel)

For image with RGB component, the Sd of block size ($B_s = 4$) for red component channel, green component channel and blue component channel is represented in Fig. 4, 5 and 6 respectively. The variation of Sd is with sorted block number.

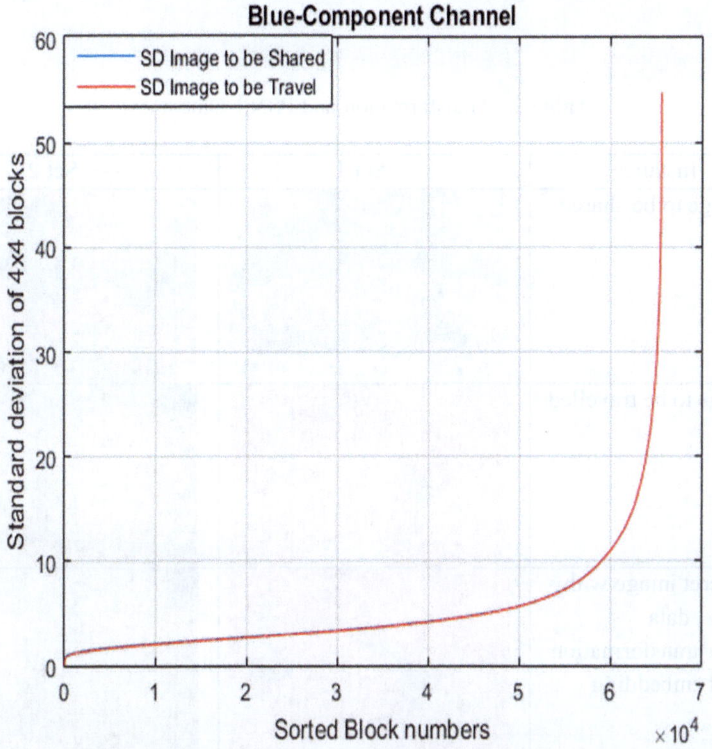

Fig. 6. Sd for images to be shared and travel vs Sorted block numbers (For Blue Component channel)

4.2 Transformation for Embedding

With both image to be shared and travel, i.e. (I_S and I_t) and block size $B_s$reversible transformation of image with varying input of factors like λ and overflow pixel value (P_{of}) along with underflow pixel (P_{uf}) (<0), the indexes(x_i) are created with critical information and blocks are arranged as per $x_{i.}$ The block modification is done by reshaping with default values of P_{of} and P_{uf}. Finally, the coloumns are rearranged into image as the secret image. Table 1 describe one instant observation for time taken as well as PSNR value.

Table 1. Analysis of time and PSNR for one instance

Instance	Execution time (s)	PSNR
1	142.30	53.4173

The process of transformation can be optimized to achieve better execution time and PSNR by adopting optimal block transformation process considering all the channel constraint. The first instance is optimized and execution time is reduced to 29.0399. The embedding process uses a secret key (k) and transmits data in the social network as hidden data image. Table 2 illustrates two sets of different images into the transition.

Table 2. Transformation and PSNR Outcomes

Instance	Set 1	Set 2
Image to be shared		
Image to be travelled		
Secret image with data After transformation and embedding		
Execution Time(s) [normal transformation/	139.7	29.1
Execution Time(s) for optimized transformation]	138.9	29.3
PSNR	53.15	53.23

For experimentation two sets of instances are used for image to be shared (I_s) and image to be travelled (I_t). The Is of set-1 is embedded with the It by using a secret key. Further, the execution time for set-1 instance is observed with normal transformation and optimized transformation as 139.7 s and 29.1 s respectively. To have further clarification in the analysis, set-2 instance is also analyzed with normal and optimized transformation which is found 138.9 s and 29.31 s respectively. From the analysis of both the intendances it is found that execution time taken by optimized transformation is very less than normal transformation. Similarly, the PSRN value of both set-1 and set 2 instances is found as 53.15 and 53.23 respectively which indicates the higher image quality.

5 Conclusion

In order to achieve secured data transmission as image, a hybrid mechanism of blocking and embedding is introduced in the context of collaborative network. The hybrid mechanism involves image data embedding (encryption) and extraction of data. The system model is conceptualized as a framework where the image to be shared (I_s) to another user or a node in a social network context. The framework provisions sampling and quantization for the digitization of image to be shared. In order to protect the content of image to be shared, another image is digitized which will travel across the network. The transformation process is optimized to achieve better execution time and PSNR by adopting optimal block transformation process considering all the channel constraint. The analysis of two set of images gives that the execution time taken by optimized transformation (29.31 s) is very less than normal transformation (138.9 s). This timing analysis represents low computational complexity of the optimized transformation. Similarly, the PSNR is computed for both the instances and that gives improvement in image quality.

References

1. Kolk, R., Razaque, A.: Scalable and energy efficient computer vision for text translation. In: 2016 IEEE Long Island Systems, Applications and Technology Conference (LISAT), Farmingdale, NY, pp. 1–6 (2016)
2. Jonaitis, D., Raudonis, V., Lipnickas, A.: Application of computer vision methods in automatic analysis of embryo development. IN: 2015 IEEE 8th International Conference on Intelligent Data Acquisition and Advanced Computing Systems: Technology and Applications (IDAACS), Warsaw, pp. 258–260 (2015)
3. Ko, T.: A survey on behavior analysis in video surveillance for homeland security applications. In: 2008 37th IEEE Applied Imagery Pattern Recognition Workshop, Washington DC, pp. 1–8 (2008)
4. Ouloul, M.I., Moutakki, Z., Afdel, K., Amghar, A.: An efficient face recognition using SIFT descriptor in RGB-D images. Int. J. Electr. Comput. Eng. **5**(6), 1227–1233 (2015)
5. Bazmi, A., Faez, K.: Increasing the accuracy of detection and recognition in visual surveillance. Int. J. Electr. Comput. Eng. **2**(3), 395 (2012)
6. López-Valcarce, R., Romero, D.: Defending surveillance sensor networks against data-injection attacks via trusted nodes. In: 2017 25th European Signal Processing Conference (EUSIPCO), Kos, pp. 380–384 (2017)
7. Ferrara, M., Franco, A., Maltoni, D.: Face demorphing. In: IEEE Transactions on Information Forensics and Security, vol. PP, no. 99, p. 1
8. Bhatnagar, S.C., Singh, N.: Assessing the impact of e-government: a study of projects in India. Inf. Technol. Int. Dev. **6**(2), 109 (2010)

9. Margaret, B.D., Kumar, R.J.V.: The socio-economic impact of e-governance: Indian scenario. Int. J. Sci. Res. Manag. **3**(1) (2015)
10. Tao, R., Meng, X.Y., Wang, Y.: Image encryption with multiorders of fractional fourier transforms. IEEE Trans. Inf. Forensics Secur. **5**(4), 734–738 (2010)
11. Yang, F.: Image encryption algorithm based on fractional fourier transform. In: 2013 International Conference on Computational and Information Sciences, Shiyang, pp. 705–708 (2013)
12. Elshamy, A.M., et al.: Optical image encryption based on chaotic baker map and double random phase encoding. J. Lightwave Technol. **31**(15), 2533–2539 (2013)
13. Zhou, J., Liu, X., Au, O.C., Tang, Y.Y.: Designing an efficient image encryption-then-compression system via prediction error clustering and random permutation. IEEE Trans. Inf. Forensics Secur. **9**(1), 39–50 (2014)
14. Li, J., Liu, H.: Colour image encryption based on advanced encryption standard algorithm with two-dimensional chaotic map. IET Inf. Secur. **7**(4), 265–270 (2013)
15. Keyvanpour, M.R., Farnoosh, M.B.: A new encryption method for secure embedding in image watermarking. In: 2010 3rd International Conference on Advanced Computer Theory and Engineering (ICACTE), Chengdu, pp. V2-403-V2-407 (2010)
16. Malakooti, M.V., Khederzdeh, M.: A lossless secure data embedding in image using DCT and Randomize key generator. In: 2012 Second International Conference on Digital Information and Communication Technology and it's Applications (DICTAP), Bangkok, pp. 236–239 (2012)
17. Fatnassi, A., Gharsellaoui, H., Bouamama, S.: A novel approach for embedding information in image cover media with security. In: 2016 World Symposium on Computer Applications & Research (WSCAR), Cairo, pp. 104–109 (2016)
18. Al-Haj, A., Abdel-Nabi, H.: Digital image security based on data hiding and cryptography. In: 2017 3rd International Conference on Information Management (ICIM), Chengdu, pp. 437–440 (2017)
19. Zhang, X.: Separable reversible data hiding in encrypted image. IEEE Trans. Inf. Forensics Secur. **7**(2), 826–832 (2012)
20. Zhang, W., Chen, B., Nenghai, Yu.: Improving various reversible data hiding schemes via optimal codes for binary covers. IEEE Trans. Image Process. **21**(6), 2991–3003 (2012)
21. Zhang, X.: Reversible data hiding with optimal value transfer. IEEE Trans. Multimedia **15** (2), 316–325 (2013)

Calculation of Robustly Stabilizing PI Controllers for Linear Time-Invariant Systems with Multiplicative Uncertainty

Radek Matušů(✉)

Centre for Security, Information and Advanced Technologies (CEBIA–Tech), Faculty of Applied Informatics, Tomas Bata University in Zlín, nám. T. G. Masaryka 5555, 760 01 Zlín, Czech Republic
rmatusu@utb.cz

Abstract. The contribution is intended to present a method of computing robustly stabilizing PI controllers for linear time-invariant systems with unstructured multiplicative uncertainty. This graphical technique is based on the application of basic robust stability condition and plotting the robust stability border pairs of P-I parameters, which subsequently leads to the robust stability region. The illustrative example is presented to show the effectivity of this straightforward approach.

Keywords: Robust stabilization · PI controllers · Multiplicative uncertainty

1 Introduction

Robustness of Linear Time-Invariant (LTI) control systems belongs among the very attractive and deeply studied control theory disciplines of several last decades. Despite the attention of many researchers, the investigation of robust control remains highly required, especially in the case of simple and easily applicable control algorithms, such as conventional PI controllers. Many analysis and synthesis problems usually start from an LTI model that incorporates some kind of uncertainty, in order to respect the difference between the ideal mathematical model and real-life behavior. This uncertainty can be included in the mathematical model in more ways. The most frequent techniques for Single-Input Single-Output (SISO) systems are represented by parametric [1, 2] and unstructured [3–6] uncertainty. Both parametric and unstructured cases can be further classified according to more specific criteria. The usually preferred sub-type of the unstructured uncertainty is the multiplicative uncertainty [3, 5], which is utilized also in this work. Another commonly used sub-type is the additive uncertainty [3, 6].

The contribution is focused on calculation of robustly stabilizing PI controllers for LTI systems with unstructured multiplicative uncertainty. In other words, the aim is to find such combinations of PI controller parameters that preserve control loop stability for all possible members of a plant family which is described by a multiplicative uncertainty model. The area of robustly stabilizing P-I parameters is obtained graphically by means of plotting the contour for $\|W_M(s)T_0(s)\|_\infty = 1$, which corresponds to

R. Silhavy et al. (Eds.): CoMeSySo 2018, AISC 860, pp. 259–263, 2019.
https://doi.org/10.1007/978-3-030-00184-1_24

robust stability border. The illustrative example is elaborated in order to demonstrate the efficiency of this method.

2 Robust Stabilization of Plants with Multiplicative Uncertainty by Using PI Controllers

A continuous-time LTI plant with unstructured multiplicative uncertainty is described by:

$$G(s) = [1 + W_M(s)\Delta_M(s)]G_0(s) \tag{1}$$

where $G_0(s)$ is a nominal plant, $W_M(s)$ represents a weight function (typically stable) which represents the uncertainty dynamics, and $\Delta_M(s)$ is the uncertainty – an arbitrary stable function that fulfills $\|\Delta_M(s)\|_\infty \leq 1$.

The weight function has to be defined in order to meet the inequality:

$$\left|\frac{G(j\omega)}{G_0(j\omega)} - 1\right| \leq |W_M(j\omega)| \quad \forall\omega \tag{2}$$

The closed-loop control system which contains a plant with multiplicative uncertainty is robustly stable if and only if [3, 4]:

$$\|W_M(s)T_0(s)\|_\infty < 1 \tag{3}$$

where $T_0(s)$ is a complementary sensitivity function:

$$T_0(s) = \frac{L_0(s)}{1 + L_0(s)} = \frac{C(s)G_0(s)}{1 + C(s)G_0(s)} \tag{4}$$

More details on systems with multiplicative uncertainty, their description and robust stability analysis can be found e.g. in [3–5].

As it has been already outlined, the controller $C(s)$ which is expected to guarantee the robust stability has to be designed so that the condition (3) is fulfilled. For the commonly used controller with conventional PI structure, the robust stability region can be easily visualized in P-I plane [7] by means of computing the robust stability border pairs of P-I parameters, i.e. the pairs of P-I parameters that comply with:

$$\|W_M(s)T_0(s)\|_\infty = 1 \tag{5}$$

Then, the internal space of the resulting region represents the area of robustly stabilizing PI controllers.

3 Illustrative Example

Assume the plant with multiplicative uncertainty, the model of which was created (from the original first order time-delay plant with parametric uncertainty) in [5]:

$$\begin{aligned} G(s) &= [1 + W_M(s)\Delta_M(s)]G_0(s) \\ \|\Delta_M(s)\|_\infty &\le 1 \\ G_0(s) &= \frac{1.5}{1.5s+1} \\ W_M(s) &= \frac{1.\bar{3}s^3 + 2.4481s^2 + 3.2\bar{5}s + 0.\bar{3}}{0.\bar{4}s^3 + 1.\bar{1}s^2 + 1.\bar{6}s + 1} \end{aligned} \tag{6}$$

Then suppose that this plant is in the feedback connection with a PI controller:

$$C(s) = \frac{k_P s + k_I}{s} \tag{7}$$

The Fig. 1 shows a 3D plot of the values of $\|W_M(s)T_0(s)\|_\infty$ for various combinations of k_P and k_I parameters for the ranges from 0 to 5 with the step 0.1. The negative parameters are intentionally excluded from the calculation.

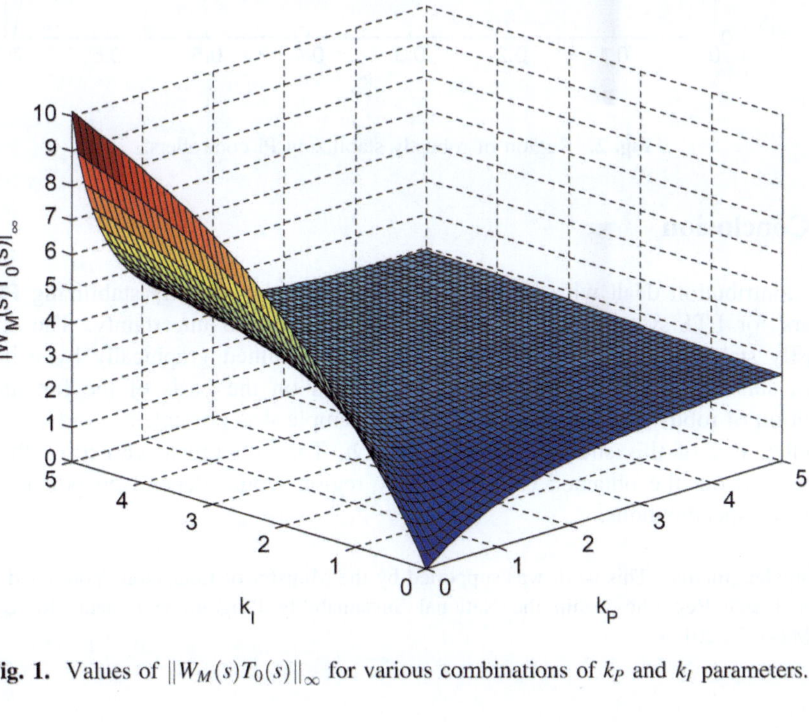

Fig. 1. Values of $\|W_M(s)T_0(s)\|_\infty$ for various combinations of k_P and k_I parameters.

Furthermore, the contour for $\|W_M(s)T_0(s)\|_\infty = 1$ is depicted in Fig. 2. Obviously, the internal space bordered by this contour represents the region of robustly stabilizing PI controllers for the plant with multiplicative uncertainty (6), because the condition (3) holds good here for all relevant pairs of k_P, k_I.

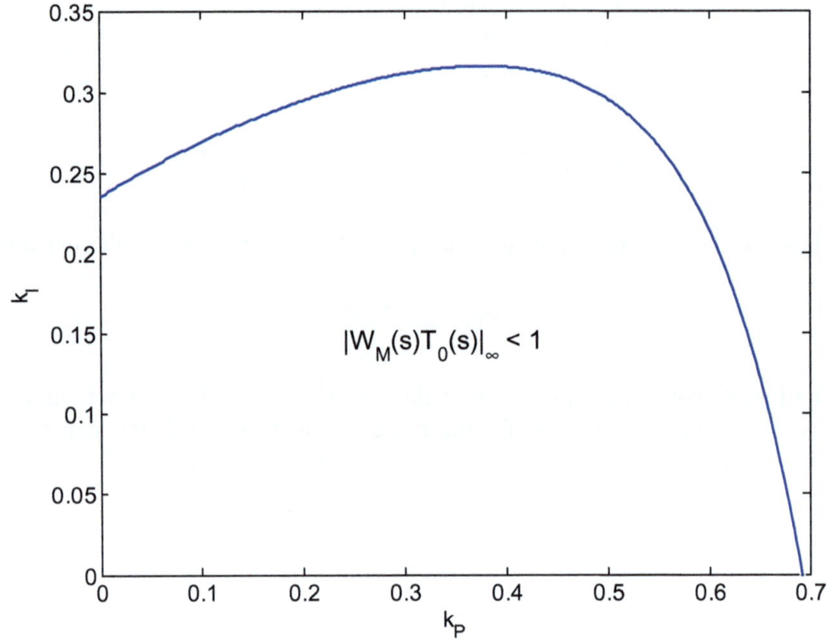

Fig. 2. Region of robustly stabilizing PI controllers.

4 Conclusion

This contribution dealt with a method for obtaining the robustly stabilizing PI controllers for LTI systems with unstructured multiplicative uncertainty. The area of robustly stabilizing pairs of P-I parameters was obtained graphically by using the robust stability border contour that was obtained on the basis of the fundamental condition of robust stability. The illustrative example was presented in order to prove the efficiency of this straightforward approach. The subsequent choice of the final controller from the obtained robust stability region would depend on potential performance specifications.

Acknowledgments. This work was supported by the Ministry of Education, Youth and Sports of the Czech Republic within the National Sustainability Programme project No. LO1303 (MSMT-7778/2014).

References

1. Barmish, B.R.: New Tools for Robustness of Linear Systems. Macmillan, New York (1994)
2. Bhattacharyya, S.P.: Robust control under parametric uncertainty: an overview and recent results. Ann. Rev. Control **44**, 45–77 (2017)
3. Skogestad, S., Postlethwaite, I.: Multivariable Feedback Control: Analysis and Design. Wiley, Chichester (2005)
4. Doyle, J., Francis, B., Tannenbaum, A.: Feedback Control Theory. Macmillan, New York (1992)
5. Matušů, R., Şenol, B., Yeroğlu, C.: Linear systems with unstructured multiplicative uncertainty: modeling and robust stability analysis. PLoS ONE **12**(7), 1–21 (2017)
6. Matušů, R., Şenol, B.: Description and analysis of systems with unstructured additive uncertainty. In: Silhavy, R., Silhavy, P., Prokopova, Z. (eds.) Cybernetics Approaches in Intelligent Systems. CoMeSySo 2017. Advances in Intelligent Systems and Computing, vol 661, pp 1–9. Springer, Cham (2018)
7. Kučera, V.: Robust controllers (Robustní regulátory). Automa **7**(6), 43–45 (2001). (In Czech)

Data Extraction from the Distorted DCF77 Signal Captured Using Low-Cost Receivers

Hubert Michalak(✉)

Department of Signal Processing and Multimedia Engineering,
Faculty of Electrical Engineering,
West Pomeranian University of Technology, Szczecin,
26 Kwietnia 10, 71-126 Szczecin, Poland
michalak.hubert@zut.edu.pl

Abstract. The DCF77 time signal is very often used in devices that need an accurate and stable synchronization to astronomic time. It has beed used since 1959, very well documented, and given to public usage. Nowadays it is quite easy to buy a complete DCF77 signal receiver that costs €3-4 and gives already demodulated digital signal to the output ready for further decoding. However, the DCF77 signal can be significantly distorted as it is modulated using the amplitude modulation (AM), that is very prone to noise, relatively high distance between transmitter in Mainflingen (Germany) and end-devices influencing the signal level together with low quality of some cheap DCF77 signal receivers. In this paper the solution to extract correct data from the distorted DCF77 signal captured using low-cost receivers is presented. As the proposed solution has low computational complexity, it can be easily implemented in low performance MCU, what is shown in this paper as well.

Keywords: Distorted signal processing · DCF77
Time synchronization

1 Introduction

The DCF77 is the timestamps transmitter for date/time reference. It is commonly used nowadays in many consumer devices, laboratories, industry, etc. as a precise time and frequency reference [8]. The main advantage of the DCF77 is in particular wide range of reception, that covers the whole central Europe. That allows reception of current time and date, that is synchronized to atomic clock, using low power devices at very low cost with the accuracy that is sufficient for many applications.

The DCF77 infrastructure is managed by the Physikalisch Technische Bundesanstalt (PTB) providing legal time and disseminating the standard frequency. The transmitter is located in Mainflingen, approximately 25 Km south-east from Frankfurt am Main working with EIRP (Equivalent Isotropic Radiated Power)

R. Silhavy et al. (Eds.): CoMeSySo 2018, AISC 860, pp. 264–275, 2019.
https://doi.org/10.1007/978-3-030-00184-1_25

of 30–35 kW. The DCF77 signal is transmitted with the carrier frequency 77.5 kHz and bandwidth of 2.4 kHz.

The system works in continuous operation with planned availability at least 99.7%. Actual availability during the years 2000–2009 was within range from 99.85% to 99.98% [1]. The DCF77 is not the only one reference global time used nowadays, some other systems are located in: China (BPC, 68.5 kHz, 20 kW), Japan (JJY, 40.0 kHz, 12.5 kW and 60 kHz, 22.5 kW), United Kingdom (MSF, 60 kHz, 17 kW), United States (WWVB, 60 kHz, 65 kW) and other countries [3].

Initially the DCF77 signal was modulated using the amplitude modulation (AM), which is quite easy to implement. Simple and inexpensive hardware can be used in the receivers. Nevertheless, the amplitude modulation is very prone to noise and distortions due to its characteristics, it also enables only the milliseconds accuracy for the process of time synchronization. That is why in 1980s the phase modulation (PM) with the Pseudo-Random Phase Shift Keying (PRPSK), using 512-bit pseudo-random code sequence, was added [9] to extend the DCF77 system and enable a higher precision and more reliable time reception. If the receiver is able to capture the phase modulated DCF77 signal, the accuracy in time synchronization can reach the level of 10 μs [9]. Additionally those two modulations - AM and PM - are orthogonal [5], what means they are both present in the DCF77 signal at the same time making possible to use two types of receivers: simple one (only for AM) which is more suitable for low-cost and low-performance devices, and advanced one (for AM and PM) that can be used in more sophisticated systems where a higher accuracy and robustness is required.

For the amplitude modulation the ASK (Amplitude Shift Keying) approach was selected. In this modulation for the set period of time (100 ms for binary "0" and 200 ms for binary "1") the amplitude of carrier is reduced to 25% of its original value. The amplitude shift keying modulation details are shown in Fig. 1. Additionally it was fixed that the start of this reduction is synchronized with the start of new second. In each second 1 bit of information is sent, so the throughput is estimated as 1 bps. The only one exception is the last - 59th second - where no reduction occurs at all, this feature is needed for the receiver's synchronization.

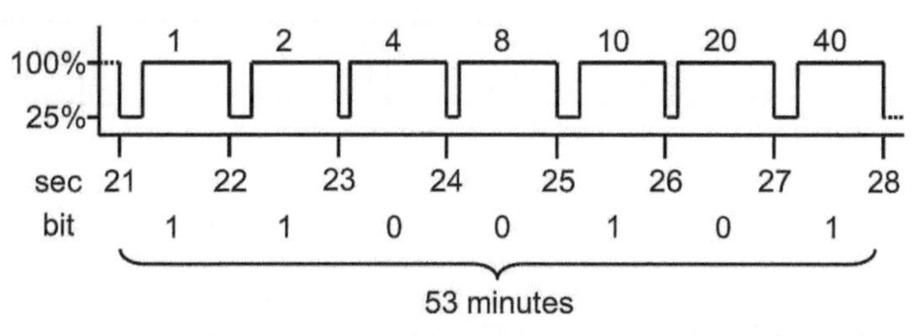

Fig. 1. Amplitude shift keying modulation detailed [10].

As it was mentioned earlier - the DCF77 signal uses also the PRPSK (Pseudo-Random Phase Shift Keying) phase modulation to increase the reliability and robustness of transmission. In this modulation a very long pseudo random noise sequence is used. This noise is made out of 512 chips, each chip lasts for 120 carrier cycles and its value is used to modulate the carrier by $\pm 13°$. The phase modulation is used 200 ms after the reduction of carrier from the amplitude modulation occurs (each second start point) and lasts for the next 793 ms. The remaining 7 ms time has no phase modulation. The details of the phase modulation used in the DCF77 are shown in Fig. 2.

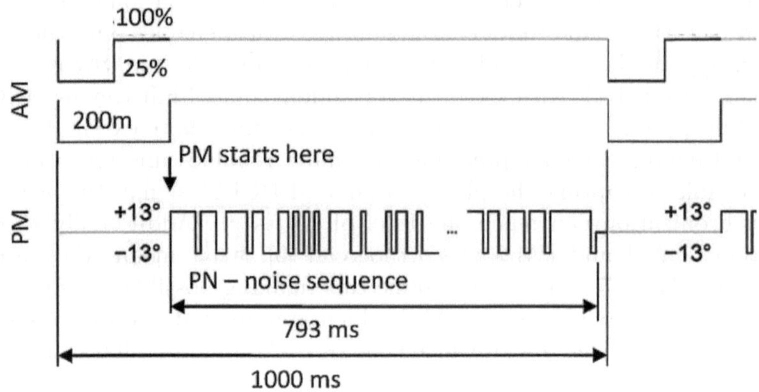

Fig. 2. Amplitude and phase modulation detailed [3].

The DCF77 frame is sent repetitively every minute and consists of 59 bits of data. Each frame holds the information of the current year, month, day, day-of-week, hour and minute. Additionally in each frame three parity bits are held to ensure the correct data decoding. The details of the DCF77 frame bits are shown in Fig. 3.

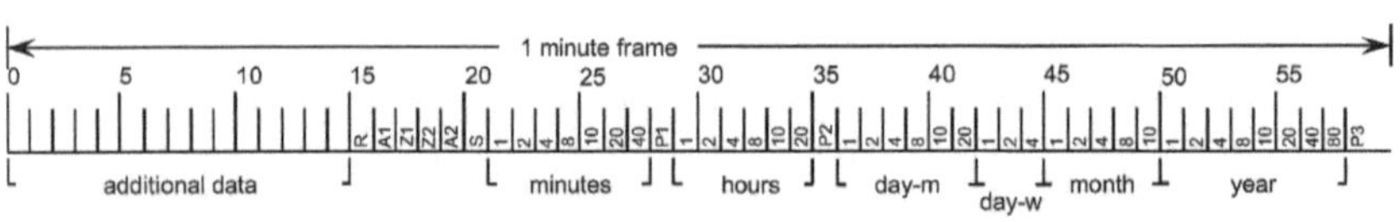

Fig. 3. The DCF77 frame detailed [10].

When using the low frequency (LF) radio wave, i.e. waves with frequency in the range 30–300 kHz, it is possible to have a service area of radius 1000–1500 km. However, the propagation of those waves strongly depends on the Earth's atmosphere. At the height of 80 Km above the sea level the ionization phenomena and the recombination of free-charged particles occur and have a large impact on propagation conditions of the LF radio waves [4]. Additionally it was verified that the propagation of the LF depends also on the time of the day: during the day the propagation is slightly worse than at night, furthermore during the short moments of sunrise and sunset the propagation becomes very bad.

Beside the influence of the Earth's atmosphere on the propagation of the LF waves, there exist many signal interferences as well (i.e. Electromagnetic Interference EMI) that come from human activities. As it is stated in [5,6] lots of the EMIs are produced from the power supplies (especially when working in switching mode) and the cathode ray tubes. All of these interferences can largely reduce the signal-to-noise ratio (SNR) at receivers, causing a bad DCF77 signal reception.

For the few years the biggest competitor for the DCF77 is the GPS (Global Positioning System). The former's work is based on LF the signalling, whereas the latter works using a completely different approach as it is the satellite based system used mainly for positioning, where the time synchronization is only a part of its functionality. Data from satellites are transmitted using the bands L1 (1575.42 MHz) and L2 (1227.60 MHz) and thus it has different propagation properties than the DCF77 that works using the LF band (77.5 kHz). The GPS system offers much higher synchronization accuracy (up to 40 ns) and provides much more services (localization, time, rescue, etc.), but has a big disadvantage - the receivers' antenna has to have clear visibility of the entire sky to be able to capture data correctly. In contrast to that, the DCF77 signal that works on LF, has the ability to penetrate through some number of obstacles and there is no need to provide the direct visibility between the transmitter and the receiver. It allows to use the DCF77 in some places where the GPS would not work (eg. in buildings). Additionally, what is also worth to notice, the DCF77 receivers (especially when using only the amplitude modulation) can be much cheaper than GPS receivers, but still provide the accuracy that is sufficient for many applications. Taking into account all advantages and disadvantages of both systems it can be concluded that both systems are complementary: they have fundamentally different architectures and signal propagation techniques thus providing various use-cases for many application.

2 Problem Description

Low-cost DCF77 receivers, besides obvious advantages (especially low price), have some number of disadvantages, among which the most important and difficult to overcome, is injecting distortions when the received signal has poor strength. If optimal working conditions will be provided: a sufficient level of the DCF77 input signal and a stable power supply, it can be expected to have the proper DCF77 pulses available at the receiver's output. In such case providing the decoding algorithm is not a complex task. Figures 4 and 5 illustrate some examples of the correct DCF77 pulses captured using low-cost receivers; depending on receiver's configuration the pulses can be positive (Fig. 4) or negative (Fig. 5).

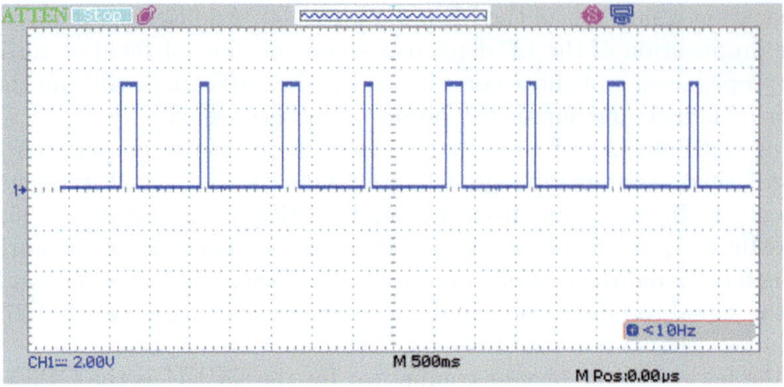

Fig. 4. A part of the DCF77 frame with positive pulses

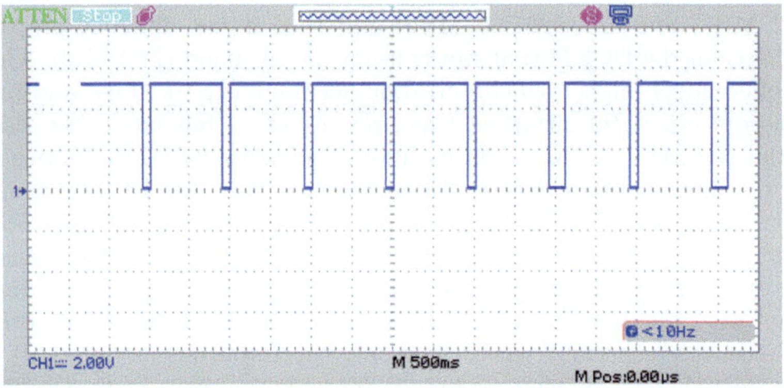

Fig. 5. A part of the DCF77 frame with negative pulses

Many of currently publicly available algorithms (for example: [2,7]) are based on the assumption that correct pulses will be provided, additionally it is required that the pulses lengths have to be in very narrow ranges. In practical solutions the pulses lengths might vary. The practically measured lengths are shown in Table 1. As it can be seen the DCF77 pulses provided by receivers in fact will not have exactly 100 ms or 200 ms duration, thus the algorithms used for the data extraction need to be prepared for such a situation.

More importantly, especially when the level of the DCF77 signal is insufficient for a correct reception, many distortions might be injected into usable pulses. Some examples of the corrupted DCF77 signals are shown in Figs. 6, 7 and 8. The types of corruptions are as follows: additional pulse that occurs between two successive pulses (as shown in Fig. 6), similar to previous one but multiple additional pulses occur (as shown in Fig. 7) and short brakes during expected pulses (as shown in Fig. 8).

Table 1. Practically measured DCF77 pulses' length for different receivers

	Logical "0"		Logical "1"	
	Receiver #1	Receiver #2	Receiver #1	Receiver #2
Pulse time (avg ± SD)	115 ± 5 ms	100 ± 10 ms	213 ± 6 ms	199 ± 9 ms
Pulse time (min)	102 ms	71 ms	200 ms	171 ms
Pulse time (max)	133 ms	129 ms	227 ms	221 ms

All of those distortions will cause the situation when simple decoding algorithms cannot be able to decode the DCF77 signal. Moreover, it is very popular to create the algorithm based on interrupts (the appropriate edge of the DCF77 signal will cause the interrupt service routine to execute to handle each pulse separately). Each time when two types of edges will be present (the expected one and the one that comes from a corrupted signal), the ISR will be executed causing to consume CPU's computing power and resources. In the extreme situation, if the received signal is very weak, a significant number of pulses might be present each second, causing the incorrect behavior of the CPU software. Taking into consideration that commonly known, simple and obvious algorithms, handle the distorted DCF77 signal very badly, a more robust algorithm proposal is needed.

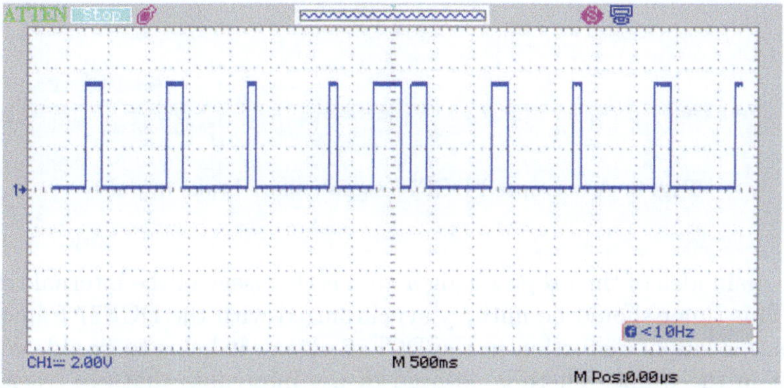

Fig. 6. A corrupted DCF77 signal: additional pulse

3 Proposed Solution

The algorithm proposed in this paper allows to correctly decode the data from the distorted DCF77 signal, captured using low-cost receivers, even if some number of corruptions occured.

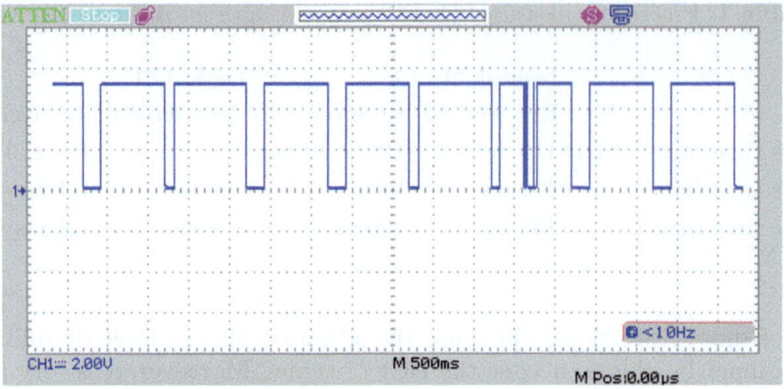

Fig. 7. A corrupted DCF77 signal: multiple additional pulses

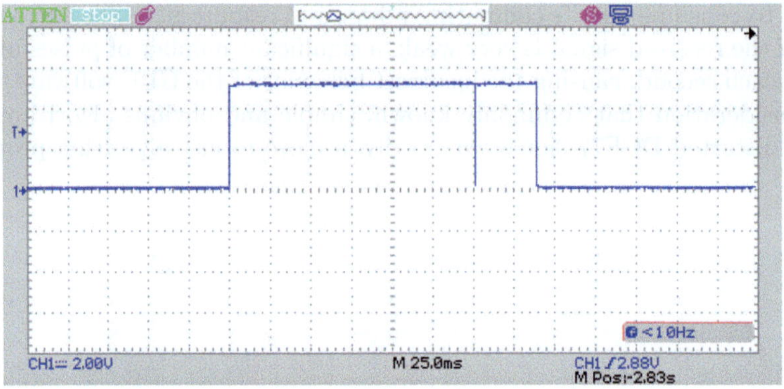

Fig. 8. A corrupted DCF77 signal: short brake in pulse

The main idea of the proposed algorithm is the usage of the internal reference timer. This timer should be initially synchronized with the DCF77 frame. Later on, even for lots of distortion in the DCF77 signal, it helps to decode correctly each bit in the frame. There is no need to use interrupts that are triggered by the DCF77 signal slopes in the proposed algorithm, thus it simplifies the system design.

The main loop of the proposed algorithm is shown in Fig. 9. This loop should be executed each 10 ms and performs the cyclic tasks depending on the current state. The first step is the synchronization of the internal timer that will be used later as the time reference for each DCF77 pulse. This synchronization is done by sampling three consecutive pulses and measuring if each slope is present at 1000 ± 50 ms. If so, the internal timer is started and synchronized with the beginning of the pulse. Since now it can be easily predicted when the next pulse should be present (each 1000 ms), and the verification if the given

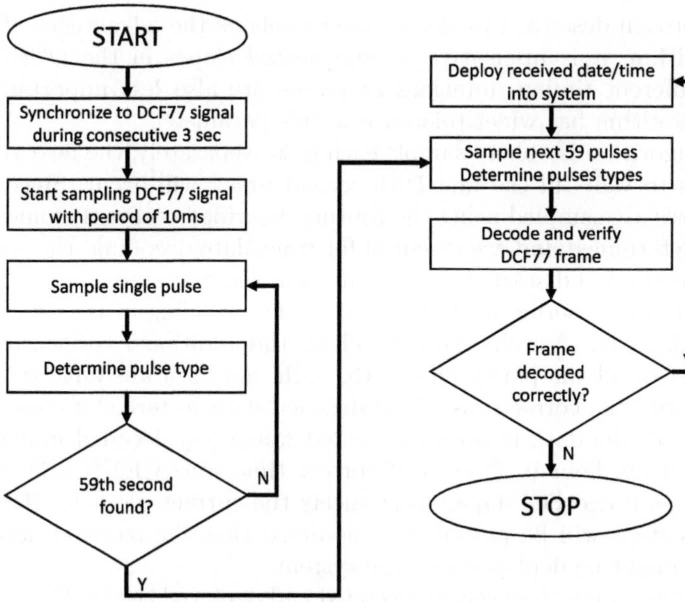

Fig. 9. The proposed algorithm for data extraction from the distorted DCF77 signal

pulse duration is 100 ms (logical "0") or 200 ms (logical "1") is present. This part of the algorithm is the only one that requires correct DCF77 signal pulses without any distortions.

Since the internal reference time is present, it is possible to sample pulses in the way that is more reliable even if those pulses might be distorted or not present at all. The sampling algorithm takes samples each 10 ms and collects them into an array. As it is known when the pulse should begin, the sampling is synchronized with this moment of time. As usable pulses should have the length of 100 ms or 200 ms, the sampling is finished after taking 30 samples (which cover the period of 300 ms).

Next, the post-processing of each sample is done to determine if the sampled pulse is the logical "0", logical "1", 59th second break (no pulse) or if it is impossible to determine the pulse type. The selection between the logical "0" and "1" is made by the analysis of the length of the pulse (in fact, the number of samples, as signal was sampled each 10 ms). If the number of samples representing the active signal for the given pulse is lower than 15 (which corresponds to the pulse length of 150 ms) it is assumed that the whole pulse represents the logical "0", otherwise it is assumed as the logical "1". If there are no samples representing the active signal - the 59th second break is assumed. If any unexpected configurations of the sampled data occur, then the type is assumed to be indeterminable.

The approach described in this chapter explains the advantages of the proposed algorithm: it is insensitive to unexpected slopes in the DCF77 signal, moreover different timing durations of pulses are also less important, as the proposed algorithm has wider tolerance of this parameter.

As the algorithm is able to sample each pulse separately, the next step of the algorithm is to wait till the first 59th second break will occur. Since that the next 59 pulses are sampled using the routine described earlier. If some of pulses cannot be determined and are essential for time/date decoding, the algorithm's loop is terminated and started once again from the first step.

After correct gathering all the 59 pulses the decoding of the DCF77 frame is performed. Next, sfter the data decoding, the additional verification is conducted. First of all the parity bits: 28th, 35th and 58th are verified. The next step is to verify the correctness of the decoded data in terms of consistency: it is verified if all the data is within expected range (eg. decoded minute should be between 0..59, hour 0..23 etc.), if correct time zone CET/CEST is present for given time, if decoded day-of-week meets the current date etc. If all of that verifications steps will be passed, it is assumed that the received date/time is correct and might be deployed into the system.

After having correctly received, decoded and deployed the DCF77 frame with date/time the proposed algorithm automatically starts to receive the subsequent DCF77 frame allowing the system to be synchronized every minute.

4 Verification

For the verification process a special setup has been prepared, consisting of: low-cost DCF77 receiver, a logic analyzer (used for gathering received DCF77 signal's waveforms for further analysis of their quality) and a micro-controller with implemented testing algorithms. For the verification purposes two algorithms have been implemented: simple decoding algorithm based on [2] and the proposed one described in Sect. 3. Both are working at the same time, and on the same input signal, additionally a number of debugging messages are available, that allow to analyze their internal states. A number of experiments have been conducted, through several days, running on the setup described earlier. During those experiments the DCF77 signal waveforms have been acquired using the logic analyzer, as well as two algorithms have been used for decoding the received data. Additionally the receiver's antenna has been moved into different places to expose the prepared setup on various receiving conditions.

The raw DCF77 signal waveforms, captured using the logic analyzer, have been used to classify the quality of received signal. In this paper four different levels have been identified:

- very good signal quality: when for the period of three concurrent minutes (180 s) signal is not distorted;
- good signal quality: when some distortions (for example: unexpected pulses) occur, but such distortions are present not more often than once per 10 s;

- poor signal quality: if distortions are occurring more than once per 10 s;
- bad signal quality: if it is impossible to determine any correct pulses or there are no pulses at all.

The acquired data have been divided into parts with different signal quality levels. On the same time it has been verified if the tested algorithms are able to correctly decode the received DCF77 signal. Those two sets of information have been compiled with each other to create comprehensive database of results, that was further post-processed.

The verification of the algorithms has been done using two parameters: Time To First Synchronization and Effectiveness of Algorithm. The former defines how long does it take since the first occurrence of 59th second break till the first correctly received and decoded date/time information. The latter defines the average percent of correctly received and decoded frames.

Table 2. Algorithms verification: average Time To First Synchronization

Received signal quality	Simple algorithm based on [2]	Proposed algorithm
Very good	1.07 min	1.06 min
Good	1.84 min	1.16 min
Poor	no synchronization	2.84 min
Bad	no synchronization	no synchronization

In Table 2 the summary results for average Time To First Synchronization are presented. The results are divided into earlier defined signal quality groups and are presented for both tested algorithms. An important fact is that in poor and bad signal quality the simple algorithm based on [2] is unable to correctly receive and decode the DCF77 frame. However, especially for the poor signal, the proposed algorithm is able to receive correct data, but generally it will take a bit more time than for the good quality signal.

Table 3. Algorithms verification: average Effectiveness of algorithms

Received signal quality	Simple algorithm based on [2]	Proposed algorithm
Very good	91.2%	93.3%
Good	46.1%	72.4%
Poor	0.0%	35.7%
Bad	0.0%	0.0%

The second parameter which is important to determine the robustness of the algorithm is the Effectiveness. Results for this parameter are presented in Table 3. Similarly as previously, the results are divided into groups of different signal quality. Simple algorithm based on [2] for poor and bad signal quality has the effectiveness of 0.0% as it is unable to correctly decode any of the DCF77 frames. The proposed algorithm is able to work with signal that has poor quality, but the effectiveness of the algorithm is reduced.

5 Conclusion

The initial version of the proposed algorithm has been implemented on the MCU and tested in practice. The results prove that the proposed approach is a better solution than the currently existing ones. The verification measurements made in Sect. 4 show that the proposed algorithm gives a much more robust way of decoding data from the distorted DCF77 signal. Additionally, what was the most important, in certain cases it becomes possible to decode the distorted DCF77 signals that come from the low-cost DCF77 receivers, whereas it is impossible using the currently available decoding algorithms. Apart of that, the effectiveness of the proposed algorithm (the ratio of correctly decoded DCF77 frames to all frames that occur during the given time) is higher for proposed algorithm. The only one disadvantage is that the algorithm is more complex, however it is not a big issue when taking into account all benefits that the proposed algorithm provides.

References

1. Bauch, A., Hetzel, P., Piester, D.: Time and frequency dissemination with DCF77: from 1959 to 2009 and beyond. PTB-Mitteilungen Special Issue **119**, 3–26 (2010)
2. Dabrowski, K.: Arduino w krotkofalarstwie. Tom 2. Krzysztof Dabrowski *OE1KDA*, Wieden (2013)
3. Depari, A., Flammini, A., Lavarini, M., Sisinni, E.: Inexpensive SDR-based long-wave radio controlled clock for time dissemination in industrial wireless sensor networks. In: 2015 IEEE International Conference on Automation Science and Engineering (CASE), pp. 125–130 (2015)
4. Dolea, P., Dascal, V.P., Cristea, O., Palade, T.: In-situ measurements regarding LF radio wave propagation using DCF77 time signal transmitter. In: 2013 11th International Conference on Telecommunications in Modern Satellite, Cable and Broadcasting Services (TELSIKS), vol. 02, pp. 449–452, October 2013
5. Engeler, D.: Performance analysis and receiver architectures of DCF77 radio-controlled clocks. IEEE Trans. Ultrason. Ferroelectr. Freq. Control **59**(5), 869–884 (2012)
6. Friedt, J.M., Eustache, C., Carry, E., Rubiola, E.: Software-Defined radio decoding of DCF77: Time and frequency dissemination with a sound card. Radio Sci. **53**(1), 48–61 (2018)
7. Klein, U.: Dcf77 library, February 2014. https://blog.blinkenlight.net/experiments/dcf77/dcf77-library/, https://blog.blinkenlight.net/experiments/dcf77/dcf77-library/. Accessed 07 May 2018

8. Niemann, L., Venzke, M., Renner, C., Turau, V.: Clock synchronization of tinyos-based sensor networks with DCF77. In: Proceedings of the 8th GI/ITG KuVS Fachgespräch Sensornetze (2009)
9. Piester, D., Bauch, A., Becker, J., Hoppmann, A.: Time and frequency broadcast with DCF77. In: 43rd Annual Time and Time Interval (PTTI) Systems and Applications Meetings, pp. 185–196 (2011)
10. Zaplata, F., Kasal, M.: SDR implementation for DCF77. In: 2013 23rd International Conference Radioelektronika (RADIOELEKTRONIKA), pp. 340–345, April 2013

A Mobile Terrestrial Surveillance Robot Using the Wall-Following Technique and a Derivative Integrative Proportional Controller

Pedro Gabriel Calíope Dantas Pinheiro[1], Maikol Magalhães Rodrigues[1], João Paulo Agostinho Barrozo[2], Jose Pacelli Moreira de Oliveira[3], Plácido Rogério Pinheiro[1(✉)], and Raimir Holanda Filho[1]

[1] University of Fortaleza, Fortaleza, Brazil
pedrogcdpinheiro@hotmail.com, {maikol,placido,raimir}@unifor.br
[2] Armtec Robotics Technology, Fortaleza, Brazil
joao.paulo@armtecbrasil.com
[3] Atlanta Tecnologia da Informação, Fortaleza, Brazil
jose.pacelli@atlantatecnologia.com.br

Abstract. This research aims to develop a land mobile robot of surveillance, whose primary task is to collect real-time videos and images, in closed environments, using the Wall-Following technique and a derivative integrative proportional controller to navigate the environment. We use ultrasonic sensors, infrared, IP camera and encoders for speed control in a closed loop. The development of the robot happened in three stages: definition, acquisition of the model and used components (mechanical structure, microcontroller, sensors, and actuators). Then, the user-robot interaction software will be defined, and the creation of the robot, of the control software and integration tests will be done. The integration test has the job of improving the programming routines of the robot and the interaction with it. Finally, the supervisory module will be described, whose interface will use the visualization and capture of real-time videos and images.

Keywords: Development · Robot · Software · Camera

1 Introduction

The resurgence of violence in the daily life of contemporary life and, at the same time, the considerable advances in science and technology have become a subject in the constant debate by civil society and governmental bodies in search of social protection, institutions and public and private assets. Given this problem, the following research question is proposed: How to develop the activity of monitoring and surveillance in internal environments with the aid of robotic technology?

Thus, this study is the development of a terrestrial mobile robot, with application directed to the monitoring of internal environments, that can be used in public, residential or private companies and institutions, to provide visualization and image

R. Silhavy et al. (Eds.): CoMeSySo 2018, AISC 860, pp. 276–286, 2019.
https://doi.org/10.1007/978-3-030-00184-1_26

capture and videos, performing mapping of the area to be covered in real time, through a supervisory system. For this, the robot will navigate autonomously and manually in internal environments, having as components: ultrasonic sensors, infrared, IP camera and encoders to perform the position and speed control of the motors.

The rest of the paper is organized as follows: Sect. 2 deals with issues theoretical foundation. Moreover, in the Sect. 3 provides a brief development, in which it is divided into two sections one describes the project as a whole, in summary, form and the other describes the robot, with its main characteristics in the development and the form that was developed being divided into three parts, are the mechanics with the structure and motors, electronics with microcontroller, sensors and camera, and programming with control and communication thereof. On the other hand, Sect. 4 describes the results the results of the mobile robot application, presents an overview of the procedures and tests adopted and their answers. Finally, in Sect. 5, the conclusions and suggestions for further work are provided [3].

2 Theoretical Foundation

2.1 Mobile Robots

Robots have different skills and competencies specific to each type of application. Given this particularity, they present themselves in a variety of ways, according to the complexity of the actions to be performed [11].

Autonomous mobile robots interact with the environment using a sensing system that allows them to pick up information and then process them, to choose the corresponding action to be performed through the middle of the drive of their actuators. They can be operated with complete autonomy or semi-autonomous, depending on the greater or lesser capacity of perception and performance in the environment, that is, the quality of the sensor system, actuators and motors, as well as the robustness of the intelligence, as it will be responsible [10]. In this paper, we present the results of the study.

2.2 Wall-Following

The Wall-Following navigation technique, which bases its orientation on wall tracking. Through this technique, it is possible for the robot to refer to the trajectory to be followed through the use of walls or the accomplishment of the contour of objects in the working environment. We can then navigate intuitively, planning the path he/she should follow indoors or indoors [5].

Navigation based on this technique allows the robot to traverse all the environment in which it is acting and to be able to return to the starting point. Another advantage that it offers in robotic applications as computational cost, well below the values referring to other techniques that are more complex, especially those that use "image processing through cameras, laser sensors and GPS (Global Positioning System)" to map the trajectory [5]. Therefore, it becomes quite attractive in the construction of mobile robot projects.

2.3 Controlador PID

A control system is one or more devices with the ability to control a particular system and is therefore specific to each of the functions it must control. According to [7], almost all electronic equipment of the modern era uses this type of control.

Proportional, Integral and Derivative control, or PID is one of the control methods that are frequently used, especially in industry. The PID control comes from the sum of three necessary control actions, each with a specific function: Proportional Action, Integral Action, and Derivative Action [4].

3 The Proposed Robot

3.1 Description of the Project

This research consists of the development of a mobile robot that uses the Arduino Mega 2560 platform based on the ATmega2560 microcontroller. For the complete development of the project, Eclipse software, a Java-based extensible free software development platform, was selected for programming the AVR in C, the software was chosen for having greater flexibility and control in the implementation of the libraries for peripheral use necessary for development [12].

The robot also has a printed circuit board in which both the Arduino board and all other sensors are connected to reduce the number of connection wires and to power the IP camera in a mobile way. As shown in Fig. 1.

Fig. 1. Sentinel Robot (Prepared by the author)

3.2 Description Do Robot Mobile

The robot uses two ultrasonic sensors HC-SR04 to its right, named as S1 and S2, which serve to guide the distance from the wall or obstacle to be circumvented. We determine the distance between the sensor and the obstacle given in centimetres using Eq. 1, where ΔT is the duration in µs of the pulse generated by the echo pin and Dist is the distance of the obstacle in centimetre (cm).

$$Dist = \frac{\Delta T}{58} \tag{1}$$

On the other hand, to avoid what is called busy waiting, in which the processor gets stuck in a loop waiting for some event to happen, the operation of the two ultrasonic sensors used in the design is entirely based on interrupts. The echo pins are connected to the external interrupt inputs of the ATmega 2560, and a timer is used to timing the events, as shown in Fig. 2.

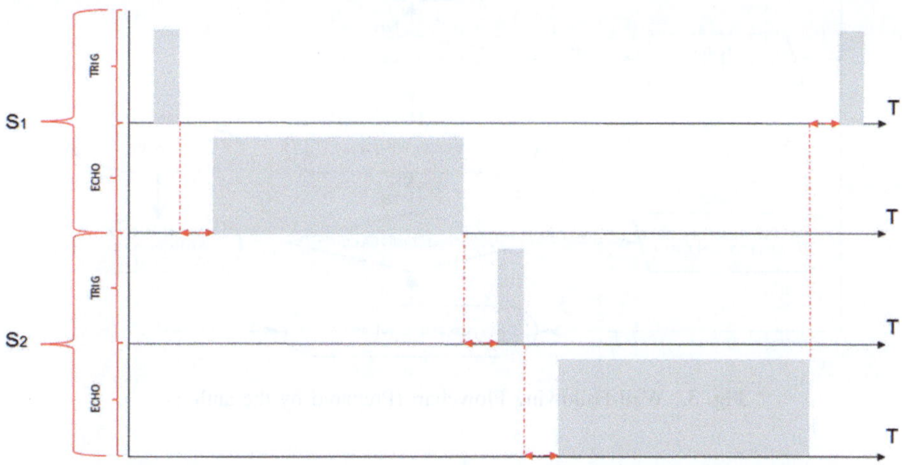

Fig. 2. Operation of the ultrasonic sensors in the robot (Prepared by the author).

Initially, the trigger pin of sensor one (S1) is set to high level, and the timer is set to generate an interrupt after a time of 10 μs. When this interrupt occurs, the trigger pin is set low, and the timer is set to generate an interrupt in 2 ms. If a rising edge is detected in the echo pin before this time, the timer is prepared to generate an interrupt in 40 ms, and the current count of this timer is saved in a variable. When a falling edge is detected on the echo pin, the current value of the timer is saved again in another variable. Subtracting this variable from the one above has the duration of the pulse generated in the echo pin that is directly related to the distance of the obstacle through Eq. 1. If the rising edge is not detected, a reading cycle of sensor two (S2).

If the process of reading sensor one (S1) occurred as expected, after detecting the descent edge on the echo pin, the timer is set to generate an interrupt in 2 ms. This procedure has the purpose of generating an interval between the end of the reading of a sensor and the beginning of the reading of the next one because it was realized that if the readings are executed immediately consecutive one sensor would interfere in the operation of the other. After 2 ms has elapsed and the timer interrupt is triggered, the trigger pin of sensor two (S2) is set to high level, and everything is processed as explained for sensor one (S1). The Wall-Following algorithm uses data from the sensors to navigate autonomously and with the developed logic it makes possible changes in the environment without the function of following walls is affected.

Figure 3 shows, from a flowchart, the reactive logic used in the mobile robot, for navigation in the operating environment, positioned at a parameterizable distance of 10 cm from the wall.

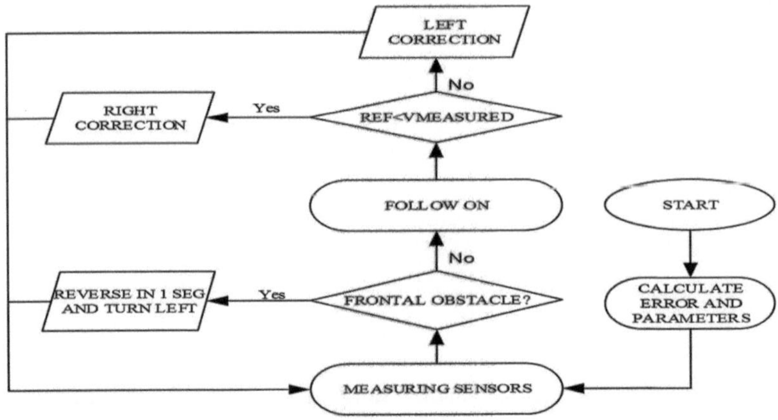

Fig. 3. Wall-Following Flowchart (Prepared by the author)

The derivative integrative proportional controller (PID) for navigation in the closed-loop speed control is carried out on-board. Thus, when receiving the data of the ultrasonic sensors, the algorithm will take the necessary actions sending proportional signals via Pulse Width Modulation (PWM) to the motors, correcting or maintaining its position. Figure 4 shows the control diagram of the mobile robot as a whole.

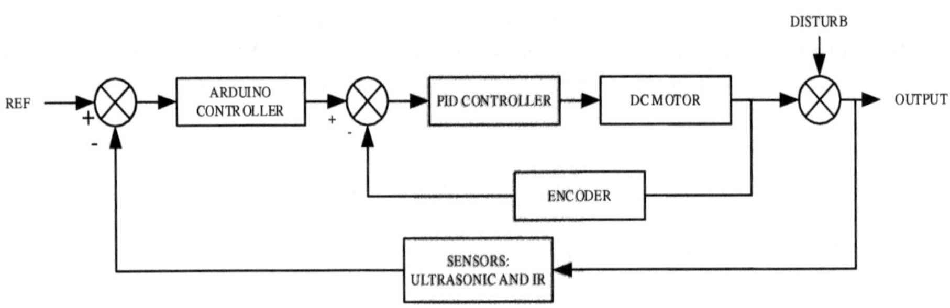

Fig. 4. Control diagram of the mobile robot (Prepared by the author)

For the implementation of the PID controller in the Atmega2560 microcontroller this be discretised. The starting point for deriving the discrete time controller is the continuous controller shown in Eq. 2.

$$U(t) = K_P e(t) + K_i \int_0^t e(t)dt + K_d \frac{d}{dt} e(t) \tag{2}$$

In Eq. 3, e(t) is the controller error and is defined as indicated in Eq. 3, where r (t) is the reference, or desired value, y(t) is the current output of the system being controlled.

$$e(t) = r(t) - y(t) \tag{3}$$

The derivative and integral present in Eq. 3 can be approximated with the expressions in Equations five and six respectively, where T is the sampling period of the system.

$$\frac{de(t_k)}{dt} \cong \frac{e(k) - e(k-1)}{T} \tag{4}$$

$$\int_0^{t_k} e(t)dt \cong \sum_{i=1}^{k} e(t_i)T \tag{5}$$

Substituting Eqs. 4 and 5 in Eq. 2 and after performing some algebraic manipulations, we arrive at Eq. 6 which represents the digital implementation of the PID controller. In this format, it can be implemented in the microcontroller.

$$u[k] = u[k-1] + \left(K_p + K_i\frac{T}{2} + \frac{K_d}{T}\right)e[k] + \left(-K_p + K_i\frac{T}{2} - \frac{2K_d}{T}\right)e[k-1] + \frac{K_d}{T}e[k-2] \tag{6}$$

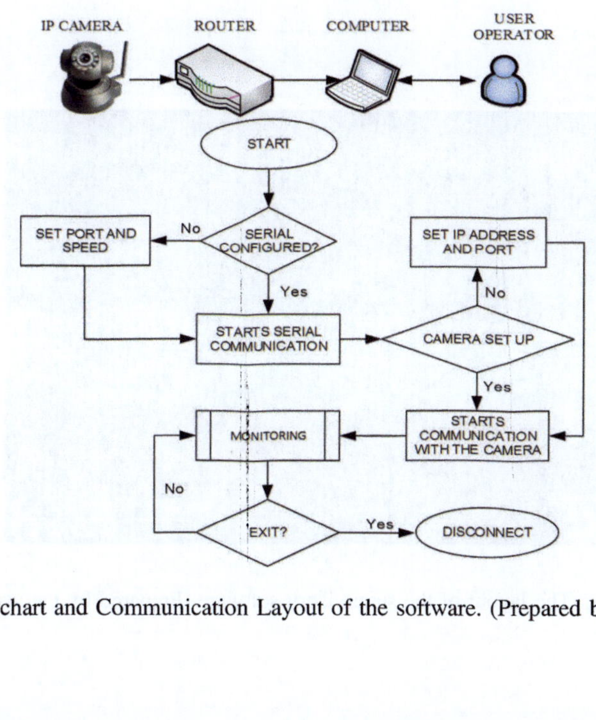

Fig. 5. Flowchart and Communication Layout of the software. (Prepared by the author).

The camera makes the video stream using the H263 codec. Provides better quality for high-resolution images and better quality in transferring audio and video over the TCP network [9]. The communication can be done with Wi-Fi 2.4 GHz or ethernet. Used for the robot will be independent of the connection of commands. The network being managed by the Wi-Fi router [8]. Figure 5 shows the layout and communication flowchart in the software developed for video monitoring.

The supervisory system used to access the camera was developed in C++ using the QT Creator platform. The developed form consists of four tabs: in the first one, there is

Fig. 6. The layout of the supervisory software (Prepared by the author).

the boot menu; in the second, the configuration menu of the camera, with the serial port used, in which the IP to be placed in that of the camera itself. The third one is the ActiveX add-in for capturing images and videos in real time. The fourth is a graphical analysis of the sensor reading [2].

Moreover, in the Fig. 6 shows the program configuration tab developed in the QT software, in which confirmation of the connected serial port can be visualized.

4 Applications of the Prototype

4.1 Computational Results

The results were obtained initially from tests performed with the sensors, in stages in which the function of each sensor was performed in isolation. After checking the functioning of the sensors in the robot, a new library was implemented in C so that they functioned properly, according to the need of navigation in an internal environment.

Furthermore, the ultrasonic sensors HC-SR04 could be read successfully, so that they worked side-by-side without interference. During the tests, it was realized that the ultrasonic sensors in parallel must have a gap of at least 2 ms between the end of the reading of one and the beginning of the reading of the other. If there is no such interval between readings, some interference occurs between them, generating inconsistent readings. Moreover, Fig. 7 shows the graph of the robot performing the correction about the reference defined in the tests as 20 cm and 30 cm demonstrating the effectiveness of the controller.

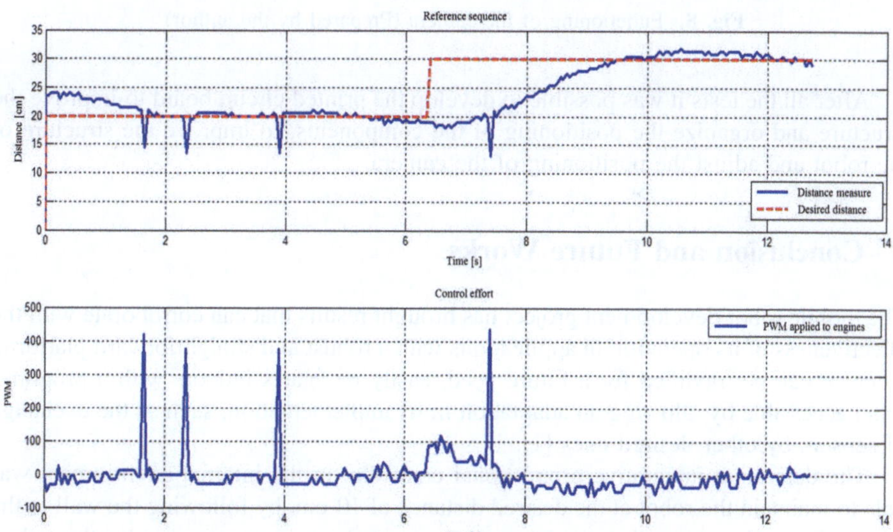

Fig. 7. Graph performed control test (Prepared by the author).

By applying the encoders coupled to the driven engines of the rover platform 5, it was possible to carry out the readings successfully. With them, it was possible to measure both the displacement and the speed of the vehicle. Using the Bluetooth module, it was possible to establish a connection between the robot and software in the computer, through which it was possible to send commands and receive measurements of the sensors.

The developed software worked expectedly, being able to view the images and videos online, provided by the camera and control the robot manually, and thus help in the surveillance and monitoring of internal environments.

The IP camera works satisfactorily by being able to transmit and capture images online through software created exclusively for this project [6]. Figure 8 shows the operation of the camera.

Fig. 8. Functioning of the camera (Prepared by the author).

After all the tests it was possible to develop the printed circuit board to improve the structure and organize the positioning of the components, to improve the structure of the robot and adjust the positioning of the camera.

5 Conclusion and Future Works

The mobile robot development project has brought results that can corroborate with the effectiveness of its operation in applications with a robust and straightforward platform, where it can be modified for a future need, easily navigates indoors with a programming accessible by allowing an adaptation in its implementation, such as the exchange of sensors by other desired ones [1].

The derivative integrative proportional controller in the internal environment was able to maintain the robot at the desired distance of 10 cm, by following the wall to the right, effectively responding to abrupt changes in the trajectory as curves that allow traversing the entire perimeter the environment.

For future work, it is believed that it is possible to perfect the controller by making the data processing remotely and include other sensors, increasing the accuracy of the robot. Real-time mapping can also be implemented to obtain two-dimensional maps and location with the printing of navigation data as coordinates and can use the robot even in outdoor environments.

Infrared sensors positioned below the robot can be implemented with the objective of detecting possible steps, avoiding falls and, consequently, damage to the robot. It is also possible to have logic for using a rechargeable base where the robot is detecting a certain percentage of the battery seeing the need to recharge it would return to its initial position.

Acknowledgements. The first author is thankful for "Coordination for the Improvement of Higher Level-or Education- Personnel" (CAPES) and the fifth author is grateful to National Counsel of Technological and Scientific Development (CNPq) via Grants #305805/2017-7. The authors would like to thank Edson Queiroz Foundation/University of Fortaleza for all the support.

References

1. Aouf, A., Boussaid, L., Sakly, A.: A PSO algorithm applied to a PID controller for motion mobile robot in a complex dynamic environment. In: International Conference on Engineering & MIS (ICEMIS), Monastir, pp. 1–7 (2017)
2. Arshad, A., Badshah, S., Soori, P.K.: Design and fabrication of smart robots. In: 5th International Conference on Electronic Devices, Systems and Applications (ICEDSA), Ras Al Khaimah, pp. 1–4 (2016)
3. Goel, A., Bhattacharya, J., Anika, G.: Robot relay: an optimal path planning for multiple client services. In: International Conference on Inventive Computation Technologies (ICICT), Coimbatore, pp. 1–6 (2016)
4. Âström, K.J., Hagglund, T.: PID Controllers: Theory, Design, and Tuning, 2 edn., 343 p. United States of America (1995)
5. Al-Mutib, K., Abdessemed, F., Faisal, M., Ramdane, H., Alsulaiman, M., Bencherif, M.: Obstacle avoidance using wall-following strategy for indoor mobile robots. In: 2nd IEEE International Symposium on Robotics and Manufacturing Automation (ROMA), Ipoh, pp. 1–6 (2016)
6. Loukil, K., Khalfa, M., Jmal, M.W., Frikha, T., Abid, M.: Design and test of smart IP-camera within the reconfigurable platform. In: 2nd International Conference on Anti-Cyber Crimes (ICACC), Abha, pp. 25–29 (2017)
7. Ogata, K.: Modern Control Engineering, 5th edn. Pearson/Prentice Hall, Upper Saddle River (2010)
8. Pinheiro, P.R., Neto, Á.M.S., Aguiar, A.B., Pinheiro, P.G.C.D.: A novel hybrid methodology applied optimization energy consumption in homogeneous wireless sensor networks. In: Wireless Sensor Networks - Insights and Innovations, 1 edn. IntechOpen (2017)
9. Purbaya, S., Ariyanto, E., Sudiharto, D.W., Wijiutomo, C.W.: Improved image quality on surveillance embedded IP camera by reducing noises. In: 3rd International Conference on Science in Information Technology (ICSITech), Bandung, pp. 156–159 (2017)

10. Siegwart, R., Nourbakhsh, I.R.: Introduction to autonomous mobile robots, 336 p. Massachusetts Institute of Technology, Cambridge (2004). http://home.deib.polimi.it/gini/robot/docs/siegwart.pdf. Acesso em 05 mar 2017
11. Lakshmanaprabu, S.K., Shankar, K., Khanna, A., Gupta, D., Rodrigues, J.J., Pinheiro, P.R., de Albuquerque, V.H.C.: Effective features to classify big data using social internet of things. IEEE Access **6**, 24196–24204 (2018)
12. Mohamed, T.B., et al.: Development of mobile robot drive system using mecanum wheels. In: International Conference on Advances in Electrical, Electronic and Systems Engineering (ICAEES), Putrajaya, pp. 582–585 (2016)

BrainMRI Enhancement as a Pre-processing: An Evaluation Framework Using Optimal Gamma, Homographic and DWT Based Methods

S. Harish[1(✉)] and G. F. Ali Ahammed[2]

[1] VTU, Belagavi, India
harishsrinivasaiah@gmail.com
[2] VTU PG-Center, Mysuru, India
aliahammed78@gmail.com

Abstract. The domain of medical imaging has been extensively studied for various parts of human bodies. Study on Brain MRI is quite helpful for the diagnosis of brain related diseases. The availability of noises and less clarity of the input Brain MRI brings ambiguity while analysis. This paper proposes optimized gamma correction, homographic filter and discrete wavelet transformation. The performance evaluation of these methods is analysed with parameters global contrast factor (GCF), contrast per pixel (CPP), contrast (C) and sharpness (S) values.

Keywords: BrainMRI enhancement · Pre-processing
Discrete wavelet transform

1 Introduction

Brain is considered as one of the cri.tical organ of human body and in case of any medical situation, it demands a sophisticated mechanism to assess its criticality situation. Out of various forms of medical imaging techniques [1], Magnetic Resonance Imaging (MRI) is one of the best method to diagnose the medical condition of brain with better granularity. Although, there are various research work being carried out towards medical image advancement of brain [2–5], still studying the precritical condition of brain using MRI is still a big challenge [6]. One of the first step of this challenge is the enhancement procedure of the images obtained from MRI of brain [7–11]. The Brain MRI enhancement is a processing aimed for various objectives of diagnosis that mainly include increasing visual perception so that the anatomy of the brain is clearly visible [12]. The core principle in the process of enhancement includes to balance the primary and secondary information that are responsible for the perceptual quality. The method of gamma-correction is evolved over a traditional system where the real-life systems are considered as a non-linear system as oppose of a linear system in the orientation of illumination intensity. A definite relationship exists for the hardware characteristics (Gamma) in terms of functional relationship as:

R. Silhavy et al. (Eds.): CoMeSySo 2018, AISC 860, pp. 287–298, 2019.
https://doi.org/10.1007/978-3-030-00184-1_27

$$I_{in} = (I_{ori})^{1/Gamma} \quad (1)$$

Where, I_{in} is pre-compensation image data, and I_{ori} is original image data value [13]. The contrast ration is improvised by the gamma correction method to achieve a better extraction of feature for classification purpose and other computer vision requirements but it require a synchronized optimization for Brain MRI input data. In this paper, initially an approach of optimization is implemented by varying value of gamma from the lower bound limit of 0.9 to upper bound limit of 1.2 and the corresponding global contrast value (GCV) is computed [14]. The situation where the fuzziness is higher in the Brain MRI inputs along with lower contrast, where it is hard for the practitioners to analyses. The optimal homographic filtering method proposed in this paper aims to improvise the contrast along with the optimization of the sharpness by means of overpowering the noises [15]. Finally, in this paper a very much optimal method suitable for the Brain MRI using different methods of the wavelet transformation is proposed. The paper is organized into different section where Sect. 2 describes some of the important work of image enhancement methods along with the methods used for MRI enhancement.

Section 2 discusses about the existing literatures towards brain MRI images followed by discussion of research problem in Sect. 3. Section 4 briefs of the system architecture of teh proposed framework as well as its associated operations followed by briefing of dataset used for analysis in Sect. 5. Section 6 discusses about the result analysis while Sect. 7 concludes the paper.

2 Related Work

An exhaustive survey study is conducted by the Harish et al. [16] with the focus on the understanding of the brain MRI enhancement techniques along with the study on classification and segmentation.In the study of Chen et al. [17], a hierarchical correlation histogram analysis is performed in order to assess the greyscale distribution with respect to degree of pixel intensity. The outcome of the study further shows that the proposed system significantly improves the adaptive contrast enhancement factor during pre-processing stage while achieving higher accuracy and lesser recognition time. Super-resolution is one of the frequently used techqniue to offer better acquisition of the information related to brain MRI. Gholipour et al. [18] have presented an unique estimation theory that is capable of reducing the error during analysing the slides of brain MRI. This technique is also adopted by Liang most recently [19]. Loizou et al. [20] have claimed that analysis of texture of the brain image could significant assists in closer investigation using frequency as well as modulation factors. The authors have used supervised learning approach to further boost up classification process. Shape-based models are considered to be another significant method to diagnose the anatomical shape of brain. According to Cerrolaza et al. [21], a hierarchical decomposition added to shape-based modelling could increase investigational speed, especially during segmentation process. Elliott et al. [22] have discussed that usage of probability theory e.g. random forest significant assists in localizing abnormality section of a brain MRI. Makropoulos et al. [23] have presented a framework that could offer better segmentation performance using expectation-maximization theory. Registration-based method is also proven to be further helpful in analysing various hidden information in brain MRI. Pilutti et al. [24] have used registration technique on the basis of multi-contrastive factor to leverage the diagnosis process registration-based method was also adopted by Marami et al. [25] who has also used Kalman filter to perform classification. Liu et al. [26] have used a fusion-based approach using labels over the brain MRI in order to assists in segmentation. According to Osadebey et al. [27], integrated usage of various operations increase the quality of investigation of brain MRI. An experimental analysis of brain MRI towards identifying specific disease was carried out by Trojachanec et al. [28].

3 Research Problem

The prior section discusses about various forms of research approaches towards analyzing brain MRI. It has been noticed that majority of the research contributions aims towards segmentation process or using certain other forms of analytical process. The primary agenda of such process is to explore some unique form of abnormality within a scope of given brain MRI. However, the biggest problem in such methodology are that very few research work has been focused towards possibility of improving the visualization of the brain MRI in the form of image enhancement. As the images are mainly considered from dataset, they are believed to be error free and hence the research work is progressed towards solving something advanced issues related to diagnosis. However, such research work may not be applicable when exposed to real-time brain MRI images as they may posses higher degree of noises, errors, and various other illumination related artifacts. Therefore, the focus toward image enhancing options is found significantly missing in existing literatures. The next section briefs of one of our research contribution to deal with this problem.

4 System Architecture of Proposed Framework

The prime motive of the proposed study is to introduce an efficient framework that is capable of performing comprehensive image enhancement for the given brain MRI image. The contribution of the proposed system is also to ensure that a universal interface system is built where it becomes feasible to perform comprehensive image enhancement of any type of brain images irrespective of an form of brain abnormalities or any form of illumination issues. Such model also aims to offer productive time to diagnose the brain MRI image more closely with less efforts.

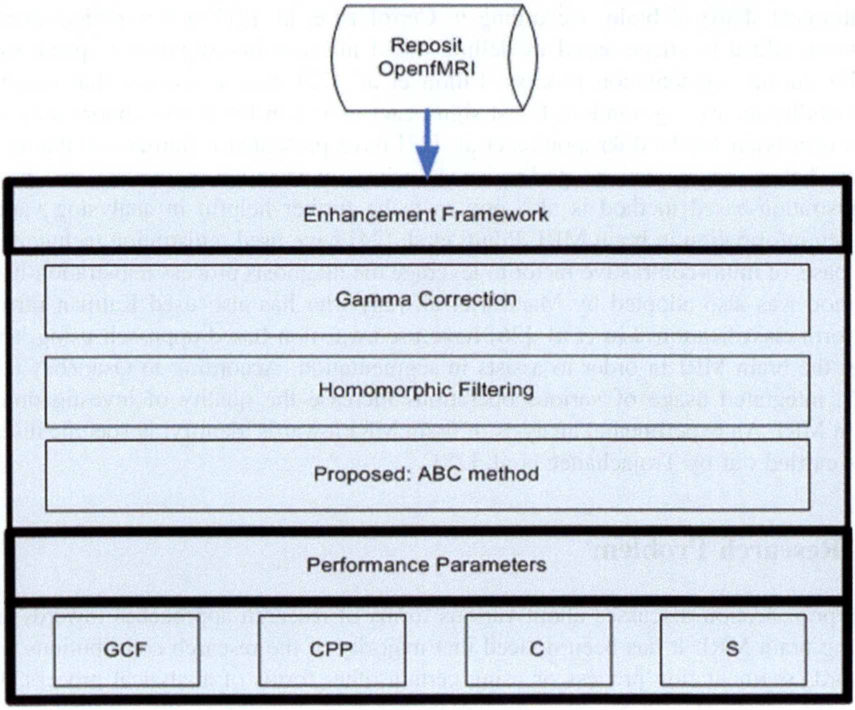

Fig. 1. Proposed framework architecture

The proposed framework architecture is shown in Fig. 1, there are three building blocks that includes (1) evaluation of BrainMRI enhancement using gamma correction, (2) evaluation of BrainMRI enhancement using homographic filtering and finally evaluation of BrainMRI enhancement using discrete wavelet transformation (DWT). The final block includes performance metrices which are: Global Contrast Factor (GCF), Contrast per Pixel (CPP), Contrast (C) and Sharpness (S) values.

4.1 Gamma Correction

Gamma correction method is adopted to benchmark the result with the proposed ABC method. In this method, the framework provisions and option to choose input BrainMRI (disk image file:D_{if}) from OPenfMRI dataset that is further goes through a process of digitization by passing the D_{if} in a combined function of sampling and quantization $f_{digital}$ and the digital pixel equivalence is stored as MxN matrix namely I_O. The computational complexity $O(D^2)$, (where D is the dimension of I_O) which is handled by the functional control of greyscale applied to I_O. The algorithm description is shown below:

Algorithm for BrainMRI Enhancement using Gamma Correction

Input: I_O
Output: I_{EN}
Start
$D_{if} \leftarrow f_{io}$(disk image file)
$I_O(MxNxD) \leftarrow f_{digital}(D_{if})$
Check for higher dimension
    If D = 3
        $I_O(MxN) \leftarrow f_{grey}(I_O(MxNxD))$
    End
Initialize gamma
$I_O(MxN) \leftarrow I_O(MxN)^{gamma}$
$I_{EN} \leftarrow 255x\ (I_O(MxN))/300$
Compute: global contrast factor, sharpness, contrast per pixel , contrast
End

The I_O obtained after grey-scaling goes through the process of gamma correction using Eq. (1) described in introduction section and finally it is being normalized by threshold factor multiplication (Th_m) of 255/300. Table 1 shows the values of GCF, CPP, C, and S for test input BrainMRI from OPenfMRI dataset.

Table 1. Values of GCF, S, CPP, C for input BrainMRI from MRI free dataset of subject AA (1): IM_00023, head spine.

Gamma	(I_O)	(I_{EN})	GCF	CPP	C	S
1	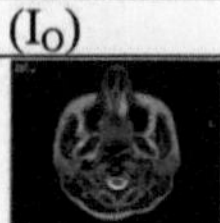	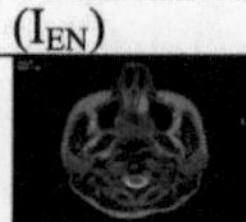	6.1799	3.9496	1.6991	5.191
1.5	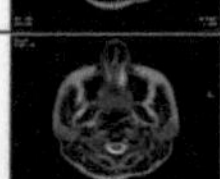	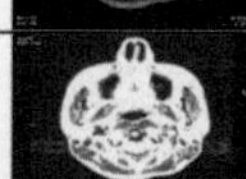	8.4829	12.3312	4.2954	10.2913
2	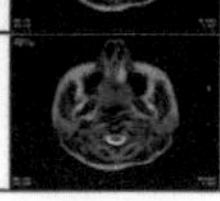		6.5822	19.1233	3.7472	16.0264

The 'GCF' shows a non-uniform pattern with the increasing value of gamma whereas the 'CPP' increases from 3.9496 to 19.1233 if gamma is 1 and 2 respectively. The value of 'C' increases if the gamma is increased from 1 to 1.5 at the same time it decreases if gamma is increased to 2 and the value of 'S' exhibit increasing pattern. Though the above result shows some acceptable visually perceptual acceptable result but require to obtain an optimal threshold of gamma to get widely acceptable visually enhanced BrainMRI, which is considered as future enhancement work.

4.2 Homographic Filtering

In order to improvise over gamma correction method, a method where brightness need to be normalized uniformly across the input BrainMRI and then further increases the contrast by removing redundant random signals which are multiplicative in nature. One of such qualifying filter is homomorphic filter which is being customized to suit BrainMRI enhancement. This is the second work along with gamma correction is provisioned as an evaluation framework to establish benchmark with the proposed ABC method. The process involved in applying homomorphic filter initially considers I_O along with the parameters α_L, α_H and the order of filter f_{order} which are represented with numerical form of metrices. Further, the process improvises the homomorphic filter and produce the enhanced image $I_{E\text{-}HF}$ in the form of a matrix consists of numerical elements. It also computed GCF and local contrast at each scale (LC). Finally, the process computed CPP, C and S. The algorithmic steps incorporated to obtained the homomorphic filter generated enhanced image are as follows:

Algorithm for BrainMRI image enhancement using Homomorphic Filtering

---

Input: I_O
Output: $I_{E\text{-}HF}$
Start
Initialize I_O, α_L, α_H , f_{order}
  Enable function enhanced $f_{homorphic}$
    $I_{E\text{-}HF} \leftarrow f_{homorphic}(I_O, \alpha_L, \alpha_H, f_{order})$
      $[]_{row\times col}$ $\leftarrow$ **Compute size**(I_O)
      Enable function $f_{create\text{-}homorphic}(row, col, f_{order})$
        for i $\leftarrow$ 1:row
          for j $\leftarrow$ 1:col
            generateH_{filter}:I_O(MxN)
          end
        end
    Compute $H_{filter} \leftarrow ((\alpha_H - \alpha_L)\times H_{filter}) + \alpha_L)$
    $H_{filter} \leftarrow 1 - H_{filter}$
    Compute log of I_O
    Perform $I_{O\text{-}FFT} \leftarrow f_{FFT\text{-}2}(\log(1+I_O)$
    Apply filter on FFT image
    Compute inverse FFT of filtered image
    Perform inverse log
    Compute normalise values
    End
  Compute GCF, CPP, C, S
  End
End

Table 2 shows the values from homomorphic filtering method for GCF, CPP, C, and S for same test input BrainMRI which is taken for evaluating gamma correction from OPenfMRI dataset.

Table 2. Values of GCF, S, CPP, C for input BrainMRI from MRI free dataset of subject AA (1): IM_00023, head spine. For $\alpha_L = \alpha_H = 0.05$, $f_{order} = 0.5, 1, 2.5$

f_{order}	(I_O)	$(I_{E\text{-}HF})$	GCF	CPP	C	S
0.5	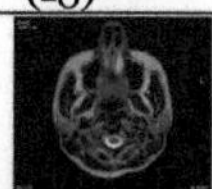	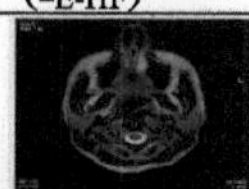	6.9531	3.0553	1.7129	5.3138
1	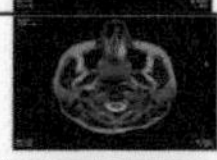	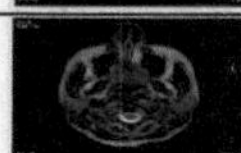	6.9531	3.0553	1.7129	5.3138
2.5	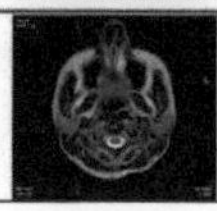	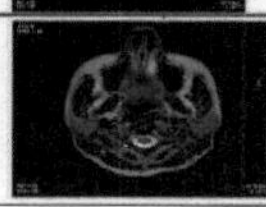	6.9531	3.0553	1.7129	5.3138

4.3 DWT Based Enhancement of Input BrainMRI

As resolution enhancement is an essential operation involved in image processing, therefore the research study performs discrete wavelet transform (DWT) numerical and functional analysis in order to sample wave-let coefficients discretely. After performing the above two processes, the image contrast enhancement is subjected to specify wavelet coefficient matrix individually. The process thereby considers different wavelet filters and provision users to apply one of them to obtain the outcome. It basically considers different wavelets such as haar, doaubechies2, symlets (5, 9), biorthogonal 1.1, biorthogonal 5.5, etc. to process DWT enhancement on the input BrainMRI image. The following analytical algorithm is designed and also subjected to assess DWT based enhancement on input BrainMRI image. It basically pro-cess single-level two-dimensional wavelet decomposition considering a specific wavelet name. The decomposition is done either with respect to the wavelet name or with respect to decomposition filters. Finally, after successful wavelet decomposition, the process computes low-low (LL), high-high (HH), high-low (HL), low-high (LH) coefficients which are basically represented with the form of a different matrix. matrix.

Algorithm for DWT based enhancement

Input: I_O
Output: I_{DWT}
Start
Initialize I_O, filter name (f_{name}), $r_{DWT\text{-}parameter}$
    Create Cell Array $\leftarrow f_{name}[1\times10]$
        for $i \leftarrow$ 1:10
            for each i
                Select specific f_{name} from Cell Array
            end
        end
        Apply $f_{dwt2}()$
        Co-eff$_{256\times256}$[] $\leftarrow f_{dwt2}(I_O, f_{name})$
        Extract [$LL_{256\times256}$, $HH_{256\times256}$, $HL_{256\times256}$, $LH_{256\times256}$] $\leftarrow$ Co-eff$_{256\times256}$[]
    Perform low-low coefficients enhancement
    $I_{DWT} \leftarrow f_{dwt2}$([LL,HL,LH, HH], f_{name})
    Compute GCF, CPP, C, S

End

The following Table 3 exhibits the visual outcomes of DWT based contrast enhancement which is performed on the test input sample BrainMRI which belong to the OPenfMRI dataset.

Table 3. Values of GCF, S, CPP, C for input BrainMRI from MRI free dataset of subject AA (1): IM_00023, head spine. For r_{DWT} = 1, 2, 3, wavelet filter: '*haar*'

$r_{DWT\text{-}parameter}$	(I_O)	Outcome of IDWT ('*haar*') wavelet filter	GCF	CPP	C	S
1	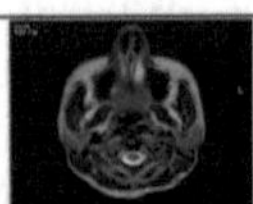	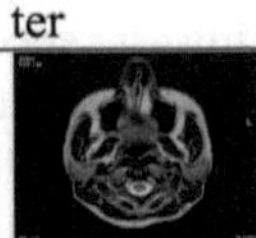	9.046	4.4748	2.6108	7.6837
2	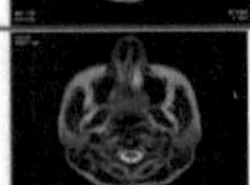	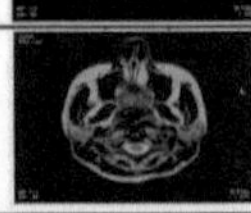	10.6615	4.9587	3.1583	9.2684
3	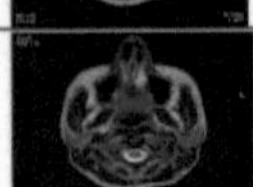	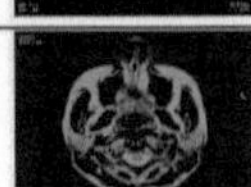	11.5899	5.4928	3.5839	10.521

The above Table 3 shows the final numerical as well as visual outcomes obtained after simulating the algorithm for DWT enhancement in a numerical computing environment. The entire simulation has been performed with respect to different values of r_{DWT}, which is a numerical DWT standard parameter. The following section further exhibits the experimental analysis which further demonstrates how the proposed system obtains optimal outcome in terms of different BrainMRI image enhancement parameters which are considered during the pre-processing step such as GCG, CPP, C, S.

5 Dataset Adopted

The dataset used in this proposed method for evaluation and apply the process of enhancement are taken from the dataset maintained by the OPenfMRI project which managed at Stanford University by the combined efforts of Centre for Reproducible Neuroscience and Poldrack along with the infrastructure support by Texas Advanced computing centre and Amazon. The funding and grants for this is obtained from the National Science Foundation, National Institute for Drug Abuse, and Laura and John Arnold Foundation [9].

6 Result and Analysis

The following Fig. 2 shows how the proposed optimization process obtained effective outcomes during BrainMRI image enhancement which is exclusively carried out during pre-processing stage.

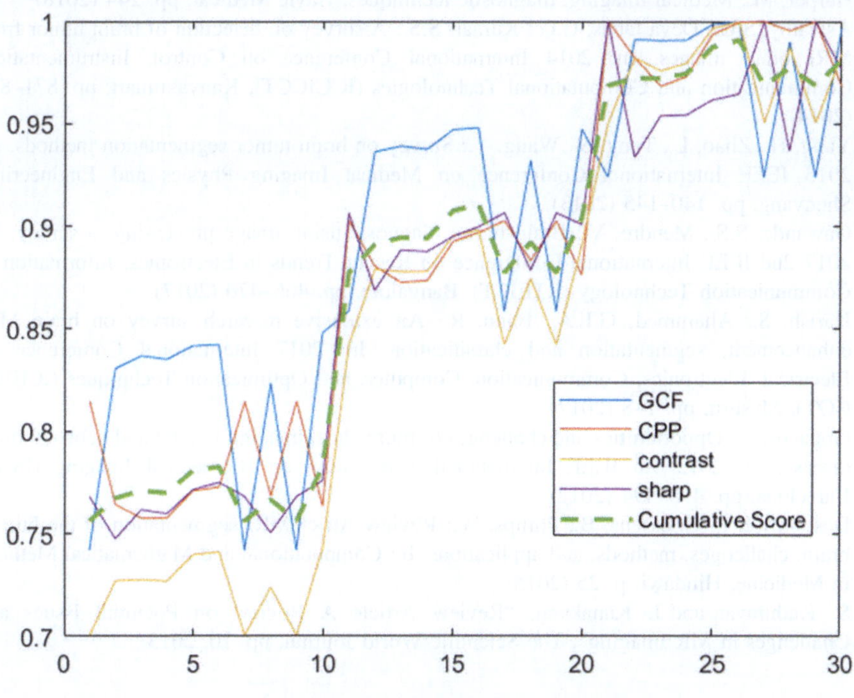

Fig. 2. Experimental Analysis

The above highlighted Fig. 2 shows the orientation of the curve of GCF, CPP, contrast and sharpness gradually increase which produce a better visual perception of the BrainMRI image with a significant cumulative score which can be further process cost-effectively for the detection of tumor. Therefore, the proposed system ensures its adaptability into futuristic BrainMRI processing and enhancement.

7 Conclusion

The proposed study presents a novel technique which offers efficient BrainMRI image enhancement by performing a cost-effective pre-processing stage. It also addresses the conventional problem pertaining to the BrainMRI images having much noisy attributes along with very less clarity, which further affects the performance of diagnostic process during detection of brain related diseases. The proposed study applied three different novel algorithms such as (1) Gamma Correction, (2) Homomorphic filtering followed by (3) DWT based enhancement which performs a lossless compression on the BrainMRI while not sacrificing the contrast level. Finally, the experimental outcomes justified the effectiveness of the proposed technique in terms of different performance metrices.

References

1. Harper, M.: Medical imaging: diagnostic techniques. Hayle Medical, pp. 244 (2018)
2. Aswathy, S.U., Deva Dhas, G.G., Kumar, S.S.: A survey on detection of brain tumor from MRI brain images. In: 2014 International Conference on Control, Instrumentation, Communication and Computational Technologies (ICCICCT), Kanyakumari, pp. 871–877 (2014)
3. Yang, H., Zhao, L., Tang, S., Wang, Y.: Survey on brain tumor segmentation methods. In: 2013 IEEE International Conference on Medical Imaging Physics and Engineering, Shenyang, pp. 140–145 (2013)
4. Gawande, S.S., Mendre, V.: Brain tumor diagnosis using image processing: a survey. In: 2017 2nd IEEE International Conference on Recent Trends in Electronics, Information & Communication Technology (RTEICT), Bangalore, pp. 466–470 (2017)
5. Harish, S., Ahammed, G.F.A., Banu, R.: An extensive research survey on brain MRI enhancement, segmentation and classification. In: 2017 International Conference on Electrical, Electronics, Communication, Computer, and Optimization Techniques (ICEEC-COT), Mysuru, pp. 1–8 (2017)
6. Gratacós, E.: Opportunities and challenges of biomedical imaging in fetal and neonatal brain disease. In: 2012 9th IEEE International Symposium on Biomedical Imaging (ISBI), Barcelona, pp. 493–494 (2012)
7. Despotović, I., Goossens, B., Philips, W.: Review ArticleMRI segmentation of the human brain: challenges, methods, and applications. In: Computational and Mathematical Methods in Medicine, Hindawi, p. 25 (2015)
8. S. Kathiravan and J. Kanakaraj, "Review Article A Review on Potential Issues and Challenges in MR Imaging", The Scientific World Journal, pp. 10, 2013

9. Chen, H.J., Zhang, L.J., Lu, G.M.: Review article multimodality MRI findings in patients with end-stage renal disease. BioMed Research International, p. 12 (2015)
10. Bahadure, N.B., Ray, A.K., Thethi, H.P.: Research article image analysis for MRI based brain tumor detection and feature extraction using biologically inspired BWT and SVM. Int. J. Biomed. Imaging, 12 (2017)
11. Oprişan, A., Popescu, B.O.: Review article intracranial cysts: an imagery diagnostic challenge. Sci. World J. 9 (2013)
12. Pekcevik, Y., Orman, G., Lee, I.H., Mealy, M.A., Levy, M., Izbudak, I.: What do we know about brain contrast enhancement patterns in neuromyelitisoptica? Clin. Imaging **40**(3), 573–580 (2016)
13. Guan, X., Jian, S., Hongda, P., Zhiguo, Z., Haibin, G.: An image enhancement method based on gamma correction. In: 2009 Second International Symposium on Computational Intelligence and Design, Changsha, pp. 60–63 (2009)
14. Matkovic, K., Neumann, L., Neumann, A., Psik, T., Purgathofer, W.: Global contrast factor-a new approach to image contrast. Comput. Aesthetics **18**(2005), 159–168 (2005)
15. Rui, W., Guoyu, W.: Medical X-ray image enhancement method based on TV-homomorphic filter. In: 2nd International Conference on Image, Vision and Computing (ICIVC), Chengdu, pp. 315–318 (2017)
16. Harish, S., Ahammed, G.F.A., Banu, R.: An extensive research survey on brain MRI enhancement, segmentation and classification. In: 2017 International Conference on Electrical, Electronics, Communication, Computer, and Optimization Techniques (ICEEC-COT), Mysuru, pp. 1–8 (2017)
17. Chen, C.-M., et al.: Automatic contrast enhancement of brain MR images using hierarchical correlation histogram analysis. J. Med. Biol. Eng. **35**(6), 724–734 (2015)
18. Gholipour, A., Estroff, J.A., Warfield, S.K.: Robust Super-resolution volume reconstruction from slice acquisitions: application to fetal brain MRI. IEEE Trans. Med. Imaging **29**(10), 1739–1758 (2010)
19. Liang, Z., He, X., Teng, Q., Wu, D., Qing, L.: 3D MRI image super-resolution for brain combining rigid and large diffeomorphic registration. IET Image Process. **11**(12), 1291–1301 (2017)
20. Loizou, C.P., Murray, V., Pattichis, M.S., Seimenis, I., Pantziaris, M., Pattichis, C.S.: Multiscale amplitude-modulation frequency-modulation (AM–FM) texture analysis of multiple sclerosis in brain MRI images. IEEE Trans. Inf Technol. Biomed. **15**(1), 119–129 (2011)
21. Cerrolaza, J.J., Villanueva, A., Cabeza, R.: Hierarchical statistical shape models of multi-object anatomical structures: application to brain MRI. IEEE Trans. Med. Imaging **31**(3), 713–724 (2012)
22. Elliott, C., Arnold, D.L., Collins, D.L., Arbel, T.: Temporally consistent probabilistic detection of new multiple sclerosis lesions in brain MRI. IEEE Trans. Med. Imaging **32**(8), 1490–1503 (2013)
23. Makropoulos, A., et al.: Automatic whole brain MRI segmentation of the developing neonatal brain. IEEE Trans. Med. Imaging **33**(9), 1818–1831 (2014)
24. Pilutti, D., Strumia, M., Hadjidemetriou, S.: Bimodal nonrigid registration of brain MRI data with deconvolution of joint statistics. IEEE Trans. Image Process. **23**(9), 3999–4009 (2014)
25. Marami, B., Scherrer, B., Afacan, O., Erem, B., Warfield, S.K., Gholipour, A.: Motion-robust diffusion-weighted brain mri reconstruction through slice-level registration-based motion tracking. IEEE Trans. Med. Imaging **35**(10), 2258–2269 (2016)
26. Liu, H., et al.: Label fusion method based on sparse patch representation for the brain MRI image segmentation. IET Image Process. **11**(7), 502–511 (2017)

27. Osadebey, M., Pedersen, M., Arnold, D., Wendel-Mitoraj, K.: No-reference quality measure in brain MRI images using binary operations, texture and set analysis. IET Image Proc. **11** (9), 672–684 (2017)
28. Trojachanec, K., Kitanovski, I., Dimitrovski, I., Loshkovska, S.: Longitudinal brain MRI retrieval for Alzheimer's disease using different temporal information. IEEE Access **6**, 9703–9712 (2018)

Efficiency Comparison of Modern Computer Languages: Sorting Benchmark

Agnieszka Bier and Zdzisław Sroczyński(✉)

Institute of Mathematics, Silesian University of Technology,
Kaszubska 23, 44-100 Gliwice, Poland
{agnieszka.bier,zdzislaw.sroczynski}@polsl.pl

Abstract. The paper surveys the execution features of ready-to-use sorting procedures in various modern computer languages/compilers. The chosen sorting functions were tested for randomly generated data sets of different size and structure resembling the lists or arrays commonly used in real life IT solutions. Our results reveal some differences between particular implementations in efficiency of sorting in terms of CPU load and execution time.

1 Introduction

The development of contemporary programming languages and compilers is driven by the rapid increase of the amount of data processed, which is often called big data computing. There are many solutions based on some kind of runtime environment in this area. Runtime environment idea has many advantages, as it simplifies multi-platform implementation and helps to increase the security, although this additional abstraction level may possibly decrease the efficiency, especially for complex calculations and sophisticated data structures. Because of that, natively compiled machinecode had been chosen for a long time as the main tool for these kinds of tasks. Nowadays, the efficiency of interpreted and runtime-based computer languages is comparable to compiled ones.

Regardless of the actual reason of that increase of efficiency, which is driven by optimizations in runtime environments and just-in-time compilation, the applications of many different computer languages vary from front-end to back-end and from UI-based to server solutions. One of the best examples here is Java [1], which become essential for enterprise-level programming despite the use of bytecode, running in special virtual machine (JVM). There is also an ongoing research on the influence of the choice of programming languages and environments on energy saving [2]. Certainly, the virtual machine idea was the proper one for multi-platform, machine independent computing, but on the other hand it has not guaranteed high efficiency for a long time.

One of the most fundamental operations in computer science is sorting, which is still the matter of extensive research [3], especially in big data processing and parallel computing [4,5]. Sorted data sets are useful for presentation purposes and reporting, giving the proper user experience of a natural order of items.

R. Silhavy et al. (Eds.): CoMeSySo 2018, AISC 860, pp. 299–310, 2019.
https://doi.org/10.1007/978-3-030-00184-1_28

Moreover, sorted data structures assure the most efficient searching algorithms, essential in many application domains. Popular non-trivial sorting algorithms, as quicksort [6,7] or mergesort [8] run in $O(nlogn)$ time in average scenario. A ready to use implementation of sort procedure is available in almost every contemporary standard library for various computer programming languages. Usually, the programmer may provide the optional comparison function or method, which allows to modify the order (from ascending to descending and vice-versa) or take into account some specific features of the data, as for example text encoding for national diacritics characters.

Because of the wide usage, well-known theoretical efficiency and implementation embedded in most of the software development environments, sorting is the ideal candidate for the programming language benchmark. The behaviour of the application testing sort procedures can reveal issues in memory management, as well as the quality of generated output machine or pseudo/byte code or the quality of interpreter/virtual machine engine for particular implementation of computer language translator [9,10].

The above considerations have motivated our research team to investigate the efficiency issue of ready-to-use sorting functions in different programming languages. We had carefully implemented sorting of arrays and collections of strings with the use of out-of-the-box procedures available in standard libraries of different programming languages of our choice. The implementations were then verified for three different data sizes in terms of the CPU load and execution time. The presented results should help in better understanding of the details of software development environments, in designing more efficient applications and avoiding some pitfalls while code reuse is needed. Moreover, we developed good benchmarks for the new or modified sorting algorithms, which can be compared in objective way with existing ones.

2 Benchmark for Sorting Algorithms

Sorting is the key task in many IT projects, involving data bases, big data analysis, business intelligence etc. Therefore for our experiments we used data structures consisting of n character strings of random length (1 to 8 characters) and random values (capital letters). This data structure, similar to ones introduced i other tests [11], resemble tables commonly spotted while processing personal information, banking information, IoT (Internet of Things) data [12], gene sets in bioinformatics [13] and every other data send over the Internet with the use of JSON or XML format [14], so it can be considered as standard example.

In this survey we present experiments held with the following computer languages and translators/compilers:

- C++ – Embarcadero C++ – Clang/LLVM-based win32 compiler (v7.30),
- mingw32-c++/gcc – win32 native compiler (v4.6.1),
- Object Pascal – Embarcadero Delphi – a native x86/win32 compiler (v32/10.2.3),

- Object Pascal – Embarcadero Delphi – LLVM-based x64 compiler (v32/10.2.3),
- Java – Oracle JVM 1.8 64bit,
- C# – Microsoft .NET 4.7 (Visual Studio 2015 Community Edition).

The translators mentioned above represent four native x86 or x64 compilers for two languages: universal C++ and Object Pascal/Delphi, used mainly for business and database desktop, UI-based applications. It is worth noting, that there are modern version among them, in particular LLVM-based ones.

The other part of the translators represent special "byte-code" compilers, which need an extra runtime environment to run their applications (JVM or .NET). We used Java 1.8 and .NET 4.7 in the experiments, because these versions are modern and widely used in production, especially for business logic and UI programming.

Despite their special characteristics, every tested language is suitable and more or less used in practice for numerical algorithms, modelling large data structures, image processing and the other complex computing.

There are some free solutions in translators test suite: Embarcadero C++, Delphi win32 (available in free Delphi Starter), Java and C#/.NET (partially open-sourced). The mingw compiler is an open-source version of gcc for Windows system. Embarcadero Object Pascal/Delphi LLVM-based x64 compiler is a commercial software available only in paid versions of the RAD Studio.

3 Test Suite Implementation

The test procedure mimicked the behaviour of the standard developer, who takes the opportunity and does not elaborate the efficiency of every available algorithm. Therefore, the test suite in conducted experiments utilized only the algorithms built-in into the standard library for the given programming language. This approach is quite common, although advanced high-performance computing projects certainly would require more comprehensive benchmarks [15].

The results of the experiments can help to distinguish the most appropriate software development solutions and to avoid common problems regarding software localization. Every test application performed two sorting procedures: using plain ASCII collation, then with the use of the specified locale. Clearly, the efficiency of localized processing was worse in every case because of computational overhead for string comparisons. An additional observation concerned the default localization, which is different for different languages. This will be discussed later in detail.

In every case, the test was developed in two stages. First, the dataset was generated with the use of random generator, then the actual sorting procedure was called. Some texts with national diacritics characters were added to ensure the proper order of the resulting dataset. The initialization of the dataset and the validation of localizations were performed before the actual execution time measurements.

3.1 Sorting Vectors in C++

The standard `vector` data structure was used to evaluate built-in sorting algorithm in C++ language (see the source code[1] in Listing 1). We used also `chrono` library with `high_resolution_clock` class for time measurements. The actual sorting procedure has been performed with the use of overloaded versions of `sort`, one for plain ASCII (commented out in the listing), and the second with extra argument – the ordering function with localization support (see Listing 2).

Listing 1. C++: implementation of sorting procedure

```
int maxv = 10000000;
setlocale(LC_ALL, "polish");
vector<string> stringv;
srand(time(NULL));

for (int i = 0; i < maxv; i++)
{
  string str = "";
  for(int j=1; j<= rand()%8+1;j++)
    str += (char)(65+(rand()%26));
  stringv.push_back(str);
}
auto start_time = chrono::high_resolution_clock::now();
//sort (stringv.begin(), stringv.end());//without locale
sort (stringv.begin(), stringv.end(), localeCmp);
auto end_time = chrono::high_resolution_clock::now();
auto time = end_time - start_time;
```

Due to some incompatibilities between versions of C++ language and available compilers, the `strcoll` function appeared to be the only portable, universal solution for localization. There was "Polish" locale used in all experiments with C++ and the other languages/environments described later. The same locale was in fact used also in the cases where the source code contained references to "default" value, because the test computer run Polish version of the Windows operating system.

Listing 2. C++: definition of the compare function

```
bool localeCmp (const string& s1, const string& s2) {
  return strcoll(s1.c_str(), s2.c_str()) < 0;
}
```

The CPU load during the experiments with sorting in C++ oscillated about 100% of one logical processor, i.e. 25% of overall computational power of the system (see Fig. 1). Therefore it can be concluded that the implementation of the `sort` standard function in C++ standard library is single-threaded.

---

[1] All listings contain only essential parts of source code, without obvious variable declarations and unnecessary function/method headers.

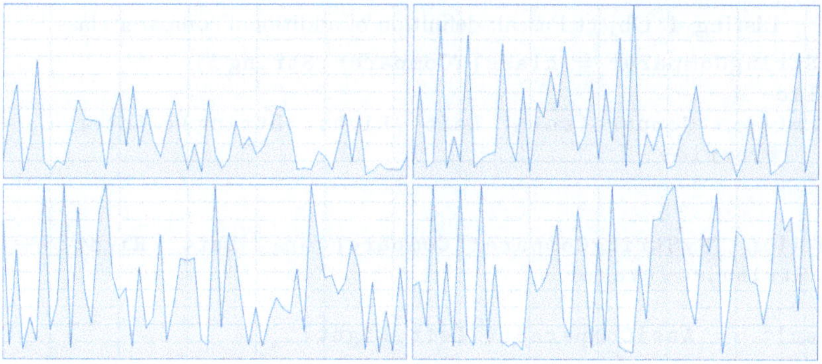

Fig. 1. C++: CPU load (0–100%) during sorting procedure for individual logical processors of the tested machine in the time span of 60 s.

3.2 Array Sorting in Object Pascal

The data structure chosen for Object Pascal implementation (see Listing 3) was the dynamic array, specific for Delphi variant of the language. Results of the other, more general tests pointed dynamic arrays as the fastest data structure for this language. To measure the time-span we used standard Windows' API: `QueryPerformanceCounter`. The actual sorting was performed with the use of generics `TArray.Sort<String>`, introduced in `System.Generics.Collections` standard library module.

Listing 3. Object Pascal: implementation of sorting procedure

```
var  tab: array of String;
maxv:=10000000;
SetLength(tab, maxv);
Randomize;
for i := 0 to maxv-1 do
 for j := 1 to Random(8)+1 do
  tab[i]:=tab[i]+AnsiChar(65+Random(26));
QueryPerformanceCounter( t1 );

//TArray.Sort<String>(tab); //without locale
TArray.Sort<String>(tab, TMyStringComparer.Create);

QueryPerformanceCounter( t2 );
QueryPerformanceFrequency( vFreq );
result:=(t2-t1)/vFreq;
```

The line invoking `Sort` method has been written in two versions, one for plain ASCII order (commented out), the second for sorting according to (Polish) locale. Special class implementing comparer is needed to sort with locale, as described in Listing 4.

Listing 4. Object Pascal: definition of additional comparer class

```
TMyStringComparer = class(TComparer<String>)
public
  function Compare(const Left, Right: String): Integer; ↩
      override;
end;

function TMyStringComparer.Compare(const Left, Right: ↩
    String): Integer;
begin
 result := AnsiCompareStr(Left,Right);
end;
```

There were two Pascal compilers used in the experiments, with different architectures (classic backward-compatible one vs LLVM-based one) and different target (32bit vs 64bit). The 64bit version performed better – the detailed results are presented later.

The CPU load while running the test application written in Object Pascal/Delphi varied, but generally stressed the processor similarly to C++. The CPU load reached 25% of the overall system power, so the sorting implementation is also single-threaded in this case. The CPU load details for the 60 s period of time are given in Fig. 2. It is worth to mention, that for majority of implementations there is a source code available, and the details of implementation could be revealed by code review as well.

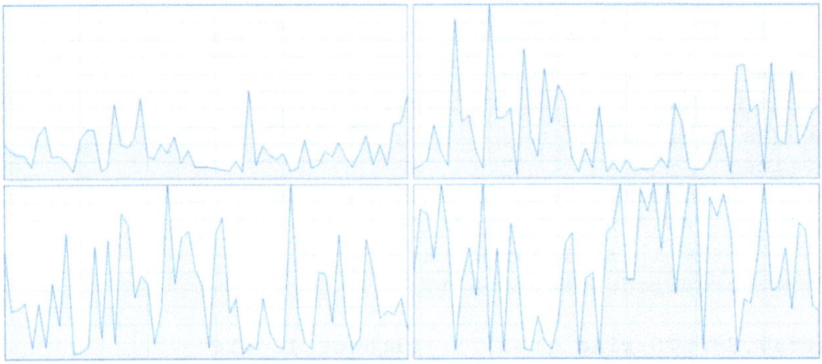

Fig. 2. Object Pascal: CPU load (0–100%) during sorting procedure for individual logical processors of the tested machine in the time span of 60 s.

3.3 Implementation of Sorting in Java

There were two implementations of sorting procedure in Java examined during our experiments. The main difference was the internal data structure used, which was a generics `ArrayList` in the first case and ordinary array in second case. The initialization of the strings for the testing was analogous, so for the brevity of the example we provide the details only for ArrayList – see Listing 5.

Listing 5. Java: initialization of the data structures and comparer

```
Random rnd = new Random();
String tmpstr;
int maxv = 10000000;

ArrayList<String> listastr = new ArrayList<String>()↩
    ;
for(int i = 1; i<= maxv; i++)
{
 tmpstr="";
 for(int j = 1;j<=rnd.nextInt(8)+1;j++)
       tmpstr+=(char)(65+rnd.nextInt(26));
 listastr.add(tmpstr);
}
Collator coll = Collator.getInstance(Locale.↩
    getDefault());
coll.setStrength(Collator.PRIMARY);
```

The source code excerpts with the calls to `sort` methods are given in Listing 6. To compute time we used standard `currentTimeMillis()` method from `System` class. There has been the default collator used for both data structures. Moreover, we tried to utilize `parallelSort` method, but for some reason it was extremely inefficient in the test environment, so we decided to omit the results for this method and plan to examine it thorough in further, future experiments.

Listing 6. Java: code invoking sort procedures

```
long start = System.currentTimeMillis();
//Collections.sort(listastr); //without locale
Collections.sort(listastr,coll);
long end = System.currentTimeMillis();
long result = end-start;
//...
String[] tabstr = new String[maxv];
//here data initialization...
start = System.currentTimeMillis();
//Arrays.parallelSort(tabstr, coll);//extremely slow
//Arrays.sort(tabstr);//without locale
Arrays.sort(tabstr,coll);
end = System.currentTimeMillis();
result = end-start;
```

The CPU load while sorting with Java (Fig. 3) was slightly higher than in two previous examples and fluctuated about 27% (regardless of data structure used), so it seems that the overhead caused by virtual machine is visible. Nevertheless, this overhead is not significant. The interesting remark concerns the results of `Arrays.parallelSort` method, which is the novelty in Java 1.8 and provides multi-threaded sorting. Indeed, the overall CPU load in this case increases over

50% (i.e. two cores are equally stressed – Fig. 4), but the final result is disappointing and greatly exceeds the others (finally we decided to omit this procedure from the summary results). Though the exact reason is unknown, it can be supposed that for the benchmark like ours the cost of synchronization eliminates profits from parallel execution.

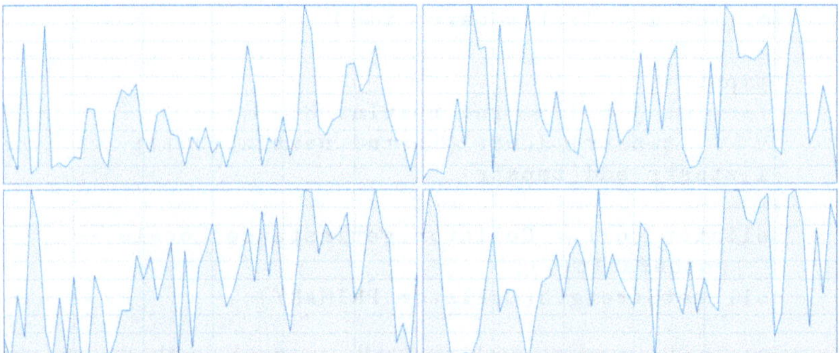

Fig. 3. Java: CPU load (0–100%) during sorting procedure for individual logical processors of the tested machine in the time span of 60 s – single-threaded sorting.

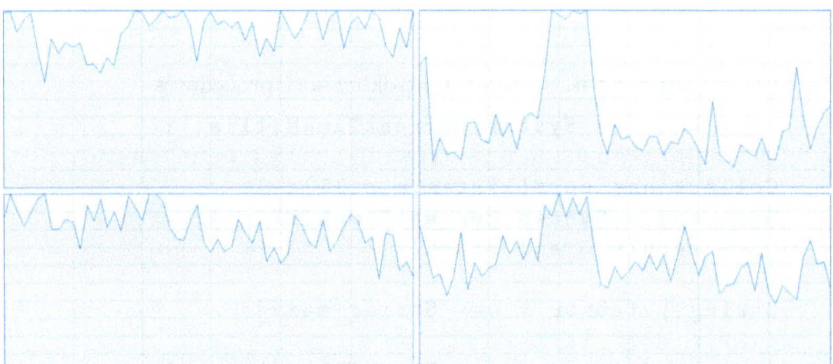

Fig. 4. Java: CPU load (0–100%) during sorting procedure for individual logical processors of the tested machine in the time span of 60 s – multi-threaded sorting.

3.4 Sorting Implementation in C#

Sorting procedure for .NET architecture has been coded in C# (see Listing 7). The data structure used in the experiment was an array, which can be dynamically sized in this language. For the performance measure we used `HiPerfTimer` class. It is worth noting, that in this case the default call to `Array.Sort` procedure with only one argument – an array to sort, invokes sorting according to current culture (i.e. collation) and it is equivalent to the following notation: `Array.Sort(tbl,StringComparer.CurrentCulture)`. To sort in plain ASCII order one has to it force explicitly: `Array.Sort(tbl, StringComparer.Ordinal)`. This is essential difference between C#/.NET and the other examined languages/environments.

Listing 7. C#: data initialization and sorting

```
Random rnd = new Random();
int maxv = 10000000;
String[] tbl = new String[maxv];
for (int i = 0; i < maxv; i++)
  for(int j=1; j<= rnd.Next(8)+1;j++)
    tbl[i]=tbl[i]+
        Convert.ToChar(65+rnd.Next(26));
HiPerfTimer pt = new HiPerfTimer();
pt.Start();
// Array.Sort(tbl); //with CurrentCulture-locale
Array.Sort(tbl,StringComparer.CurrentCulture);
// without locale:
// Array.Sort(tbl, StringComparer.Ordinal);
pt.Stop();
//result=pt.Duration;
```

In case of sorting in C# the overall CPU load was again about 27%, what can be explained by virtual machine (.NET environment) overhead. Apparently, the sort procedure itself is single-threaded. The results of C# prove, that the extra abstraction layer in application development does not have to be a reason for the efficiency degradation (Fig. 5).

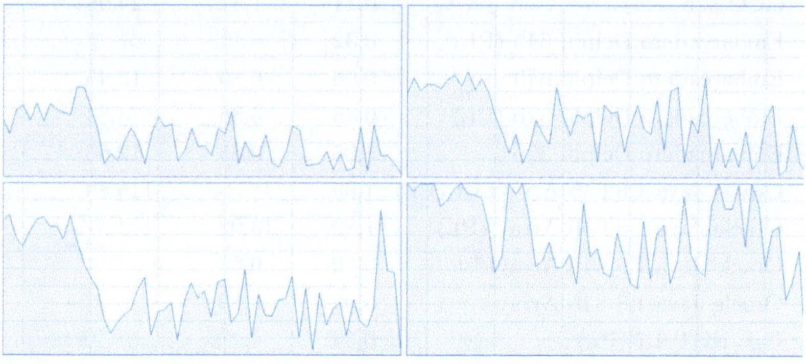

Fig. 5. C#: CPU load (0–100%) during sorting procedure for individual logical processors of the tested machine in the time span of 60 s.

4 Experimental Results

All the test applications operated at 64bit dual-core (4-threaded) Windows 10 Pro (build 1803) machine, with 16 GB of RAM. The tests were performed for datasets of length 100000, 1000000 and 10000000. There were the same circumstances while every test, especially the operating system had been restarted every time and no secondary applications were operating. The test have been repeated 5 times for every dataset size. Although the dataset was randomly generated each time, the results are repeatable and very close to each other.

Table 1 presents the average execution time of the sorting tests. The best results for ASCII collation had the code in Java, C#, than native Object Pascal compilers, and the differences between that three solutions were rather insignificant. For the national collation (Polish in our examples) the best results gave 64-bit Object Pascal code and C# again, but Java performed the worst. C++ test applications executed in disappointing time, probably because of the extra conversions needed.

Table 1. The average results of the sorting for different compilers/environments and three sizes of data [in seconds]. The fastest solutions taking into account national collation marked in green, the slowest in red. The fastest solutions for plain ASCII sorting marked in yellow.

Language/translator	Data size		
	100 000	1 000 000	10 000 000
Embarcadero C++ 7.30 32b (PL)	0,88	9,51	105,24
Embarcadero C++ 7.30 32b	0,51	5,33	57,32
GCC 32b (PL)	0,72	9,11	109,85
GCC 32b	0,11	1,81	24,98
Embarcadero Delphi 64b (PL)	0,32	**4,26**	**56,50**
Embarcadero Delphi 64b	0,06	0,86	12,45
Embarcadero Delphi 32b (PL)	0,43	5,86	73,83
Embarcadero Delphi 32b	0,05	0,85	12,60
Oracle Java 1.8 64b ArrayList (PL)	1,85	**15,44**	116,89
Oracle Java 1.8 64b Array (PL)	1,82	13,92	**118,07**
Oracle Java 1.8 64b ArrayList	0,09	0,68	**7,62**
Oracle Java 1.8 64b Array	0,03	0,89	**8,69**
C# .NET 4.7 (PL)	0,34	**4,51**	**58,38**
C# .NET 4.7	0,05	0,77	11,15

Of course the implementation of sorting differs slightly for every tested solution and there are different data structures used as well. But the goal of our comparison was to compare the sorting capabilities available through standard libraries of the particular language/environment. So the proposed implementations tend to mimic the behaviour of the programmer willing to develop the sorting task with the minimal effort and time-cost.

There were two monitoring applications used to determine CPU load during the tests. The first was the built-in Windows' Task Manager, the second was CPUID HWMonitor[2]. In general, the results obtained from these two sources were consistent.

5 Summary

The test implementations of sorting in different computer programming languages illustrate the common approach of calling procedures built-in into standard libraries and process the standard data structures, generally list-alike collections or dynamic arrays.

It was shown that solutions based on some kind of virtual machines and intermediate code (Java, .NET) can reach efficiency of machine compiled executables. Actually, the observed differences concerned more the language and its translator itself than particular architecture.

On the other hand, the results of experiments prove that classical solutions can still be efficient, especially when it comes to business desktop applications, for which they are designed. This case concerns, for example, Object Pascal used in Embarcadero Delphi [16]. This shows that the proper selection of software development tools can help to create more comfortable user experience in complex applications.

One of the key factors here is the difference between collation used for default sorting procedures. While the most of implementations does not take into account national order of characters, .NET defaults to `CurrentCulture` collation, which is certainly much slower. Observations of this kind can be essential for the software engineering projects integrating many different programming languages or while source code reuse.

Although the presented experiments provide new insights, valuable for contemporary software developers dealing with the choice of the software tools for a sorting (and similar) tasks, the problem deserves further examinations, especially by extending the list of computer languages tested, comparison with self-written code and multi-threaded execution.

References

1. Hylock, R.: UPC: large-scale memory efficient Java primitive collections. JSW **11**(3), 251–271 (2016)
2. Couto, M., Pereira, R., Ribeiro, F., Rua, R., Saraiva, J.: Towards a green ranking for programming languages. In: Proceedings of the 21st Brazilian Symposium on Programming Languages, p. 7. ACM (2017)

[2] https://www.cpuid.com/softwares/hwmonitor.html.

3. Kokot, M., Deorowicz, S., Długosz, M.: Even faster sorting of (not only) integers. In: International Conference on Man–Machine Interactions, pp. 481–491. Springer (2017)
4. Slagter, K., Hsu, C.-H., Chung, Y.-C.: An adaptive and memory efficient sampling mechanism for partitioning in mapreduce. Int. J. Parallel Program. **43**(3), 489–507 (2015)
5. Bingmann, T., Eberle, A., Sanders, P.: Engineering parallel string sorting. Algorithmica **77**(1), 235–286 (2017)
6. Hoare, C.A.: Quicksort. Comput. J. **5**(1), 10–16 (1962)
7. Woźniak, M., Marszałek, Z., Gabryel, M., Nowicki, R.K.: Preprocessing large data sets by the use of quick sort algorithm. In: Knowledge, Information and Creativity Support Systems: Recent Trends, Advances and Solutions, pp. 111–121. Springer (2016)
8. Woźniak, M., Marszałek, Z., Gabryel, M., Nowicki, R.K.: Modified merge sort algorithm for large scale data sets. In: International Conference on Artificial Intelligence and Soft Computing, pp. 612–622. Springer (2013)
9. Sestoft, P.: Microbenchmarks in Java and C#. Lecture Notes, September 2013
10. Costa, D., Andrzejak, A., Seboek, J., Lo, D.: Empirical study of usage and performance of Java collections. In: Proceedings of the 8th ACM/SPEC on International Conference on Performance Engineering, pp. 389–400. ACM (2017)
11. Marszałek, Z.: Performance test on triple heap sort algorithm. Tech. Sci. **20**(1) (2017)
12. Sroczyński, Z.: Internet of things location services with multi-platform mobile applications. In: Proceedings of the Computational Methods in Systems and Software, pp. 347–357. Springer (2017)
13. Gruca, A., Sikora, M.: Data-and expert-driven rule induction and filtering framework for functional interpretation and description of gene sets. J. Biomed. Semant. **8**(1), 23 (2017)
14. Bier, A., Kapczyński, A., Sroczyński, Z.: Pattern lock evaluation framework for mobile devices: human perception of the pattern strength measure. In: International Conference on Man–Machine Interactions, pp. 33–42. Springer (2017)
15. Vogels, W.: Benchmarking the CLI for high performance computing. IEE Proc. Softw. **150**(5), 266–274 (2003)
16. Sroczynski, Z.: Human-computer interaction on mobile devices with the FM application platform. In: R. M., P. P. (eds.) Internet in the information society. Insights on the information systems, structures and applications. Academy of Business in Dabrowa Gornicza Press (2014)

Input Output Data Converter for the Math Engine in an Expert System

Simon Barkovskii[1], Larisa Tselykh[2(✉)], and Alexander Tselykh[1]

[1] Department of Information and Analytical Security Systems, Institute of Computer Technologies and Information Safety, Southern Federal University, Nekrasovskii, 44, 347922 Taganrog, Russia
kharitonov.simon@yandex.ru, ant@sfedu.ru

[2] Chekhov Taganrog Institute, Rostov State University of Economics, Initsiativnaya, 48, 347936 Taganrog, Russia
l.tselykh58@gmail.com

Abstract. This study offers a simple and inexpensive tool for semi-automatic control of computational algorithms with different input/output data structures. The proposed approach to data recognition is based on creating generation rules using data structure sample. Our contribution lies in proposing a built-in semi-automatic engine for recognizing and transforming input/output data (MMC Tool) does not require special end-user queries; there is no needs special end-user skills; the confirmation procedure is carried out in a user-friendly view. The proposed MMC Tool is an effective solution for automation of plugging and switching math modules running in a single system without having to resort to third-party software developers.

Keywords: Expert systems · Input output data · Data parser · Matrix data

1 Introduction

Year by year, the demand for knowledge-based information systems and expert systems (ESs) for management tasks in socio-economic systems is increasing. At the same time, their complexity also increases. The development of such systems stimulates the growth of the number of software applications used to solve various tasks. Various specialized computing applications, included to the product ESs, provide pre-processing data preparation for the inference subsystem. Computing applications can form computation chains. The emphasis in the area of automation of the ES knowledge acquisition module includes the joint operation of various computational algorithms and the provision of processing large amounts of information. This information forms a data flow transferred from one module to another, and shared between different applications. The dynamic nature of applications and variability of the product line of modules generates the compatibility problem of formats, types, and structures of input/output data. The difficulty in switching math modules is that in most cases such modules can process the input data in a pre-determined form only. This makes it necessary to develop special utilities for data transformation. Repeated requests addressed to manufacturers mean additional costs for the ES user organization and

R. Silhavy et al. (Eds.): CoMeSySo 2018, AISC 860, pp. 311–322, 2019.
https://doi.org/10.1007/978-3-030-00184-1_29

additional difficulties in adapting to new services for end-users. Therefore, the development of a mechanism in the ES structure that would automate the process of replacing and plugging-in new applications/modules without changing the internal operating logic of the plugged-in math algorithm and would minimize requirements to end-users is a topical issue.

Our research is focused on solving the problem of automating plugging in computational modules. In this article, we propose a control module of input/output data of various computational algorithms united in a single cluster, with an engine for semi-automated extraction, recognition and transformation of these data structures. The proposed engine is geared towards data recognition and transformation. These data are matrices that are given in a different form and structure (semi-structured and unstructured data, arrays, unnamed lists, etc.). Control is performed at the physical level using report files. Our approach to matrix recognition is to create generation rules using a data structure sample. The control process does not require end-user intervention, except for the final approval of recognized data.

2 Related Work

The authors in [1] proposed the MATHSCOUT package to be used as a preprocessor that mechanically converts output data from the Gaussian program into the input data for MATHEMATICA. The conversion is based on finding keywords and phrases in data consisting of named lists and tables. A set of commands is provided using special encoding methods. This approach is stipulated by the output data of the Gaussian program, being only one type of output data in the form of numeric values of the characteristics computed by the program. This method is not applicable if the output data have another type, for example, matrices or sets of numbers (as in clustering programs).

In [2–4], the problem closest to our problem was considered – recognition and transformation of data to ensure compatibility of interconnected specialized software applications.

In [2] a data formatting solution for Geoscientific computing based on programming through the interpreted Output Input Language (OIL) was proposed. Recognition of input files uses the OIL program library, which can be filled by other users. Thus, if the OIL program for any format is not created, then the file cannot be read. It is also assumed that OIL will be used by end-users without the intervention of professional application developers. This requires a certain level of expertise from end-users, both in the field of programming, and in specific areas.

The article [3] examines the problem of analyzing the results of computer modeling, both in the process of one data flow and the flows of several data sets. The authors propose to control the flow of files and data at the stage of creating raw files. The main idea is based on the specifications of the three types of user SQL-queries based on the predefined domain attributes in each data element. The proposed solution is not applicable in the absence of such attributes, and also requires professional skills of users.

Yang, et al. 2018 [5], proposed an original solution in the form of an e-infrastructure for the integrated management in a specialized field. The proposed approach uses a web browser to control data by the user. Data processing and control is carried out by means the file system and the NoSQL database. At present, MatCloud can extract only predefined characteristics. In MatCloud, data control is implemented at a logical level, with data being converted to *json* format, so the engine is not available to end-users.

The need to develop an automatic engine to control external structured, semi-structured and unstructured data from various sources for business analytics is noted in [6]. Nath et al. offered the Semantic Extract-Transform-Load (SETL) engine for data semantic transformation based on the combination of Semantic Web and Data Warehouse technologies. Semantic Web provides data in Data Warehouse in XML, XML Schema, RDF, RDF Schema, OWL, CSV and several other formats. The extraction of the corresponding data from several heterogeneous data sources is performed through the extraction function using the R2RML mapping language. The definition of customized mappings from relational data into the Resource Description Framework is performed by the user.

The authors in [7] proposed the visual dataflow programming language VisualTPL for generating report understandable to end-users. The basic idea is in the top-down decomposition; when large and complex structured content is divided into simpler layouts to arrange the content in a certain way. The authors used special extraction functions based on canonical tabular data. The engine they offer cannot be used for raw data recognition. VisualTPL includes programming techniques, therefore working with VisualTPL can be difficult to users not familiar with them.

The authors in [8] developed the TabbyXL console application to bring the tables to their canonical relational form. Their method is based on recognition of tagged documents in such formats as Excel, Word or HTML, functional and structural analysis of the logical structure of the table and its further transformation. Cells are defined as a set of specific attributes, such as Location, Style, Content, and Annotation. The method includes the modification of cells in the process of canonicalization. This type of modification can incorrectly change the contents of the cells in the tables, which may lead to incorrect interpretation of the data. For other data structures, data need to be preconfigured into a canonical table format.

3 Methods

3.1 The Framework of Math Module Control

A typical solution (including ESs) used in decision support software for socio-economic systems, includes a hard-coded computational complex. This complex can include various algorithms in the form of subprograms or individual applications. These algorithms can receive data both from the system and from each other. Computations, as a rule, are organized as follows: the algorithm is rigidly fixed in the system, and the user only needs to follow it. Replacing or updating specific algorithms that make up the chain of computations, is either possible through general updating the software as a product line, or requires the development of special utilities.

Such solutions are very costly, and require high level of professional skills from users, not only in the subject area, but also in programming and mathematics, which small and medium size organizations may find difficult to afford. Even for large organizations, this can be a big burden. Therefore, it is necessary to develop a simple and inexpensive tool that provides automatic or semi-automatic interaction and compatibility of computing programs.

To illustrate the problem to be solved, consider the example in [9] that presents the ES includes the computing module "Math engine" consisting of several computing applications. The architecture of this ES is discussed in more detail in [9]. Issue of the compatibility of different applications within the ES module, and the transfer of data between it and other modules of the ES is solved by including the additional "Math module control tool" (MMC tool). The set of computations is performed by the "Math engine". In particular, a chain of computations is performed, the result of which is the identification of antecedents and consequents passed to the "Inference engine" module. Applications that perform computations can be replaced with updated versions and supplemented with other programs of algorithms. The applications of the "Math engine" module use the input/output data of each other and other ES modules.

The plugged-in math algorithms can have a predefined specification, both as input data and output data, so to use them one must, first, recognize the structure of source data, and secondly, transform them into the required form. "MMC Tool" provides the following compatibility functions for input/output data in the context of the problem being solved (Fig. 1):

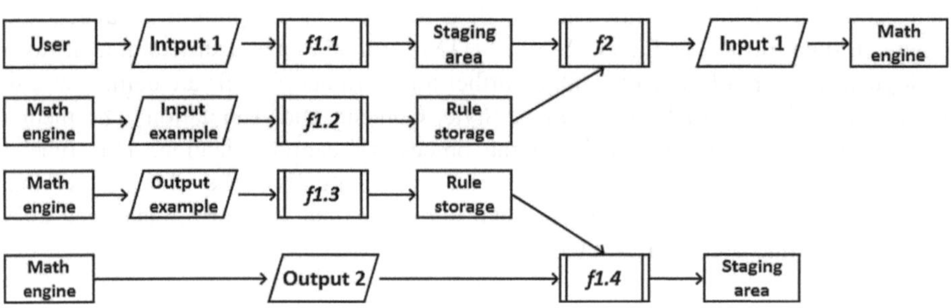

Fig. 1. The scheme of compatibility functions of input/output data.

- Function 1.1 (*f1.1*): recognition of input data for the "Math engine" (Input 1);
- Function 1.2 (*f1.2*): recognition of example input data of the math algorithm;
- Function 1.3 (*f1.3*): recognition of example output data of the math algorithm;
- Function 2 (*f2*): transformation of output data (Output 1) into the type of input data for the "Math engine";
- Function 3 (*f1.4*): recognition of output data from "Math engine" (Output 2) to be transferred to other modules.

The above functions are realized in the "MMC Tool" module, which includes the following components: (1) "Module selector"; (2) "Data parser": "Input recognizer", "Input rules generator", "Output recognizer", "Output rules generator"; (3) "Data generator"; (4) "Rule storage"; (5) "Staging area". Using the "Module selector", the end-user selects a math algorithm in the "Math engine". The input/output data (f1.1 and f1.4) are recognized by the "Data parser". The recognition function creates the generation rules using the input/output data sample provided by the application of the selected algorithm. The recognition process is carried out by two subprograms: "Input recognizer" and "Output recognizer", which perform the function of input/output data recognition, respectively. The "Data generator" realizes the data transformation function (f2). In the "Data generator", the source data are transformed into the required data type based on the generation rules gotten through recognizing the input data (f1.2). The "Rule storage" stores the created generation and recognition rules of data from math algorithms, which allows not to carry out these operations anew. The "Staging area" is intended for storing intermediate data with the result of the system operation at a certain stage.

To realize of the recognition and transformation functions in the "MMC Tool", the following conditions for plugging-in the math algorithm in the "Math engine" module must be met. The application package must include the following files: (i) executable console application; (ii) input data in txt format; (iii)'output data to the console or as a text document; (iv) input/output data sample; (v) input/output data notation. Notation includes the rules for generating specific mandatory fields in the input data, which must be generated to ensure the operation of the math algorithm. Notation can also include the rules designed to simplify the recognition of the math algorithm output data.

The working process of the MMC Tool, shown in Fig. 2, uses the following services in ES that implement the data flow:

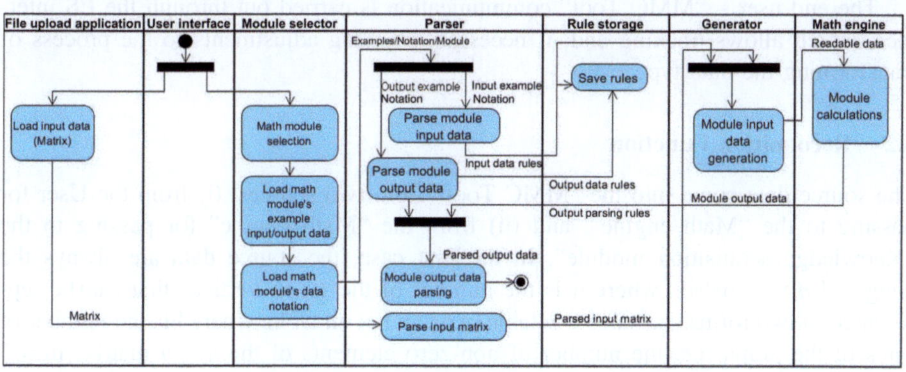

Fig. 2. The plug-in scheme of the math algorithms.

– User interface. ES's interface, which allows the user to interact with all application components.
– File uploads application. The component loading the user's source data.

- Module selector. ES component is realizes the possibility of selecting a math algorithm, loading the input/output data sample and notation for the math algorithm.
- Data Parser. At the stage of selecting a math algorithm, the Data Parser is responsible for processing of input/output data samples from a math algorithm and creating the generation and recognition rules. At the stage of completing computations by the math algorithm, the Data Parser recognizes the output data of this algorithm.
- Rule Storage. The Rule Storage component is designed to store the created the generation and recognition rules.
- Data Generator. The Data Generator performs the transformation of input data for the math module in accordance with the generation rules created earlier.
- Math Engine. The Math Engine starts a computational process by a user-selected algorithm.

4 Results

4.1 The Transformation Cycle of Input/Output Data

As we mentioned earlier, to solve the problem of compatibility of input/output data it is necessary implement the functions of data recognition and transformation. The transformation cycle involves successive transformations of the data structure passed from internal modules and applications to the “MMC tool”. As input data, the “MMC tool” takes the input/output data sample and the user input data. As the control element, the “MMC Tool” takes the notation of the input/output data from notation file. An example of successive transformations of $n \times n$ adjacency matrix into the required data type for the math algorithm is shown in Fig. 3.

The end user – “MMC Tool” communication is carried out through the ES interface, which allows tracking and, if necessary, making adjustments to the process of transforming the data type.

4.2 Recognition Function

The source data come into the “MMC Tool” from two sources: (i) from the User for passing to the “Math engine”, and (ii) from the “Math engine” for passing to the “Knowledge acquisition module”. In the first case, the source data are always the weighted $n \times n$ matrix, where n is the number of the graph vertices that can be represented in two forms – as an $n \times n$ table or as an l $\times$ m table, where l is the number of links of the graph, i.e. the number of non-zero elements of the $n \times n$ matrix; m = 3 represents the columns in which the output vertex, the input vertex, and the numeric value of the non-zero element of the matrix are set. The initial matrix can come in the “MMC Tool” in two ways: as an excel table loaded directly into the ES, or as built in the ES itself. In addition to the matrix data, the source data can include various explanatory texts and other matrix characteristics (for example, metrics, eigenvalues, etc.). In the second case, the source data are the result of the computation from the “Math engine”. These data can also have numeric and non-numeric information.

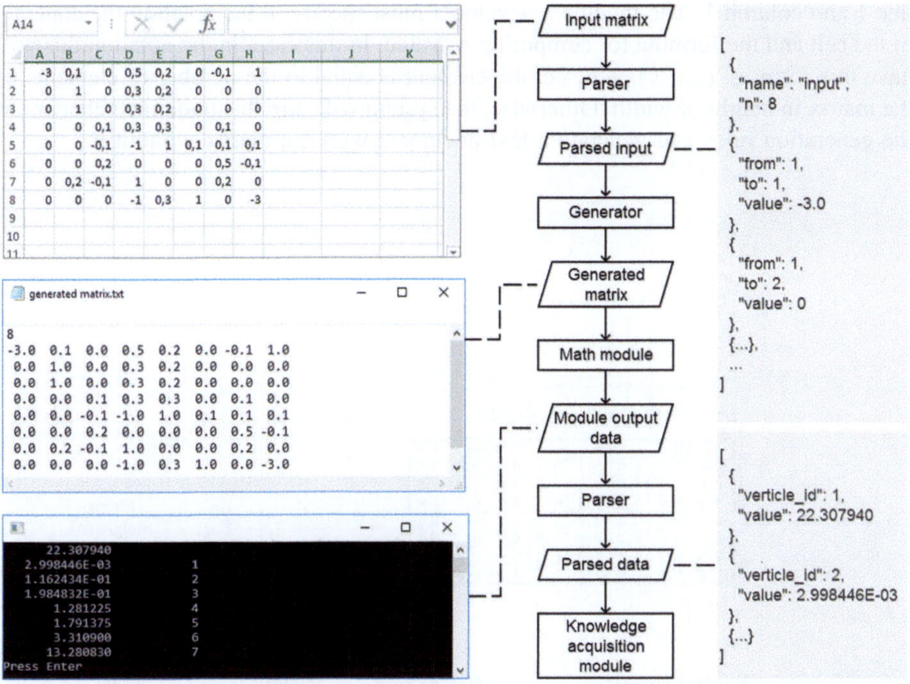

Fig. 3. Example of the transformation cycle of data type at different stages of ES's work with the math algorithm.

The process that provides the recognition function includes the following steps:

Step 1. Data processing using the sample enclosed in the application package loaded into the "Math engine". Identification of the redundant information. Here the identification of all numbers from the text document occurs.

In Step 1 the source data are cleared of additional and redundant information, and a file is created with a set of numerical source data.

Step 2. Creating the generation and recognition rules by the "Input rules generator" and "Output rules generator". The rules for transforming the source data are used to form a file with a certain type of data.

In this step, the identified areas of numerical values are analyzed to construct a matrix of the required form for the selected component of the "Math engine". For floating-point numbers, the presence of decimal separators "." and "," will be checked. Then the separator between the numbers is detected – the system will go through all the lines of the document. After detecting the separators, a grid of matrix elements of $n \times n$ dimension is constructed. All elements that do not fall into the set of matrix cells are considered undefined elements. Such elements should be specified in notations in a special way by the developer, which would allow creating a rule for generating this element. For example, in Fig. 4, an undefined element is the element in line 1 and column 1. The module's developer must specify in the notation the contents of the cell and the formula for computing its value. In this case, in the notation this will have this form: `e [1; 1] = n`, i.e. the element is equal to the number of elements of the matrix in height or width. Otherwise, the system will skip this undefined element in the generation rules and generate a text document with the matrix without it.

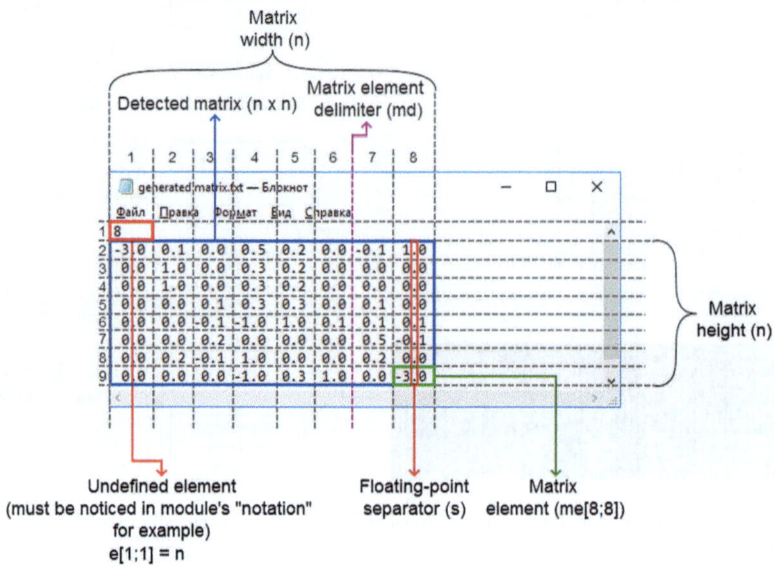

Fig. 4. The technique of recognizing a sample of input data from the math module.

Thus, the rules for generating input data have the following attributes:

`S = ".”` – decimal separator.

`E[1;1] = n;` – the equation of the element in cell (1;1), equal to the height of the matrix.

`Md = "TAB”` – the type of the matrix elements delimiter.

`MatrixBeginDrawLine = 2;` – the line number in the document from which to start the process of drawing the matrix.

These rules will apply to the original user matrix for its transformation.

In Step 2 a set of rules is created for transforming the source data submitted by the user, to the required form for the math algorithm.

Step 3. Writing rules for generating and recognizing input/output data into the “Rule storage”.

4.3 Transformation Function

The data transformation process is implemented in the "Data generator" following these steps:

Step 1. Feeding the recognized user data in *json* format into the "Input converter".

Step 2. Invoking of the generation rules for transforming of the input data from the "Rule storage" and feeding them to the "Input converter".

Step 3. Transforming the recognized *json* format user data into the corresponding data type using the input data generation rules for the math algorithm. Using the `{Md = "TAB"}` attribute, the grid of the matrix consisting of cells is drawn, where the tab delimiter is the separator. Using the `{E[1;1] = n}` rule, the "Data generator" will record a specific element in the cell in the first column of the first line (the element specified by the developer in the notation). `{MatrixBeginDrawLine = 2}` will indicate number of line to begin to fill the matrix transformed from *json* format with the separator `{S = "."}`..

Step 4. Approval of the transformed end-user input. The created data is presented to the end-user for final confirmation of the correctness through the user interface. There is possibility to manually change the generation rules and start the process of transforming the first data type again.

Step 5. Feeding of the transformed data into the math algorithm

Note that by correcting the generation rules, we can transform the recognized user data not only into the $n \times n$ table, but also into the $l \times m$ table, where m = 3 represents the columns in which the output vertex, input vertex, and the numeric value of a non-zero matrix element are set.

The "Extraction-Recognition-Transformation" processing of user data shown in Fig. 5 uses the following data:

1. User input data in matrix form, submitted in ES in *excel* format.
2. Recognized and successfully validated user matrix, recorded in *json* format.
3. The rules for transforming the matrix in *json* format into the required form for the math algorithm.
4. Preview of the data fed to the math module. The user can confirm the created data for further work with the math module.
5. The generated document in *txt* format with the input data for the math module in the required form.

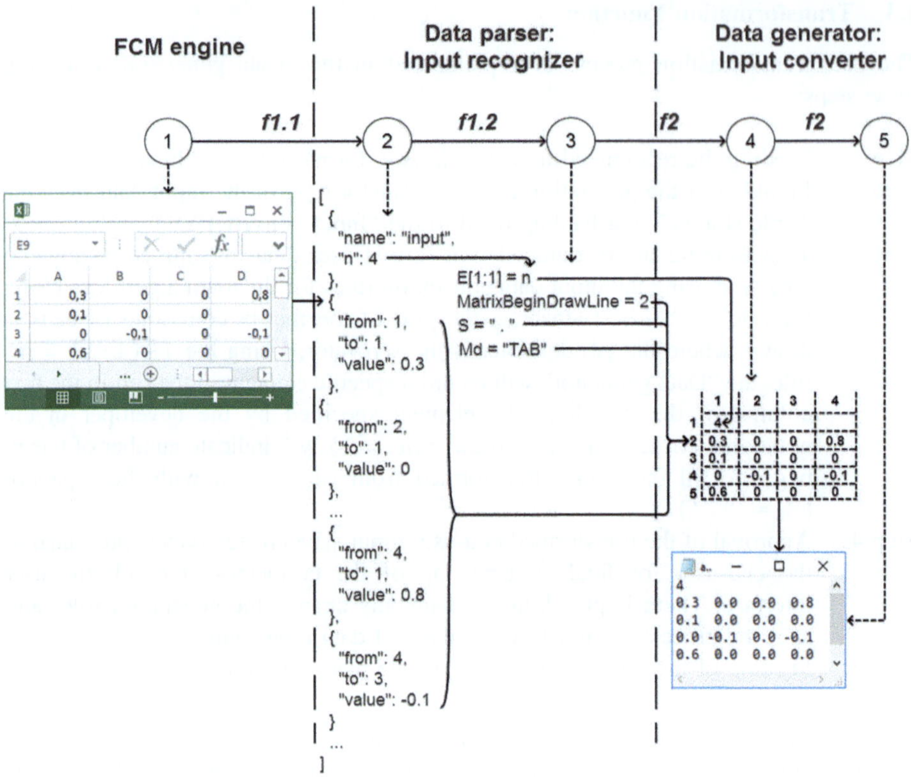

Fig. 5. Transformation of the user input data into the input data for the math module.

5 Discussions

The proposed solution is evaluated on the following: productivity of the MMC Tool and its value for the ES; end-user friendliness; support of the MMC Tool; performance problems of the MMC Tool; future of the MMC Tool.

The MMC Tool solves the automation problem of plug-in and compatibility of various computational algorithms. The following automation tasks were solved: (i) recognition of the input/output data of math algorithms; (ii) transformation of the input/output data. Two types of restrictions were identified: (i) for programs of math algorithms: inavailability of open source code and heterogeneity of the input/output data, and (ii) for users: lack of professional education in programming for end-users and minimizing their involvement in these processes.

To overcome the limitations of the first type, the recognition of the input/output data can occur at the physical level, i.e. based on transforming the data from the file system. In this case, the files do not allow for support specialized queries, as well as indexing the content of the files. The choice of control at the physical level does not require the involvement of professional programmers and domain experts at each change of the computational algorithm, which significantly reduces the costs for user organizations. Addressing the file system does not require the open code of the computational algorithm programs; however, it requires tools for recognizing non-predefined data types. Our solution is to obtain additional information in the form of the following files: (i) input/output data sample to recognize the structure of the input/output data, and (ii) notation for recognizing additional undefined information. The request is made to the technical support provided by the developers of the computational program.

Another problem caused by the first type of limitations was that the output files of the algorithms may contain many different tables and individual data. In our case, the input/output data are primarily matrices and semi-structured tables. Therefore, our first task was to detect the matrix, the second was to find additional components in the data.

Math algorithm uses the square matrices as the input data. It was the starting point for solving the problem of recognition of the input data. The proposed process of recognizing the input data involves the identification of numeric values in a document, their verification, making up an array of values, and writing it into a square matrix in *json* format based on several templates. The task of transforming the input data involves the presence of a sample of input data in the math algorithm package. This problem is solved by creating generation rules for transforming the data in *json* format into data suitable for the operation of the math algorithm.

In general, we propose a semi-automatic system that implements a plugging-in and choice of third-party math applications within the ES. Simplifying the plugging of computational algorithms without requesting notation is the next step in this study. A simple and intuitive interface that would allow the user to create notations for connecting math algorithms independently is an obvious solution to this problem.

The MMC Tool allows the users to call different computational algorithms simply by selecting the appropriate one. One of the main advantages of working with the MMC Tool is that all the steps in the "Extraction-Recognition-Transformation" process can be easily supported and implemented with a simple interface that does not require programming skills from the end-user.

6 Conclusion

This study proposes a simple and inexpensive tool for semi-automatic control of computational algorithms with different input/output data structures. The proposed approach to data recognition is based on the development of generation rules using data structure samples. Our contribution lies in proposing a built-in semi-automatic engine for recognizing and transforming input/output data (MMC Tool) does not require special end-user queries; there is no needs special end-user skills; the confirmation procedure is carried out in a user-friendly view. As no big data processing is required, the execution time is non-critical.

The proposed MMC Tool is an effective solution for automation of plugging and switching math modules running in a single system without having to resort to third-party software developers. The automatic data transfer from application output files to workflows opens up new opportunities for the use of ESs by managers of socio-economic systems who do not have sufficient professional competencies in programming and mathematics.

Acknowledgements. This work was supported by the Russian Foundation for Basic Research [grant number № 17-01-00076].

References

1. Barnett, M., Capitani, J.: The MATHSCOUT mathematica package to postprocess the output of other scientific programs. Comput. Phys. Commun. **177**, 944–950 (2007). https://doi.org/10.1016/j.cpc.2007.07.009
2. Khan, K., Akhter, G., Ahmad, Z.: OIL – output input language for data connectivity between geoscientific software applications. Comput. Geosci. **36**, 687–697 (2010). https://doi.org/10.1016/j.cageo.2009.09.005
3. Silva, V., de Oliveira, D., Valduriez, P., Mattoso, M.: Analyzing related raw data files through dataflows. Concurr. Comput. Pract. Exp. **28**, 2528–2545 (2016). https://doi.org/10.1002/cpe.3616
4. Silva, V., Leite, J., Camata, J., de Oliveira, D., Coutinho, A., Valduriez, P., Mattoso, M.: Raw data queries during data-intensive parallel workflow execution. Future Gen. Comput. Syst. **75**, 402–422 (2017). https://doi.org/10.1016/j.future.2017.01.016
5. Yang, X., Wang, Z., Zhao, X., Song, J., Zhang, M., Liu, H.: MatCloud: a high-throughput computational infrastructure for integrated management of materials simulation, data and resources. Comput. Mater. Sci. **146**, 319–333 (2018). https://doi.org/10.1016/j.commatsci.2018.01.039
6. Nath, R., Hose, K., Pedersen, T., Romero, O.: SETL: a programmable semantic extract-transform-load framework for semantic data warehouses. Inf. Syst. **68**, 17–43 (2017). https://doi.org/10.1016/j.is.2017.01.005
7. Chen, W.-K., Tu, P.-Y.: VisualTPL: a visual dataflow language for report data transformation. J. Visual Lang. Comput. **25**, 210–226 (2014). https://doi.org/10.1016/j.jvlc.2013.11.003
8. Shigarov, A., Mikhailov, A.: Rule-based spreadsheet data transformation from arbitrary to relational tables. Inf. Syst. **71**, 123–136 (2017). https://doi.org/10.1016/j.is.2017.08.004
9. Tselykh, A., Tselykh, L., Vasilev, V., Barkovskii, S.: Expert system with extended knowledge acquisition module for decision making support. Adv. Intell. Syst. Comput. **680**(2), 21–31 (2017). https://doi.org/10.1007/978-3-319-68324-9

Adaptive Software System for Optimization of the Admission and Management Process for Doctoral Students

Pavel Zlatarov(✉), Galina Ivanova, Desislava Baeva, and Diana Antonova

University of Ruse, 8, Studentska Str., Ruse, Bulgaria
pzlatarov@uni-ruse.bg

Abstract. The preparation of a doctoral student and a successful thesis is considered an achievement – not only for the doctoral student and their scientific supervisor, but also for the institution that has taken responsibility in developing a young scientist. Doctoral student education is a complex activity, which includes administrative services, discursive social practice, document turnover, etc. The creation of a software system aimed at tracking the development of the doctoral students themselves and their scientific interests is one of a series of initiatives that the University of Ruse has taken to optimize the organization activity related to Ph.D. student education. The base theoretical concepts forming the foundation in the build process of a software system are outlined in this article, as well as their successful practical implementation into a well-functioning software resource. An analysis of the results of a poll intending to survey the opinions of the users in their work and experience with the system is also included.

Keywords: User centered software system · Doctoral students Process · Adaptive software system · Information system · Admission

1 Introduction

Modern educational institutions offer a variety of Internet-based information platforms that aim to aid the process of under- and postgraduate student admission and education. This is a matter of prestige and of educational process quality, and is a powerful factor determining the organization's degree of competitiveness. The management, control and decision-making involved in the education of specialists in the doctoral educational and scientific degree requires timely and operational access to student-related data by administrative personnel and students alike. Its optimization and modernization has been achieved by creating an automated information and reference system, allowing the collection, storage, processing and quick access to information related to the educational and scientific development of the doctoral student [1].

The increased aim towards quality improvements in the operational provisioning of the post-graduate educational process in the University of Ruse, the collection, systematization and control of trusted information required for the evaluation and

R. Silhavy et al. (Eds.): CoMeSySo 2018, AISC 860, pp. 323–332, 2019.
https://doi.org/10.1007/978-3-030-00184-1_30

improvement of the activities involved in planning of the educational and administrative work, the need of maintaining an effective feedback channel and coordination of the interaction between scientific supervisors and students are all strong motivators in the creation of a specialized software platform for education management for doctoral students in University of Ruse. By contrast, compared to most popular information systems that primarily focus on satisfying the needs of administrative personnel, the University of Ruse team aimed to analyze and implement the basic stages and procedures of admission, education and dissertation defense, while focusing on the doctoral students themselves, and on the harmonization of their interests with the needs of the administrative system.

The aim of this article is to outline the basic theoretical concepts that have laid the foundation of the development of the software system and their successful practical implementation into a well-functioning software resource.

2 Theoretical Concepts Used in the Development of the Software System

1. The scientific mentorship is notably important for the doctoral student as well as the whole scientific community and plays a high-priority role in the design and development of the system structure [2]. This is the reason that, when determining the **relationships** between the different roles in the software model, **a complex approach**, which should take into account a series of characteristics – of the socialization, the scientific environment and the scientific mentorship of doctoral students, needs to be applied.
2. The most important factor for the successful application of a software product is its user acceptance. The user-centered software design (UCD) guarantees that the goals and needs of the end users lie in the center of the product development. UCD requires constant and continuous control and evaluation of user activity and perfection of the designer concepts and prototypes [2, 3]. The appropriate functional scope and the agile methodologies used to determine how useful can the developed software be for the end user [5].

The process of building applications using the user-centered design approach (UCD) is described by the 13407 ISO-standard - Human–centred design processes for interactive systems (ISO 1999). The standard describes UCD as an iterative process, consisting of five steps and building upon the following principles [6]:

- Active user participation and understanding of their requirements and tasks;
- Appropriate distribution of the functionality between user and system;
- Iteration of the design solutions;
- Multidisciplinary design teams, [6].

3. A doctoral student's individual curriculum possesses powerful regulatory elements, but also gives students the ability to actively participate in the process of building, assessing and correcting their own educational path. The maintenance of each

component composing the individual curriculum was determined to be a key factor for building the software system during the preliminary assessment.

The descriptions of the processes analyzed during the preparatory stage of the creation of the software system require documents to be presented as a logically ordered aggregation, in which the execution of activities required by legal and regulatory acts are described. This exposes the causal links in the organizational process as a whole, the main activities that compose it, the main process integrators (such as methodologies, documentation, formats, procedures, etc.), and the ways to monitor, control and evaluate the effectiveness of the learning management system.

4. Undoubtedly, there exists a need to ensure unrestricted access to sources and resources consisting of teaching materials, normative acts and administrative forms. In the modern information environment, this is done through the development of a specialized library fund.

Its prospective planning and systematization in a quantitative and thematic aspect, as well as by type of documents, enhances the possibilities to obtain all the documents that would be needed by the potential users of the library.

3 Practical Realization of the Conceptual Solutions

The design and development of the target information system of the University of Ruse imposed actions on the following components of the system:

1. **Definition of user target groups** and their access rights. The following user types exist: an administrator with exclusive rights to administer and manage the information in the web application; doctoral candidate; scientific supervisor; visitor; and an administrative officer/employee.

As a main user type, the focus is placed on doctoral students. Their functions and interactions have been formalized (see Fig. 1).

2. **Interface and functionality** - When building an efficient web-based system for the organization of the educational process, attention should be paid not only to the educational content and learning activities but also to the design of the computer interface. When the interface is difficult to use, it can interfere with the efficiency of the targeted activities. While it is a critical aspect, the interface design as an element of software systems is often overlooked, even though it helps to establish visual communication between the user and the hidden application logic in the product. The "effective visual communication" category contains essential factors, rules and models for creating and operating the graphical user interface. Compliance with the four principles of interface screen design (objects are arranged in a logical order, buttons are visible, labels and symbols are clear, buttons are grouped according to their functions and frequency of use) guarantees easy orientation of the users and motivates them to work with the software product. Formal models and visual metaphors have been used to make understanding more effective, since the interface is based on already known concepts. The information presentation is in accordance

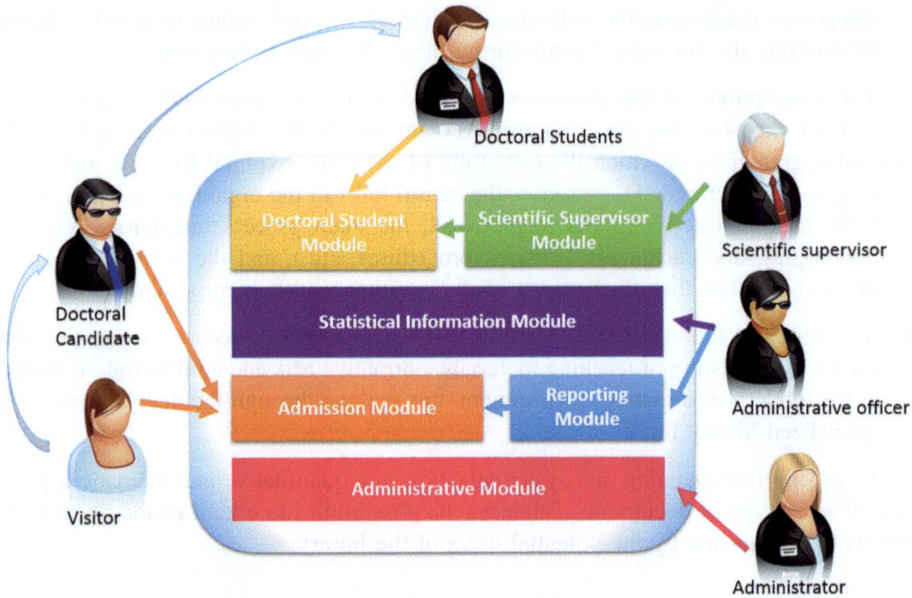

Fig. 1. Organizational scheme of the "Doctoral Student" software platform

with the contemporary requirements of the technical means of visualization and fits into the overall design of University of Ruse's website. The application allows all users to use it on mobile devices thanks to the responsive design - the content is automatically adapted to the screen size of the device that is used for access.

The end result is a finished web-based system that is publicly available on the Internet and provides information on the announced vacant places for doctoral studies in the University of Ruse. It provides for the collection of proposals by the research units and their generalization for new competitions. Creating a competitive environment, fair competition and monitoring are key factors in the selection of young scientists. This requires **transparency** in conducting the complex procedures for the selection of doctoral students, ensuring publicity and quality information about their progress. The doctoral candidates participate in their preferred competition by completing an online application form. After the examination and approval of the applications (see Fig. 2), the system provides candidates with information and all the necessary materials for the examination procedures. Certain documents, such as the enrollment order, are automatically generated with the data of candidates who passed the exams successfully pre-filled; emails are sent to the approved candidates. By logging into the application, the admitted doctoral students have the opportunity to track each stage of their education.

Maintaining a detailed database provides quality information service to all users of the system. Any piece of information about accredited doctoral programs at the University of Ruse has been made easy to access, along with the corresponding

Fig. 2. A sample screen from the system, ensuring transparency in the admission process

structure of professional fields, specialties and areas of higher education, as well as their relationship with the units responsible for the training.

Doctoral students' profiles belong to different professional fields. Storing, updating and retrieving information about doctoral students, their current status and the activities they perform in connection with their PhD degree is ensured, as well as information about the courses they are planning according to their individual curriculum, taking into account and according to the group curriculum (see Fig. 3). The profiles for the various forms of doctoral studies are differentiated as follows: full-time, part-time, free.

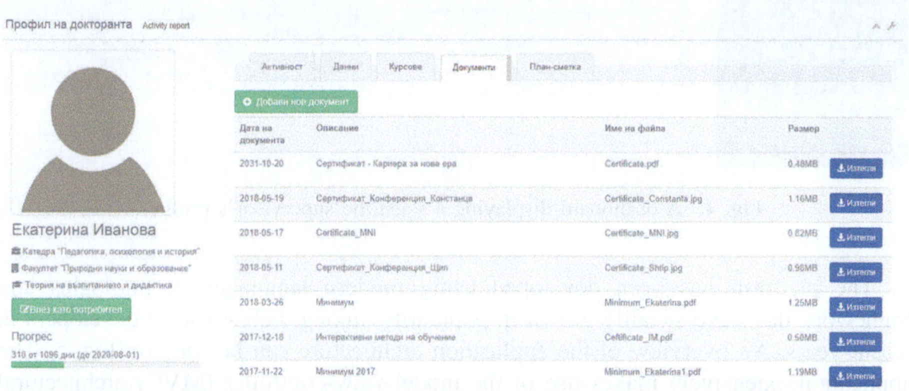

Fig. 3. A profile of a doctoral student, containing administrative information

The systems allows for the generation of dynamic reports of a different nature - for doctorate courses in professional fields, faculties, as well as reports used for reporting and control by the Ministry of Education and Science, and operator-chosen formats.

An important functionality that is embedded in the software system is the support of scientific supervisors and research units in managing the process of training and scientific development of PhD students. A dynamic portfolio of the supervisor is maintained, which summarizes data for all PhD students they supervise.

Interdisciplinarity is a modern paradigm for new creative approaches to building research teams. A common practice in interdisciplinary topics is that one PhD student should have two scientific supervisors (or consultants), this being provided and included as a functional option in the platform.

It is a global practice that a vast majority of scientists are to independently manage the documentation related to their scientific work [8]. Products such as Google Scholar, Microsoft Academic are primarily focused on the storage and analysis of scientific publications. The web-based system supports a number of ways to track the individual path of young scientists learning. This includes creating an archive of individual administrative documents, articles, training certificates, quarterly and annual individual reports, a financial plan account, and so on.

In order to ensure the academic mobility of doctoral students, the software also supports a credit system; credits are accumulated at each stage of the individual curriculum, and stored in the database (Fig. 4).

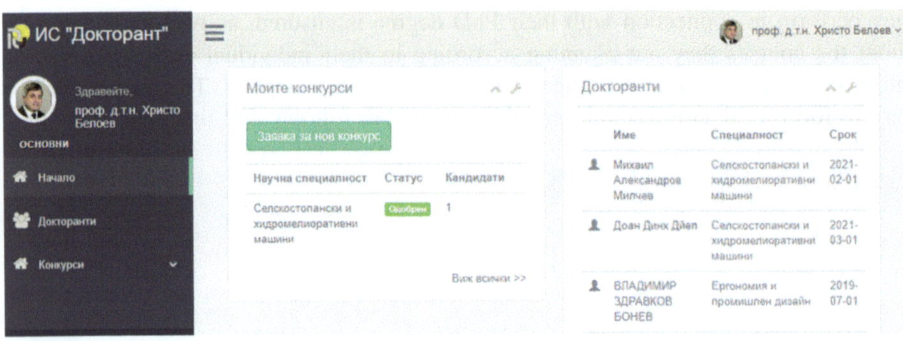

Fig. 4. A dashboard displaying a scientific supervisor's profile

The platform has been developed using modern languages, technologies and framework that have steadily grown in popularity among web software developers in recent years. An overview of the application architecture can be seen on Fig. 5. The application extensively makes use of the model-view-controller (MVC) architectural pattern, which, among other benefits, means that the application logic is completely separate from the business logic and data access. The application has been developed using PHP 7 and the Laravel framework, making use of most of Laravel's core features and having been upgraded to newer versions of the framework several times. A search server has been added to automatically index searchable items and facilitate search throughout the application, making processing of search requests much faster compared to directly searching in the database. The application has low memory and disk space requirements and can be hosted on any of the three popular operating systems (Windows, macOS and Linux), and most popular web servers (Apache, Microsoft IIS, Nginx, etc.).

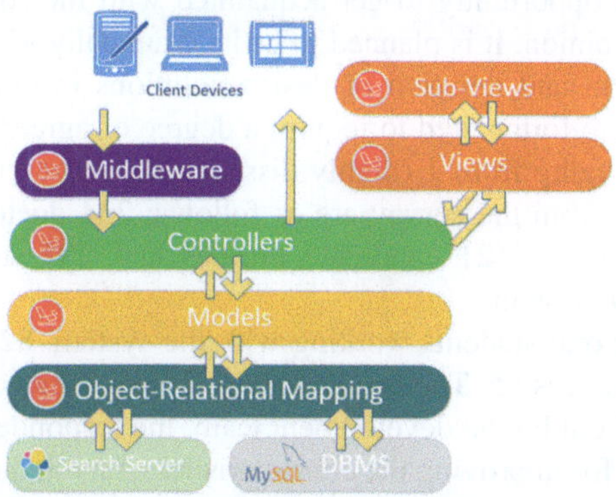

Fig. 5. An overview of the application framework and data flow between modules

According to the European Commission's decision of 2013 [9], through the initiative to open education, training institutions have the opportunity to analyze their organizational models and to determine what changes are needed to turn the challenges into opportunities. "Opening up education" means implementing the digital revolution in education. "Digital technologies allow people to learn everywhere, at any time, on any device and with the help of everyone". This requires the creation of opportunities for organizations, teachers, and students to use digital technology and content as well as to teach and learn in innovative ways. In line with the initiative, University of Ruse's Ph. D. platform provides access to all electronic curriculum materials from the group curriculum courses as an integral part of the Virtual Library for doctoral students included in the system of the University of Ruse's Doctoral School.

4 A Survey Measuring the Satisfaction of Doctoral Students from Working with the Platform

Wixom, Todd et al. examine the characteristics of a quality software product and classify them into several groups: system quality, quality of information, service quality, utility, ease of use, expected results, organizational factors [9]. They point out that these specific conditions are derived from the integration of the factors mentioned in the literature on consumer satisfaction.

Although these characteristics have a common relevance, the importance of each of them depends on a particular system and application direction.

To determine the degree of user satisfaction from the software system, an anonymous online survey was developed and positioned in social media. The questionnaire is composed of question groups generated according to the criteria of what system qualities are concerned: a group referring to the characteristics of the software system; a group referring to its functional capabilities; and a group about access to specialized information. The survey was conducted in April 2018, after the newly admitted doctoral students had the opportunity to get acquainted with the software system and to express a competent opinion. It is planned to be held annually as a feedback tool in the process of management and iteration of design solutions in building the specialized interface. Constructs are formulated to require a degree of agreement, assessed on a 5-degree Likert scale ranging from 1 (highly dissatisfied) to 5 (very satisfied).

The data obtained from the survey are as follows: 234 doctoral students' profiles (120 of which are active), 121 candidate profiles and 204 supervisor profiles are registered in the Ph.D. system.

Of 120 active doctoral students working with the system, 82 responded; scientific supervisors - 14; employees - 5. The results obtained are presented in Fig. 6. As a result of the analysis carried out by the development team, the respondents' comments are an objective prerequisite for improving the characteristics of the indicators with the lowest (end) estimates. The dominant opinion is "very satisfied" (73–77% of respondents) about the capabilities offered by the software.

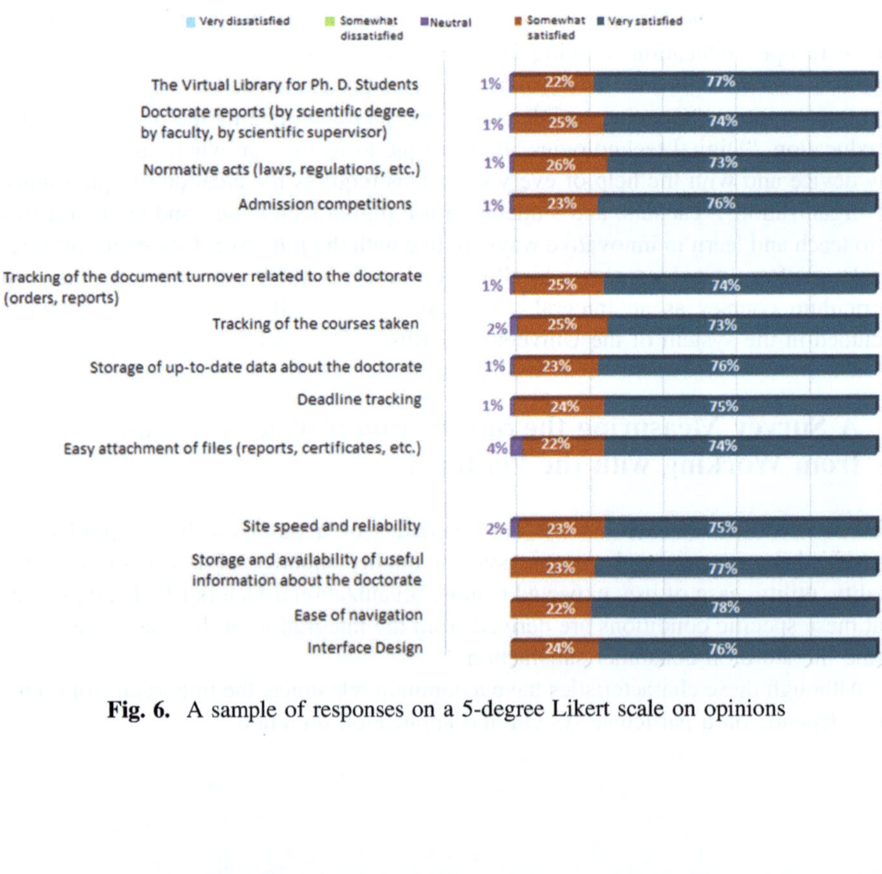

Fig. 6. A sample of responses on a 5-degree Likert scale on opinions

Feedback with users and the ongoing provision of information about their motivation to work with specialized software is of great importance for the quality control of the activities performed.

According to the analysis of respondent groups, the following summaries are made:

- A large number of respondents believe that a well-organized interface and easy access to the system make it easy to work with it and are a prerequisite for good service.
- The majority of the users of the doctoral platform are happy with the operation of the software resource. They recognize the ability to use the electronic administrative services that it offers as an advantage.
- The attitude towards the possibility of online access to up-to-date information - normative documents, learning resources, etc. – is definitively positive.
- During the patching of the system, the recommendation made by the respondents to pay attention to the creation of quality tools for internal communication among the different groups of users will be taken into account, which would increase the control over the work of the participants in the doctoral education process and would lead to a more rigorous compliance with the deadlines in the implementation of the individual curriculum.

5 Conclusions

According to the Bologna process, doctoral programs are the "third degree" of higher education, and at the same time the beginning of scientific growth (after bachelor and master degrees), with these characteristics being an important tool for the development of the "society of knowledge" [7]. The proposed approach to using technology to track and organize their study place PhD students at the center of educational process management.

The web-based platform for doctoral students has been tested in real conditions, validated and implemented for use at the University of Ruse in 2016, with subsequent upgrades and updates, according to the requirements of the legislative and regulatory base. It ensures transparency in the enrollment of the students in a doctoral program, which is in line with the Standard for doctoral students (adopted on 23.01.2018 by the Academic Council of the University of Ruse) and serves as a solid basis for the established system of the Doctoral School of the University of Ruse, [7].

As an innovative tool for managing and reporting the teaching and service activities of doctoral students at the University, the system synchronizes the actions of the individual research units responsible for doctoral studies, the general university faculty development department, the scientific supervisors and doctoral students.

Through the platform, the individual learning pathway of the young scientists is tracked and a synergistic effect is achieved - the doctoral students have access to all the electronic materials for the courses in the group curriculum, to the Virtual Library for doctoral students and to the current legislative documents from the legislative base, related to the development of the academic staff and the competitive procedures.

The theoretical concepts behind the development of the software system, their practical implementation, and the view of the users show that the designed and built web-based platform is a successful software educational resource that responds to and satisfies the user requirements of the target groups and stakeholders.

Acknowledgements. The study was supported by contract of University of Ruse “Angel Kanchev”, № BG05M2OP001-2.009-0011-C01, “Support for the development of human resources for research and innovation at the University of Ruse “Angel Kanchev”. The project is funded with support from the Operational Program “Science and Education for Smart Growth 2014–2020” financed by the European Social Fund of the European Union.

References

1. Ivanova, G., Baeva, D., Zlatarov, P., Ibryamova, E.: Conceptual model of a software system for organizing and tracking the admission and training of PhD students. In: Education, Science, Economics and Technologies, Pedagogy and Quality of Education, vol. XIII(3), Bourgas, pp. 96–100 (2017). ISBN 13126121
2. Fogel, G.: A Model of Learner-Centered Approach and Best Practices in Distance Education (2016)
3. Kottmann, A.: Reform of Doctoral Training in Europe, Reform of Higher Education in Europe, pp. 29–43. SensePublishers (2011)
4. Fox, D., Sillito, J., Maurer, F.: Agile methods and user centered design: how these two methodologies are successfully integrated in industry. In: 2008 Agile Conference (AGILE2008), 4–8 August 2008, Toronto, ON, CA, pp. 63–72 (2008)
5. Lindström, H., Malmsten, M.: User-centric design and agile development: rebuilding the swedish national union catalog. Code4lib, Issue 5 (2011). ISSN 1840-5758
6. ISO, ISO 9241-11: 1998 ergonomic requirements for office work with visual display terminals (VDTs) - Part 11: Guidance on usability (1998)
7. Standard for Ph.D. Students in University of Ruse (2018)
8. Barbara, H., Wixom, P., Todd, A.: A theoretical integration of user satisfaction and technology acceptance. Inf. Syst. Res. **16**(1), 85–102 (2005). ISSN 1047-7047. McIntire School of Commerce, INFORMS
9. Zlatarov, P., Ivanov, G., Baeva, D., Ibryamova, E.: A user-centered software system for admission optimization and education process management for Ph.D. students. In: The 14th International Scientific Conference eLearning and Software for Education, Bucharest (2018). ISBN 10.12753/2066-026
10. European Commission. Opening education with new technologies. http://ec.europa.eu/education/policy/strategic-framework/education-technology_en
11. Iliev, Y.: New incentives and challenges for doctoral education in Europe, Infosvet 1: Forming and Building the Modern Young Scientist: Approaches and Practices (2017). ISSN 2367-9417

The Production and Simplification of Evidence - Enhancing Trust and Costs Reduction on Court

Miguel Matos, Duarte Duque, Joaquim Silva, and Irene Portela(✉)

Instituto Politécnico do Cávado e do Ave, Barcelos, Portugal
{mmatos,dduque,jpsilva,iportela}@ipca.pt

Abstract. To solve the problem of slow/costly and inefficiency of the e-Justice system, the TESTOR project consists of improving legal certainty, expeditious dispute resolution, and costs reduction. The objective is to simplify the production of evidence in order to be used for legal purposes. For that, the evidence (any type of digital document/file) is digitally signed by all the stakeholders (may include a solicitor/lawyer) and submitted to an online platform that creates a proof of evidence, stores and manages permissions access to the evidence. Enabling to follow and connect the proof, that can be any digital content, with the civil process. The proof, with the full range of problems that are associated with it, has to be the nerve center of the system and Justice. The material facts must be proved and this is crucial to establish the proof of value - the value of what is provable.

Keywords: Proof of evidence · Digital signatures
Long term validation · Cryptography · Law

1 Introduction

The Decree-Law No. 76-A/2006, of 29 March, the number 1 of article 38, assigns responsibilities to solicitors performing "simple recognition and special mentions, in person and by similarity, authenticate private documents and certify translations of documents pursuant to notarial law." The number 2 of the same article confers probative force to acts of recognition, endorsements and certifications held by solicitors, since, as explained in paragraph 3 of article 38, are carried out by registration in a computer system [10].

In this context, the Ordem dos Solicitadores e dos Agentes de Execução (OSAE), the Portuguese Chamber of Solicitors, has developed an application of "Online Register of Solicitors" called ROAS[1], which allows these professionals, the registration of acts of recognition, authentication and certification, through platform authentication for a user/password mechanism assigned by the Chamber of Solicitors.

[1] https://www.solicitador.org/roas2/login.jsp.

R. Silhavy et al. (Eds.): CoMeSySo 2018, AISC 860, pp. 333–340, 2019.
https://doi.org/10.1007/978-3-030-00184-1_31

The legality and certification of documents deposited in the ROAS, is ensured by the use of paper-based authentication seal sticker vignettes/number of the vignette, which is issued to each solicitor and sent to the solicitor's professional domicile, by registered letter.

In accordance with the provisions of the "Regulamento de Publicidade e Imagem dos Solicitadores e Agentes de Execução" (Regulation 786/2010), the authentication of documents composed of more than one page is performed using a single seal, placed on a staple that unites the pages in the document. In the authentication process, the supplicant must also sign or initial the seal, writing the date on which the signing or initialing was affixed. This procedure transforms a "private" document in a "particular authenticated" document, but not giving it the force of "public document" as for example with the notarial deed. The document signed by this professional can be fraudulently altered, that is, without the consent of its author or the parties involved (in the case of legal transactions). Are here concerned the values of authenticity, integrity and non-repudiation. On the other hand, there is the possibility of a document be destroyed or damaged, and this possibility increases as time passes, and documents can be exposed to corrosive effects, causing irreparable damage on the content and form of the documents.

The digital information is presented in a number of formats, mapped into binary code in computer storage media, and their access and preservation places to the new "curators" of contracts/wills/statements for future memory, the challenge of "irreversibility and their incorruption". This information, encrypted, secure and verifiable comes to convert the action of Solicitors and agents of enforcement to authenticators agents, and plenipotentiaries certifiers of the powers that the legislature assigned to them, the role of real justice auxiliaries.

The main objective of TESTOR is to provide a proof of evidence for any type of digital document which can be legally accepted and indisputable on law processes. The digital document is stored on an online platform with access restrictions, working TESTOR as a trusted archive service.

In this paper is discussed the similar work, the technologies available and the TESTOR approach presenting the security requirements, the system architecture and how is produced the proof of evidence.

2 Related Work

Digital document certification services are an enabler of e-Government and are present in some countries e.g. Hungary [6] and Estonia [11]. These services make use of digital signatures and require a public key infrastructure for digitally sign files/documents, working as an e-Notary. Although electronic notarizations have been legally available in many states for more than a decade, the state of Virginia (USA) expanded electronic notarizations to allow their notaries to also legally perform remotely based notarizations through a secure webcam session [12]. Allowing notaries to be able to electronically notarize documents from virtually anywhere at any time.

The e-Court system from Estonia also shares some use cases from the TESTOR project. Estonia's e-Court is a system for managing all court procedures. The solution has been implemented for Estonian courts under the Ministry of Justice and paperless proceedings have been held since 2015. Once a claim is entered via a public portal and confirmed, the workflow engine delivers the necessary data to the portal for the allocated judge. Meanwhile, the judge and other participants can submit further items of evidence electronically, answer questions and even involve legal representatives and lawyers in the process [13].

On TESTOR project is allowed not only certify documents (as a notary) but also follow and connect the proof, that can be audio, video or any other digital content, with the civil process. The aggregation of the document with a time stamp is a core functionality and the time stamp needs to be preserved for an undetermined long time.

The constant technological evolution and the increase of the computational capacity of the electronic devices, cause the cryptographic algorithms to become obsolete and lose some of the capacity for protection of digitally signed content. When archive the digitally signed documents is the objective, the digital signature needs to be a Long Term Validation (LTV) signature. An LTV signature acts by applying time stamps with metadata integrated into the content of the document, thus ensuring its preservation and content unchanged over an extended period of time [5] enabling the archive of the content [2].

An alternative to digital signatures is blockchain technology. Blockchain can preserve the digital documents with a time stamp with the secure, distributed and open database technology [4]. For that, a hash function is applied to the signed document and added to the blockchain, the hash at blockchain is also sent to the client. To prove the integrity and authenticity of the submitted document, a hash function is applied to the document in question and seen if it is present on the blockchain, being present is proven the authenticity and integrity of the document.

Documents are digitally signed by all the stakeholders and may include a solicitor or a lawyer, the signature is performed with the smart card (citizen's card) [7] of each stakeholder. Portugal offers the service Chave Móvel Digital (CMD) which allows the citizen to perform, by his own will, a qualified electronic signature with the password chosen by him and his security code [8]. The CMD service to being allowed to sign needs the association of a mobile phone number with the civil identification number (NIC) for a Portuguese citizen or passport number for a foreign citizen. The CMD registry can be done online or in person in the service.

3 TESTOR Project

The actual platform (ROAS) is not functional at this moment, because it's impossible to store digital documents and metadata ensuring levels of scalability, reliability and availability of "evidence" in the system.

Evidence that directly proves a fact, without an inference or presumption, and which in itself, if true, conclusively establishes that fact.

In a civil action, the requisite degree of proof is a preponderance of the evidence. The plaintiff must show that more probably than not the defendant violated his or her rights. However, each alleged fact must be proved separately, as must all the facts necessary to reach a judgment for the plaintiff (the person filing a lawsuit) or for the prosecution (the "people" or "state" represented by the prosecutor). The defendants in both civil suits and criminal trials need not provide absolute "proof" of non-responsibility since the burden is on the plaintiff. An adequate evidence, adequate proof legally presented at trial, or not, fulfilling a legal responsibility is a sufficient corroboration.

3.1 Security Requirements

Identification of the requirements of the Digital Certification System to be adopted for the signature and submission process, and for the migration process (if appropriate).

Communication Security. The data communicated between the client (browser) and the server cannot be visualized by a third entity nor can be changed in the communication, passing the solution by the encryption of the data. The solution involves the use of the Transport Layer Security (TLS) security protocol [1]. This protocol must be used between the browser and the server and between the server and possible integrations with other systems, such as an external digital signature service.

Protection of Data. The stored data will be metadata about the submission and the file(s) submitted. Submission metadata and user data will be stored in a relational database and documents in a file system where the database indexes the location of submitted documents. There are measures to ensure the security and privacy of data, solutions to data recovery, data access and infrastructure are addressed. Access to information and content of submissions is only authorized to those who have permission for such access, and the permissions are defined in the act of submitting the documents. All access to the data is recorded. For this, all access to the contents of the submissions in the database is registered in the database. The data present in the database is encrypted.

Technology Migration. The creation of future formats for reading a text, audio, video files, among others, can make obsolete the existing file formats in the system, making the old documents submitted in a possible invalid proof or difficult to read.

To guarantee that the digital documents that are being stored today will still be readable by the computers and programs that will be used in the future in order to mitigate the problem of technological obsolescence and ensure continued access to digital information, various strategies for preserving digital information were pointed out, such as emulation, format migration and encapsulation. It can be concluded that the following set of services is sufficient to implement automatic preservation procedures (in this phase, exclusively based on format migration) that operate transversally on collections of digital objects [9]:

- **Service of format identification** - fundamental for obtaining a map of the formats that constitute the collection of objects to be preserved. Through this information, decisions can be taken as to the best preservation strategy to take;
- **Obsolescence notification service** - service required to ensure the automation of preservation processes. This service permanently monitors the current technological context and determines the level of obsolescence of a given format and the risks associated with its conservation. In a situation of imminent technological breakdown, it informs the preservation system that should initiate a preservation intervention;
- **Format conversion service** - service responsible for the execution of preservation actions;
- **Quality control service** - converters are not all the same and conversions are not always perfect. This service is responsible for assessing the quality of a conversion and subsequently, the converter or converters used;
- **Conversion strategy selection help service** - an object, format or class of objects can be converted to a large number of different formats. A migration can be accomplished using a multitude of conversion tools. This service allows identifying, among all the options recognized by the system, which guarantees the highest level of satisfaction of the preserver.

3.2 Architecture

The trier of fact, whether a judge, must review the evidence presented, evaluate it, and determine if it meets the standard of proof. The weight of evidence is based on the believability or persuasiveness of evidence.

The information recorded in the database must meet a number of requirements to be conferred probative value, namely: maintain consistent registration systems; and information must be kept inviolate, consistent, auditable and removable. The database should be based on PostgreSQL, since: it is an open source database, free of licensing; there is no risk of the information stored in the database is compromised, for license compliance verification audits; and it is available on multiple platforms (Linux and Windows).

The interface with the users should be guaranteed through a web browser, and one should be careful to ensure maximum compatibility with browsers with greater diffusion.

This activity should be developed by all web services support access system metadata, and must themselves ensure the authenticity, privacy, and integrity of data exchanged between client and server. In order to prototype validation and evaluation of its suitability for subsequent transposition into a real environment, it should be subjected to tests, in which will participate in a pre-selected group of Solicitors. During the test phase, the system users will use the system proposed here, in parallel with the current ROAS platform.

As Fig. 1 describes, paper documents should be scanned. Using a smartcard (Citizen's Card), the solicitor digitally signs the document. Any documents in digital format, including videos, may be signed. Documents are sent to the platform through a web interface, using web services.

The documents received by the platform, which are accompanied by descriptive metadata, are re-signed (with a time stamp) by the platform itself. After the signature procedure, the documents (the original and the signed by the platform) and their metadata are stored in a database. The Solicitor, as well as other entities of the judicial system (who are granted access to the document), may access the metadata and its documents through a web interface. Interfaces may also be created for connection to other information systems.

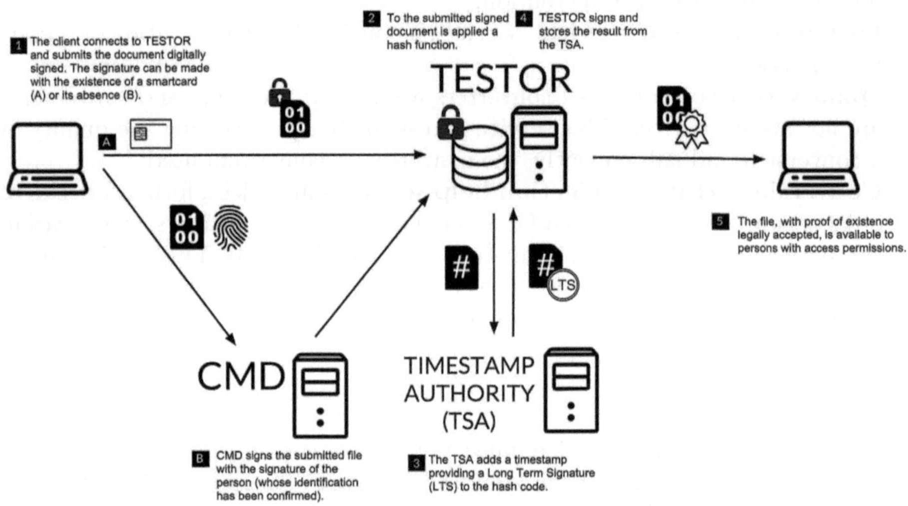

Fig. 1. System architecture.

TESTOR will make use of smart cards, to stakeholders sign digital documents; digital signatures, the online platform TESTOR sign each submitted digital document; and an external service for a long-term validation signature which includes the timestamp. The signature with smartcard also may be done with the Portuguese public service Chave Movel Digital (CMD) [8], this service allows the signer to not have the smart card when performing a signature, CMD needs a previous adhesion where the smart card is registered and when a signature is needed a message with a secure token is sent via SMS that is needed to sign the document on CMD service.

3.3 Proof of Evidence

As Fig. 2 presents, a hash function is applied to the submitted and previous signed (by the stakeholders) document, resulting in one hash code. Then the

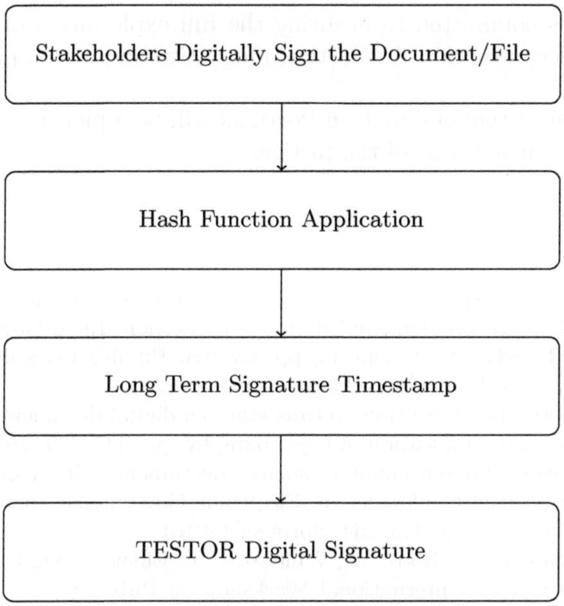

Fig. 2. Generation of proof.

hash code is sent to a recognized external authority that adds the timestamp [3] and digitally signs with a long-term validation signature [5], providing a legally recognized digital signature, and the central entity signs and delivers the result to the client as proof of evidence. The central authority key pair is unique and used for all documents to be signed. This key pair may include a certificate by an external recognized authority.

The client with the indicated result and the document equal to the submitted document can verify its submission, even if the central entity removes the data referring to its submission. When the authenticity of the date and time of submission is challenged, the proof is made through the date data present in the long-term signature. When the document's authenticity is challenged the proof is made by comparing the application's hash function to the document in question with the hash present in the signed result, the hash's not being equal the documents are not the same.

4 Future Work

The vision for the project TESTOR is eminently practical and aims to simplify procedures, providing security, speed, prevent and avoid legal disputes in court.

The TESTOR prototype will be used in a proof of concept. We hope that this project will generate new opportunities for social and professional applications in the field, with significant consequences in the economy, in the e-Justice and finance.

The OSAE is committed to ensuring the full exploitation of the prototype as a public service, a service of e-Justice, innovative and with a professional and social focus.

The e-Justice system of OSAE in Portugal will be a pioneer, and his use will be the basis for a new phase of the justice.

References

1. Dierks, T.: The transport layer security (TLS) protocol version 1.2 (2008)
2. Dumortier, J.: E-government and digital preservation. In: E-Government: Legal, Technical and Pedagogical Aspects, pp. 93–104. Publicaciones del Seminario de Informatica y Derecho, Universidad de Zaragoza (2003)
3. Haber, S., Stornetta, W.S.: How to time-stamp a digital document. In: Conference on the Theory and Application of Cryptography, pp. 437–455. Springer (1990)
4. Ølnes, S.: Beyond bitcoin enabling smart government using blockchain technology. In: International Conference on Electronic Government and the Information Systems Perspective, pp. 253–264. Springer (2016)
5. Ansper, A., Buldas, A., Roos, M., Willemson, J.: Efficient long-term validation of digital signatures. In: International Workshop on Public Key Cryptography, pp. 402–415. Springer (2001)
6. Kiss, P.J., Kiss, J.K., Klimkó, G.: Electronic document certification service: an enabler of e-Government uptake in Hungary. In: International Conference on Electronic Government and the Information Systems Perspective, pp. 276–286. Springer (2016)
7. Vasconcelos, A.: The Portuguese interoperability framework applied to the Portuguese citizen card project. In: Presented at the OECD Workshop on Digital Identity Management (IDM), 9 May 2007 (2007). http://www.oecd.org/dataoecd/36/9/38573902.pdf. Accessed 24 Apr 2018
8. Chave Móvel Digital (2018). https://www.autenticacao.gov.pt/a-chave-movel-digital. Accessed 24 Apr 2018
9. Ferreira, M.: Preservação de longa duração de informação digital no contexto de um arquivo histórico, pp. 183–184. Doctoral's thesis, Universidade do Minho (2009). https://repositorium.sdum.uminho.pt/handle/1822/9563. Accessed 24 Apr 2018
10. Decreto-Lei n. 76-A/2006. In: Diário da República n. 63/2006, 1 Suplemento, Série I-A de 2006-03-29 (2006)
11. eRik, Estonia e-Notary. http://www.rik.ee/en/other-services/e-notary. Accessed 24 Apr 2018
12. Virginia Electronic Notary, Secretary of the Commonwealth Kelly Thomasson. https://commonwealth.virginia.gov/official-documents/notary-commissions/enotary/. Accessed 27 Apr 2018
13. Estonia e-Court, e-Estonia. https://e-estonia.com/solutions/security-and-safety/e-court/. Accessed 4 May 2018

Author Index

R. Silhavy et al. (Eds.): CoMeSySo 2018, AISC 860, pp. 341–342, 2019.
https://doi.org/10.1007/978-3-030-00184-1

MIX
Papier aus verantwortungsvollen Quellen
Paper from responsible sources
FSC® C105338

If you have any concerns about our products, you can contact us on
ProductSafety@springernature.com

In case Publisher is established outside the EU, the EU authorized representative is:
Springer Nature Customer Service Center GmbH
Europaplatz 3, 69115 Heidelberg, Germany

Printed by Libri Plureos GmbH
in Hamburg, Germany